Although this account deals exclusively with military units, the difficulties described here can develop in any large organization or social movement. The author suggests that high morale is something that cannot be coerced; it can only be won by providing rank and file members with whatever they believe to be worthwhile.

Tamotsu Shibutani is Professor of Sociology, University of California, Santa Barbara.

the derelicts of company K

A Sociological Study of Demoralization

by Tamotsu Shibutani

University of California Press
Berkeley • Los Angeles • London

University of California Press
Berkeley and Los Angeles, California

University of California Press, Ltd.
London, England

ISBN 0-520-03524-0
Library of Congress Catalog Card Number: 77-79237

Printed in the United States of America
Designed by Dave Comstock

1 2 3 4 5 6 7 8 9

contents

List of Tables and Illustrations vi
Preface vii
Acknowledgments xiii

1. The Problem of Group Solidarity 1
2. The Challenge to Nisei Loyalty 19
3. The Social World of Nisei Servicemen 71
4. The Initial Break in Discipline 101
5. The Disposition of Surplus Personnel 169
6. Impetuous Reactions to Degradation 221
7. Fluctuations in Company Morale 286
8. Anarchy at the Replacement Depot 364
9. Demoralization as a Social Process 411

Glossary 437
Bibliography 441
Index 447

tables and illustrations

Tables

1. Nisei Casualties Listed in the *Pacific Citizen* 59
2. A-9-3 Training Schedule for the Week of April 22 131
3. Student Personnel of Company K (October, 1945) 199

Plates

1. Evacuation, U.S.A., 1942 46
2. A Relocation Center 48
3. Insignia of Sergeants in the U.S. Army, 1945 120
4. Company K and the "Regular" School 201
5. Rikoran 291
6. Scenes from Rikoran movies 292
7. Transportation facilities were limited 366
8. All brothels were off limits 377
9. Annex of the 4th Replacement Depot 390

preface

Although this book is about soldiers, it is not a saga of heroes. It is about some servicemen who never entered a combat zone. They led brief and undistinguished military careers; for the most part their only contribution to the war effort was the inflation of statistics on the number of men under arms. Compared with the enormous sacrifices rendered by millions of others in World War II their experiences seem trivial, for the little known incidents recorded here hardly affected the outcome. Why, then, should we be concerned with an obscure company of garrison troops—a tawdry backwash of a gigantic war?

The aim of this study is twofold. It is a chronicle of one of the more disorderly units in United States military history, and one objective is to learn just how it deteriorated to this point. An account of what happened to a single company, however, is not the sole purpose of this inquiry. The second objective is the formulation of sociological generalizations concerning the process of demoralization. Like so many other soldiers, these men were forced into roles they did not seek, caught in momentous events they could not understand, and compelled to cope with disagreeable situations as best they could. Most of their reactions were characteristically human. Most of the behavior patterns described have been observed before and are likely to develop again in many other historical contexts. The record is to be taken, then, as a case study that will serve as a vehicle for a broader investigation of the breakdown of collective undertakings.

The most serious impairment in the fighting capacity of the United

States armed forces occurred during the latter phases of the Vietnam war, especially during the prolonged program of disengagement. It became apparent that most American soldiers no longer believed in the war, and many openly flaunted peace symbols. Observers agreed that Americans fought well, but only if they were attacked. In isolated operations many merely went through the motions of patrolling, deliberately avoiding contact with the enemy. Further fighting seemed pointless, for the men were convinced that nothing would change when they departed; they were not willing to risk their lives for nothing. Enlisted men questioned orders; if the explanations were unsatisfactory, in some cases they flatly refused to obey. Career officers, commissioned and noncommissioned—scored as "lifers"—and the young draftees did little to hide their disdain for one another. Many rumors developed over the "fragging" of unpopular officers, and the Department of Defense subsequently disclosed that 551 such incidents had occurred between 1968 and 1972, leaving 86 dead and over 750 injured (Cortright, 1975:43–44). Officers stopped making patriotic appeals, and NCOs did not even bother to pay lip service to the reasons given by the government for the American presence in Vietnam. The use of drugs that obviously would impair combat effectiveness became widespread. One official estimated that over half the troops were using drugs illegal in the United States, and in August 1970 the army established an amnesty program in which heroin addicts could report for treatment without risk of punishment (*Newsweek*, 1971). Although none of the disturbances in Company K even remotely approximated these incidents in importance, there were many remarkable similarities, and the availability of a detailed record provides an opportunity for a careful analysis of phenomena that have thus far proved difficult to comprehend.

Nor is the problem of maintaining adequate morale confined to military organizations. Demoralization can develop in any organization or community, seriously disrupting its operation. Scandals have rocked hospitals in large cities, where unnecessary agony and even death have been attributed to the carelessness and neglect of a dejected staff. In recent decades the Catholic church has lost the services of thousands of once dedicated priests and nuns, and other denominations have encountered difficulty in recruiting competent clergymen. The chronically unemployed in urban ghettoes have on occasion become so despondent that they have refused to participate in programs designed to improve their lot. Policemen and firemen, overworked and underpaid, have angrily walked off their jobs, leaving large communities unprotected. Many insurrections have succeeded not so much from the adeptness of revolutionaries, but from the inability of the privileged to mobilize their resources to defend their interests. Indeed this inquiry should not be regarded as a study in military sociology. Although it deals exclusively with army units, none of the men involved were subjected to the intense, sustained, nightmarish stress of combat; and most of the problems discussed are of the sort that arise in any formal organization—a school system facing student restiveness or a union confronted by a wildcat strike.

Despite the generally acknowledged importance of the subject, most explanations of demoralization are inadequate. A frequent reaction of authorities confronted by such breakdowns is to blame "a small criminal element," "psychopathic malcontents," "outside agitators," "inadequate facilities," or "misunderstandings of policy"—some antecedent event or condition that is regarded as the "cause." Poor leadership is a frequent explanation, and local officials are sometimes reprimanded for "permissiveness," "unnecessary harshness," or "poor communication." Such explanations are simplistic and stereotyped, and it is not surprising that remedial measures based on them so often fail to rectify the situation. Explanations of demoralization by social scientists are generally less vindictive, but there is considerable disagreement among them. Sociologists who account for variations in morale in terms of "causal factors" are only citing some preexisting condition; they too are using the simple "cause" and "effect" mode of explanation of daily discourse.

This study differs from most previous investigations of morale in that attention has been focused on the subjective experiences of the demoralized men themselves. Many disorders have been studied from the outside—statistics of the damage done, the number of people who defected, opinions of officials and observers. Much of this account deals with the beliefs and sentiments of enlisted men, not because they are more accurate than those of officers, but because their views make the behavior of common soldiers understandable. By concentrating on the manner in which their perspectives are constructed and reaffirmed in daily conversations, it becomes possible to record the manner in which a sequence of events is viewed by men in intimate contact, independently of official pronouncements. Seemingly trivial preoccupations have been described at length, for we can understand what people do only when we grasp the manner in which they see their world. This picture is often quite different from that of outsiders. Violence, for example, always appears ugly to observers, especially to the victims and their friends. When seen from the standpoint of its perpetrators, however, the same deeds appear quite appropriate, sometimes even ennobling. Thus, much of this book becomes a record of the reactions of ordinary men to extraordinary events.

A question arises immediately over the dangers of generalizing from a single case, and the risks involved should not be underestimated. A solitary case is presented so that intimate details about life in the barracks can be examined and analyzed. It is the availability of such particulars that makes possible the formulation of the kinds of generalizations that have seldom been entertained in previous studies of morale—hypotheses about the kinds of beliefs and sentiments that underlie widespread insubordination or retaliation against an organization. Hypotheses about the manner in which such beliefs arise can be constructed only when detailed data are available on the interchanges among the persons involved. Once such hypotheses have been formulated, they may subsequently be tested in studies that focus on specific items. Some of the findings of this study are embarrassingly obvious. They have been spelled out in painful detail precisely because they differ from currently accepted explanations of demoralization.

Although this book reports a sociological investigation, its format differs from what has become customary for research monographs. It appeared to me that the subject matter—the demoralization of a single unit in an otherwise efficient organization, the eruption of minority group violence, the collective adjustments made in anarchic situations—would be of interest to a much larger audience. Hence, all technical matters of interest primarily to sociologists have been confined to chapters 1 and 9 and to the analytic summaries at the end of the intervening chapters. Chapters 2 through 8 contain a record of what happened to the men who made up the core of Company K, along with the historical background necessary for those unfamiliar with the plight of Japanese-Americans during World War II. The events reported took place at the military installations named on the dates cited; only the names of the individuals involved have been disguised. Since some uncommon words have been used repeatedly—from military terminology, the argot of American soldiers in World War II, Hawaiian pidgin, common Japanese expressions, and Nisei slang—a short glossary has been appended.

Since more than three decades have passed since these data were collected, questions will inevitably arise concerning the delay in preparing this book. When my field notes were initially organized and transcribed in 1947, I encountered two difficulties. Because of deep emotional involvement in the events, the intensity of which I could not appreciate at the time, I was unable to understand many things that now appear obvious. The passage of time has made possible a more candid appraisal of what happened. In addition, I discovered several serious gaps that could be filled only by checking military records, but many of the necessary documents were still classified. By the time they became available, I was already committed to several other projects that could not be set aside for a task of this magnitude. The long delay was unintentional, but the overall consequence has been beneficial.

To be sure, some of the materials are clearly dated. The armed forces of the United States are no longer segregated in the manner described, and many of the practices that the men found so obnoxious—reveille, KP, frequent inspections—have been eliminated. Hawaiian Nisei are no longer so concerned with the linguistic problems that plagued so many of them during World War II, and younger Americans of Japanese ancestry are rapidly losing the cultural traits that account for many of the distinctive reactions described. But the delay has turned out to be a blessing, for it is now easier to view the materials more dispassionately. What happened to Company K is no longer of political consequence to anyone, but many current disorders of a similar nature may prove difficult to comprehend because of our personal involvement. Accounts of comparable events in contexts that no longer elicit emotional reactions may enable us to view them from a more detached standpoint.

acknowledgments

Prior to my induction into the armed forces in 1944, I spent more than two years as a research assistant in the University of California's Evacuation and Resettlement Study—as a participant observer in the evacuation of the Japanese from the Pacific Coast, in the Tule Lake Relocation Center, and in the resettlement program in Chicago. Had I not gotten into the habit of keeping a detailed journal of each day's events during those turbulent years, I doubt that I would have been able to conduct fieldwork under the circumstances described in this book. For the research training I received and for her forbearance and understanding, I am deeply indebted to the director of that study, Professor Dorothy S. Thomas.

The field notes were mailed almost daily to friends and relatives, who kept them for me. From time to time I prepared progress reports, and Professor Herbert Blumer, then at the University of Chicago, kindly took time off from his heavy wartime duties to make critical comments and suggestions. Others whose support and assistance during this period proved so important include Tomika Sollen, Robert and Hanny Billigmeier, and Koso and Emily Takemoto. The transcription and initial organization of the field notes were made possible by a predoctoral training fellowship from the Social Science Research Council.

Once the notes had been organized, many gaps became apparent; and in the course of filling them I incurred many additional debts. Relevant information concerning the movies enjoyed at Fort Snelling was provided by various executives of the Toho Kabushiki Kaisha, who not only answered my questions but arranged for permission to reproduce some of the photographs used in this book. I wish in particular to express my appreciation to Mr. Rin Masutani. George Matsuoka played a key role in making these arrangements possible. Many of the questions that had arisen about Hawaiian AJAs (Americans of Japanese ancestry) were answered during a visit to the University of Hawaii. For their part in providing indispensable background material I wish to thank the faculty members of the Department of Sociology—especially Bernhard Hormann, George Yamamoto, and Douglas Yamamura. Official records of the units studied were consulted at the Military Personnel Records Center, General Services Administration, in St. Louis. My work there was greatly facilitated by the generous help of the staff—in particular Phillip C. Chrisman, Michael T. Vranesh, and Herman Henderson. Additional materials were drawn from two archives of the University of California—from files of the Evacuation and Resettlement Study on the Berkeley campus and of the Japanese-American Research Project on the Los Angeles campus. All of the postwar trips were made possible by grants from the Research Committee of the Santa Barbara campus.

Obviously this study could never have been made without the cooperation of the men themselves. They tolerated my perpetual note taking, responded candidly to my questions, volunteered information, warned me of impending events that might be of interest, and covered for me whenever fieldwork required my unauthorized absence from formations. Most members of the various platoons to which I was assigned knew of the study and contributed in some way, but I am especially indebted to the following members of Company K: Henry Nakamura, George Norikane, Harry Shinozaki, Harry Tsutsui, and Fred Yamashiro. Part of the account would have been quite one-sided were it not for data about the difficulties faced by the Officer Candidate School (OCS) men, and I wish to acknowledge the assistance of the following members of Company A: George DeVos, Herbert Passin, and Gerald D. Worth. I am especially grateful to Mr. Worth for preparing a detailed account of the stabbing incident.

I have benefited immeasurably from the suggestions and encouragement of those who were kind enough to read the manuscript, and I wish in particular to thank Sandra Gettman, Tetsuden Kashima, Yukiko Kimura, and Anselm Strauss. After the manuscript was submitted to the University of California Press it underwent substantial transformation. A major revision was based on a penetrating but sympathetic critique prepared by Frank Miyamoto. Further rewriting resulted from detailed comments and suggestions on style by Marilyn Schwartz. I am deeply indebted to Grant Barnes not only for arranging for both reviews but for showing me the many other ways in which an astute editor can guide an author toward the latter's objectives.

Without the generosity of all these people this book would not have assumed its present form, and I wish to express my gratitude to them.

1.«» the problem of group solidarity

The brawl started on a small streetcar—called the "dummy line"—that shuttled for about a mile from the gate to the interior of the fort. It was crowded with soldiers returning from pass, hurrying back to quarters before bed check. In the pushing and scuffling one soldier, whose foot had been trampled, objected loudly and shoved the offender. He found himself confronted by an infuriated man, who cursed him as a "dirty Haole" and muttered to his companions that such creatures should be "fixed." At the end of the line the three Caucasian soldiers aboard nervously made for their barracks at a brisk pace. They were overtaken at a firebreak and quickly surrounded by the enraged man and his friends. Others stopped to watch.

As the protagonists faced one another, the irate man screamed, "You white! You t'eenk you better dan me! Look at me! Yallow! But me, I no scared of you! If you so good, why you no heet me?"

The outnumbered men declined to fight, exclaiming repeatedly that no offense had been intended. As they started to run, they were set upon by the gang. One man was beaten to the ground and, as he lay helpless, was kicked repeatedly. As some onlookers began to intervene, the other two succeeded in breaking away. It was not until they approached their company area, however, that one of them realized that he had been stabbed in the back.

This incident took place on 17 October 1945 at Fort Snelling, Minnesota, at that time a training center for the Military Intelligence Service Language School. It was there that selected troops received specialized instruction for combat intelligence duties in the Pacific theater.

Over 90 percent of the trainees were Nisei, Americans of Japanese ancestry, all but a handful enlisted men. Most of the others were Caucasian, nearly all of them officer candidates.

As soon as the incident was reported, the duty officer, accompanied by military police, rushed directly to the barracks of Company K. The men were awakened and ordered to display all their clothing, including dirty laundry, and other personal belongings on their bunks; one by one they were required to hold out their hands for a knuckle inspection. The stabbing was but part of a general pattern of disorder and violence that had broken out at the fort during the late summer and autumn of that year, and Company K had been credited with most of the trouble.

Some Symptoms of Demoralization

The men of Company K forged a formidable record of discord. Wherever they went disturbances ensued. At times discipline broke down to such an extent that the men themselves wondered jokingly whether they were still part of a military organization. They were astonished that so few of them had been court martialed for misconduct, insubordination, and protests that sometimes bordered on mutiny. More than one visiting officer asked in dismay, "Where do soldiers like these come from?"

What is even more surprising is that the record of Company K presents such a stark contrast to the exemplary performance of most Nisei troops in World War II. Since they were identified with the enemy by so many, persons of Japanese ancestry were acutely conscious of their tenuous position in American life, and they grasped at every opportunity to demonstrate their loyalty. In the European theater the 442nd Regimental Combat Team fashioned a distinguished battle record in some of the most grueling fighting that took place in France and in Italy. Nisei troops also played an important part in the Pacific, where they interrogated prisoners, translated captured documents, and monitored Japanese broadcasts; they were subsequently described as the "eyes and ears" of the Allied command. Even in garrison duty they had won the esteem of their comrades and officers for their bearing and dependability. On learning of the slovenly appearance and troublesome conduct of Company K, many Nisei combat veterans cursed bitterly those who threatened to tarnish a reputation secured through such costly sacrifices; some were almost in tears, wishing that the company could somehow be disowned. In the Minneapolis-St. Paul area Company K men were feared by other Nisei, roundly condemned as hoodlums, and in some instances ostracized.

Why the radical deviation of this group from the general pattern set by other Nisei soldiers? When asked to explain themselves, Company K men would list their "reasons": at the outbreak of war they had been accused of disloyalty, even though there was no evidence of it; they had been evacuated from their homes on the Pacific Coast and incarcerated without trial; just as they were becoming reestablished in the Midwest or East they had been drafted, in most cases while their parents were still behind barbed wires in relocation centers; once in the army they had been subjected to discriminatory treatment. While their grievances were not unjustified, the same objections could have been raised by virtually all Nisei drafted in the continental United States.

One readily discernible contrast was in morale. Most Nisei units were characterized as having high morale. Company K was generally acknowledged to be demoralized, both by observers and by the men themselves. It had been labeled as a "fuck-up outfit" and scorned. Most of the men were disconsolate; they felt disinherited and sometimes referred to themselves as the "orphans of the war." This leads, then, to the initial question: What is *morale?*

Morale is a popular concept. The term refers to something that is recognized intuitively as important. Much has been written about it, but authors disagree over what morale is, for popular concepts are notoriously vague in reference. For some it is the frame of mind of an individual—his dedication, eagerness, or willingness to sacrifice. For others it is a social phenomenon—collective enthusiasm or the persistence of a group under adverse conditions. Popular concepts often contain implicit value judgments. Although this has never been demonstrated to be the case, morale is frequently assumed to vary along a unidimensional scale—from high to low. Even casual observation and reflection suggests that we are dealing with a complex phenomenon, one that is neither clearly identified nor well understood.

In this inquiry morale is regarded as one of the attributes of groups—the degree of effectiveness with which the recognized goals of joint enterprise are pursued. In virtually all of the component transactions that make up the structure of any group, there is a differentiation of tasks among the participants, and the accomplishment of avowed objectives requires cooperation. We are concerned with the *quality* of coordination and integration of the contributions of those who are involved in some kind of common endeavor (Barnard, 1938:55–56). We are also concerned with *persistence* in the pursuit of collective goals—the capacity of a group to maintain its integrity and steadfastness of purpose until its objective is attained (Wirth, 1941:426). Thus conceived, it is obvious that morale is of importance in innumerable contexts: the success of a band of pioneers struggling in the wilderness, the league standing of athletic teams, the survival of religious cults facing ridicule and persecution, the achievement of a production unit in a factory, the resilience of protesting demonstrators confronted by armed troops, or the doggedness of passers-by who try to save a drowning child. Morale is important in any enterprise that requires sustained cooperation for success; it is for this reason that it may be decisive for those involved in any kind of contest.

Although sociologists have yet to develop adequate tools for the analysis of the expressive component of society, it would appear that every organized group with continuity of personnel is likely to develop a distinctive *style of performance*, much like the personality of each individual. Each group acquires a characteristic stance—a distinctive way of perceiving and defining events, predispositions to respond in particular ways to frustrations, and typical ways of getting things done. A distinction may be made between structure and style. Group structure refers to the accepted and agreed upon procedures for accomplishing the goals of enterprise. In formal organizations such as an army, norms are explicitly stated in regulations; what is to be done is clearly prescribed, as are the penalties for those who do not comply. But style does not depend on this institutional framework. In their study of U.S. Army Air Force units in World War II, for example, Grinker and Spiegel (1945:25–26) note that after a crew had been in combat for some time it took on a definite color, making it distinct from other similarly structured groups. One could recognize the "hot outfit" with its impressive record of tactical efficiency and accomplishment; the "snafu outfit" with its discontent, bickering, and chronically poor performance; the "carefree outfit" with its exuberance, dash, and recklessness; as well as the sedate, colorless unit marked by sober dependability. All units in a given military organization are governed by identical regulations, and in time of war objectives are agreed upon to the point where they can be taken for granted; but style of performance varies from one group to another. Furthermore, such differences are definitely related to proficiency in combat. Similarly, within the same factory productivity rates vary from one section to another; and in large prisons, riots erupt more frequently in some cell blocks, while others only lend reluctant support.

An organized group is characterized as having high morale when it performs *consistently* at a high level of efficiency, when the tasks assigned to it are carried out promptly and effectively. In such units each member is likely to contribute his share, willingly doing what he believes to be worthwhile and assuming that his associates will do their part. When necessary, the men help one another, without even being asked. Mutual encouragement is commonplace, and those whose zeal is exemplary are singled out for praise. The few who do not share the prevailing orientation feel pressures to comply; those who fail repeatedly to live up to expectations are scorned as "slackers," and efforts may be made to expel them from the group. The successful completion of each transaction occasions no surprise; it is the expected thing. Members of such groups usually place a high evaluation on themselves. They often develop a strong sense of identification with each other; they develop pride in their unit, become conscious of its reputation, and take pleasure in displaying emblems of belonging to it. It must be remembered, however, that such performances may be executed in several different ways. Units marked by sober dependability can perform just as effectively as those marked by enthusiastic outbursts of maximum effort, but their styles are discernibly different. High morale, then, is a label of approval.

If an organized group develops a style that results in consistently desultory and sluggish performance, it is characterized as having low morale. Such units are often marked by constant bickering, factionalism, and reluctant compliance with regulations that results more from a fear of punishment than from a sense of duty. The continuous vituperation of such men is not to be confused with the good-natured griping frequently heard among rank-and-file members of large organizations. For example, a study of the U.S. Army during World War II reveals that a rather substantial proportion of enlisted personnel, especially in the lower ranks, expressed dissatisfaction over their fate (Stouffer et al., 1949: I, 82–104). As long as military service is accepted as legitimate, however, those who complain assume that the various hardships must of necessity be endured. But in low morale groups serious doubts arise about the avowed goals, and those who remain conscientious are dismissed cynically as "suckers." Absenteeism may become commonplace, and a sense of futility and despair may be felt and voiced. Members of such groups usually place a low estimate on themselves; there is no group pride, and some try to transfer elsewhere. Some seek solace in esoteric interests; others become preoccupied with hedonistic pursuits or lose themselves in alcohol or drugs. When confronted by adversity, such groups are more likely to disintegrate, as each person becomes preoccupied with saving himself. Thus, any group that develops a style of performance not compatible with the type of endeavor required encounters difficulties and is likely to be labeled as being deficient in morale.

Some social scientists have contended that tenacity in the face of adversity is the most unequivocal index of high morale (Lasswell, 1933: 640; Blumer, 1969:108). Indeed, it is when confronted by severe stress that some high-morale groups achieve peak performances. The emotional climate becomes one of grim determination, even in the face of grave danger. Attention becomes focused on the task at hand, and all resources are mobilized. Everyone pitches in, doing whatever he can. The division of labor develops spontaneously, as various participants improvise whatever is necessary to get the task done. Any call for help is heeded at once. Obstacles and setbacks are viewed as temporary challenges to be overcome, and the participants regroup to grapple with various problems with renewed zeal.

When faced with similar adverse conditions, however, some low-morale groups collapse. Activity becomes disorganized, as the participants work at cross purposes. Very little is accomplished. Since each group has a unique style of performance, each flounders in its own way. What is of special interest, however, is that in spite of these differences the symptoms of breakdown are astonishingly similar; they have been observed over and over in diverse historical contexts. Evasion of duty becomes commonplace. As authority deteriorates, local officials are no longer able to enforce regulations. Increasing numbers refuse to obey, even when threatened with punishment. Since disciplinary action against everyone is impracticable, many formal norms are simply ignored. During the latter part of the Vietnam intervention, for example, American soldiers shirked dangerous missions; "search and destroy" was relabeled "search and avoid." This practice became so widespread that Viet Cong representatives announced at the Paris peace negotiations that Communist forces had been ordered not to molest Americans unless they were provoked (Heinl, 1971:31). Such disorganized situations are often marked by a widespread sense of frustration, occasional outbursts of hostility, and egoistical individualism. Those who are trapped in them rely increasingly on informal arrangements among themselves. Factionalism becomes pronounced, and further disruptions result from the disunity. Different individuals and cliques move in separate directions, and coordination of their respective efforts becomes impossible. It is precisely this type of failure in collective adaptation that is often referred to as *demoralization*. The term refers to a breakdown of collective effort that results from the unwillingness of people to continue performing their duties.

When a group becomes demoralized, not only is efficiency impaired, but in some instances the participants actively engage in obstruction. The men get together among themselves to mount campaigns of resistance and transgression. Inefficiency becomes intentional and organized, as in the deliberate slow-down tactics of sullen factory workers. When authorities try to make corrections, they often encounter open defiance. In demoralized army units blatant disregard for dress regulations and military courtesy become commonplace. Such efforts interfere further with the attainment of organizational goals and compound the difficulties.

Although violence is not a necessary part of demoralization, it frequently erupts. Where differences in class, religion, or ethnic identity have already led to divisive factionalism, animosities may be intensified. Unrestrained by law, some of the disgruntled may band together into gangs that launch campaigns of terror. For example, since the 1960s the school boards of several American cities have reported open defiance of regulations, rampant truancy, rape, arson, assaults on fellow students, intimidation, extortion, vandalism, and even homicide. In 1972 alone, 541 teachers were assaulted in New York City, and in Detroit such attacks occurred at the rate of about 25 per month. Administrators and teachers have been virtually helpless, for policemen cannot be stationed everywhere (*Time*, 1973). Riots may break out. Toward the end of the Vietnam war several hostile outbursts occurred in the U.S. armed forces. In October 1970 some 400 Negro and Caucasian soldiers in Chu Lai engaged in a shooting spree after a brawl at a service club. On the night of 12 October 1972 a riot erupted on the attack carrier U.S.S. *Kitty Hawk*, following the questioning of a Negro sailor concerning his involvement in a fight at the enlisted men's club in Subic Bay. Apparently under the impression that their fellows were being harassed unfairly, marauding bands of enraged Negroes attacked the others, in some instances pulling sleeping men from their berths and beating them with fists, chains, wrenches, metal pipes, fire extinguisher nozzles, and broom handles. They were heard to shout: "Kill the son of a bitch!" "They're killing our brothers!" "Kill the white trash!" By the time order was restored, 47 men had to be treated for serious injuries; three required medical evacuation to shore hospitals (Cortright, 1975:120–21).

Embittered members of demoralized units sometimes turn against the organization itself. Their retaliation may take several forms—verbal expressions of indignation, petitions, picketing, demonstrations, sabotage, defection, or outright rebellion. Collective protest consists of more than mere convergence of the spontaneous reactions of outraged individuals. The participants communicate and thereby reinforce one another's feelings of bitterness; whether they merely jeer in unison or engage in violent reprisals, their reactions involve some measure of organized coordination. In Vietnam, where many soldiers came to regard their national leaders and officers as their true enemy, such retaliation assumed incredible proportions. A mass movement, involving both soldiers and civilians, was launched in 1967 and operated with varying degrees of success until 1972. More than 300 underground newspapers were published, and arrangements were made for aiding deserters, protest demonstrations, strikes, and law suits (Cortright, 1975:50–105). When some official is singled out as being particularly odious, schemes may be contrived for vengeance. In Vietnam bounties were raised by common subscription for the heads of various officers and NCOs. After the costly assault on "Hamburger Hill" in 1969, *GI Says*, an underground newspaper, openly offered a bounty of $10,000 for the death of the colonel who had ordered and led the attack (Heinl, 1971:31). Sabotage is another form of retaliation against the organization. Unexplained damage to equipment and intentionally bungled repair work become commonplace. In 1971, the U.S. Navy reported 191 investigations for sabotage, 135 for arson, and 162 for "wrongful destruction" of property (Cortright, 1975:123–26). Even when men in demoralized units are unable to accomplish anything else, some of them may mobilize at a high level of efficiency for retribution. The impact may be considerable. The protest movement in Vietnam significantly altered national policy by limiting options.

Although nothing so momentous was at stake, Company K had all the symptoms of a demoralized group. The most common form of resistance was evasion of duties, but opposition often went well beyond mere recalcitrance. Shouting obscenities and insults from ranks was commonplace. Violence erupted, and a brief reign of terror developed. Local officials, though held personally accountable for the accomplishment of military objectives, found it impossible to maintain order; at times the men became so unruly that the officers simply left them to their own resources. Situations approaching anarchy, though largely ignored in sociological literature, are of considerable interest; to understand how human society hangs together it is instructive to examine situations in which it is falling apart. The availability of this detailed, day-to-day record provides an opportunity to find out just how such conditions develop. The first objective in this inquiry is to ascertain just how this particular unit became demoralized.

The Autonomy of Primary Groups

On learning of the disturbances in Company K, administrative officers at Fort Snelling were astonished and outraged. They had done the best they could under difficult circumstances for a useless lot of soldiers; compared with other Nisei the sacrifices asked of these men seemed minimal. Furthermore, even if their lot was not an enviable one, did they not care about the reputation of the Nisei as a whole? These officers reacted much as many administrators do when confronted by such disorders. They were confused and resentful. They punished the company commander. They scolded the men. They made some precipitous changes. When these remedial measures failed to restore order, they concluded that the men were hopeless.

But what else could these officials have done? Intelligent remedial measures are predicated on a comprehension of what is wrong. Responsible authorities cannot deal effectively with demoralization without an adequate explanation of it. It has already been noted that effectiveness of performance cannot be explained in terms of group structure. Even when regulations are explicitly stated and clearly understood, the manner of adherence is always voluntary; those who desire to excel may do so, and evasion of some kind is usually possible. Soldiers can be ordered to perform with enthusiasm, but it is generally possible to distinguish between an ostentatious display of ebullience and spontaneous, whole-hearted endeavor. Nor does quality of performance depend entirely on the skill of the personnel. Just as a symphony orchestra consisting of the finest musicians may perform poorly, teams composed of athletes acknowledged to be the most gifted sometimes "choke up" in a tight pennant race and give way to others who had not been considered seriously as contenders.

The second objective in this inquiry is to construct an explanation of the process of demoralization. What are some of the conditions under which the performance of a unit becomes consistently ineffective? What are the conditions under which a group breaks down under pressure? What are the conditions under which resistance and obstruction develop? What are the conditions under which disgruntled men mobilize to retaliate against their organization? We are dealing with complex phenomena, but we can take at least a few steps toward a better understanding of what is involved.

In their studies of morale sociologists have focused their attention on primary groups. Ever since publication of the study of the Hawthorne plant of the Western Electric Company, students of industrial organizations have been concerned with the informal norms that develop among those who work together in close contact. Careful observation of a room of fourteen workers engaged in wiring banks of telephone equipment revealed that there were organized patterns of behavior that differed from company regulations. Some workers were frequently helped by their colleagues; others were not. Games were played regularly at lunch time that included some workers but excluded others. A number of common understandings were shared: don't be a "rate buster" by working too quickly; don't be a "chiseler" by working too slowly; if one is a straw boss, be a "regular guy"; don't be a "squealer." The workers had their own conception of what constituted a "fair day's work," a production standard that fell below the level that management regarded as desirable and below what several of the workers were capable of doing. The faster workers simply stopped working earlier than the others; those who were slower were taunted to keep up (Roethlisberger and Dickson, 1939:379–548). Furthermore, pioneering studies of emotional climates that sometimes envelop such units have pointed to the importance of interpersonal relations and especially to the importance of local leadership (Lippitt, 1939; White and Lippitt, 1960).

One widely entertained hypothesis in military sociology is that high morale depends on the personal obligations that soldiers develop in primary groups (Ardant du Picq, 1921). In their analysis of the fighting effectiveness of the German army in World War II, Shils and Janowitz (1948) conclude that high morale in the *Wehrmacht* did not depend so much on dedication to Nazi ideology as on the extent to which its primary groups met the soldier's organic needs, offered him affection and esteem, supplied him with a sense of power, and adequately regulated his relations with authority. Whenever primary group life was disrupted —by separation, breaks in communication, loss of leadership, incompatible ethnic groups, diversity of age, or depletion of personnel—steadfast endeavor was more likely to collapse. Some elite units, such as the *Waffen SS*, were of course characterized by strong ideological commitments, but most German soldiers were more concerned with protecting their immediate associates and complying with their expectations. Although the authors cite other variables, variations in morale are explained mainly in terms of the capacity of primary groups to avoid disintegration. In their subsequent writings each author has elaborated on this position. Janowitz

(1959:65–73) accounts for the relative solidarity of primary groups in terms of such variables as weapon systems that require close cooperation, manner in which such units are formed, homogeneity of personnel, seriousness of threats faced, and local leadership. Shils (1957) has distinguished three kinds of primary groups, including one that is based on adherence to lofty goals. Several other studies have reiterated the importance of these small local units in morale (Marshall, 1947; Mandelbaum, 1952; George, 1971; Lang, 1972:71–77), and in this study an effort will be made to push this line of inquiry a bit further.

A primary group is a constellation of individuals who know one another on a personal basis, who see and treat one another as unique individuals. Common examples include the family, adolescent gangs, cliques of executives, small bands of outlaws, squads in military organizations, patients in a chronic ward of a hospital, or people who are imprisoned together. Such groups are generally small, and contacts are frequent and sustained for a considerable period of time; otherwise it would not be possible for the members to get to know one another's idiosyncrasies and problems, aspirations and interests. Members of primary groups generally partake together in a wide variety of transactions. Their interaction is not specialized; they exchange stories, become familiar with each other's foibles and virtues, learn when to joke and when to be serious, know what subjects to avoid when certain individuals are present. In doing so many different things together, they learn of many facets of one another's personalities; the more diversified the transactions, the more rounded the knowledge they have of each other. Members of such groups generally form a strong sense of mutual identification.

Even within the standardized structure of an army, the size and composition of primary groups vary, and the boundaries of coteries are often difficult to ascertain. Their perimeters may or may not coincide with the boundaries of organizational units. The borders of primary groups are set by the limits of sympathetic identification and effective communication. The composition of each coterie may change over time. There may be a falling out among friends, and newcomers may become more fully accepted as others become better acquainted with them. Furthermore, many primary groups overlap, some members being involved simultaneously in several. Nisei soldiers' contacts crisscrossed organizational boundaries. Most members of each platoon were acquainted with one another, but each man knew others outside it. Previous contacts had been established in the segregated communities on the Pacific Coast and in Hawaii, in the relocation centers, in schools, and in previous military assignments. The unit of analysis depends on the range of reciprocal obligations established by people who know one another as individuals.

Although the key attribute of primary groups is the reduction of social distance among its members, the extent to which personal reserve is relaxed varies from one group to another and from one individual to another within each group. The network of interpersonal relations that develops in each primary group is unique, and the degree of intimacy and mutual identification attained varies. To the extent that the members identify with one another, however, they become sensitive to each other's feelings. With increasing familiarity each person lowers his guard, becomes more candid, and reveals more of his inner secrets. A primary group, then, becomes a significant orbit of responsiveness. Precisely because they know one another so well, they can more readily appreciate each other's interests. Human beings are most responsive to the expectations they impute to those with whom they identify, for one can readily appreciate how a disappointed friend feels and is struck by feelings of guilt. Thus, the informal norms that develop in a primary group are very difficult for most of its members to violate (Cooley, 1909:23–50). It should be emphasized, however, that all members of such coteries do not necessarily like or approve of one another. But they can understand each other as human beings.

Since military organizations are so highly institutionalized, one might wonder if there is any room left for informal activities on a person-to-person basis. The responsibilities of those holding each position are explicitly specified, and the proper procedures for the performance of all required tasks are described in minute detail in manuals, directives, and specific orders. Commands are passed down a clearly defined hierarchy, and nothing is left to the judgment of those of lower rank other than the details of local implementation. Furthermore, many activities that would be left to the discretion of individuals elsewhere are specifically prescribed—such as hours of sleep, time for eating, circumstances for using the toilet, frequency of shaving, choice of underwear, and even the site of a toothbrush that is not in use. Penalties are prescribed for deviations and in most instances are immediately imposed. Yet there are many interstices between the regulations. Since soldiers are assigned to units on an impersonal basis, common membership in a squad is a matter of chance. Yet those who are thus thrown together realize that they must cooperate in many common transactions and quickly become acquainted. Furthermore, it is widely known that men who have endured such service together often get to know one another more intimately than they do members of their own family. They share hardships and inconveniences; they undergo many grueling experiences together and learn to admire one another's courage and endurance. Probably most important, one's own platoon is the only place where he is known as a specific human being—something other than a serial number and an entity addressed elsewhere as "soldier."

What is of decisive importance is that most primary groups operate independently of the larger contexts in which they are located. *Each member of any large organization or community tends to perceive his surroundings from the standpoint he shares with his immediate associates*, an outlook that becomes a filter through which he interprets a succession of situations. Each primary group develops its own perspectives and its own informal norms. Thus, the same event would be viewed in somewhat different ways by members of each primary group.

The interpretations that emerge in each primary group, whether or not they support organizational goals, develop *independently* of formal announcements and exhortations. All large organizations have at least two sets of communication channels. The formal channels are clearly delineated, and all orders and official announcements come through them. But each primary group has its own communication network, and the members evaluate any item of information they receive by pooling their own intellectual resources. If external compliance is required, they may pretend to accept the official position while actually questioning it. Official pronouncements are not necessarily rejected; on the contrary, they are usually accepted at face value. Men become suspicious only when a statement strikes them as being implausible. Only when they have some reason to distrust formal sources, when they impute foul motives to officials, do they become skeptical and seek information elsewhere. It is in such circumstances that rumors emerge.

All formal organizations—armies, business corporations, governmental agencies, large hospitals, universities—are made up of a multitude of interrelated and overlapping primary groups at all levels—cliques of high-ranking executives, friendship circles in middle management, coteries of clerical workers. Quality of performance can vary even in organizations with common customs and standard regulations and may fluctuate considerably over time within the same unit precisely because each primary group is to some extent autonomous. The mobilization of collective effort occurs largely on a person-to-person basis. Enthusiasm, reluctance, defiance—all emerge in personal contacts. The manner in which a situation is defined is often as much a reaction of the participants to one another as it is a reaction to the event itself. Whenever people are working together as a unit, the extent to which the participants are committed and the manner in which each contributes to the undertaking give the group as a whole its characteristic style (Hocking, 1941). As Barnard (1938:286) pointed out, formal structures establish the general pattern of activity, but informal structures determine its vitality. To understand what happens in a demoralized unit would require getting at its component primary groups to ascertain just how the participants view the situations in which they are involved.

Collection and Analysis of the Data

The record of Company K deviates markedly from that of other Nisei units in World War II. The most frequently used procedure in sociological research would call for matching the company personnel with a sample drawn from other Nisei units in terms of such categories as average age, level of educational attainment, original place of residence, average intelligence, religion, or proportion with parents still in relocation centers. This has not been done, for there is no reason to believe that such an analysis would be productive. These men were drawn from the same manpower pool as all other replacements for the 442nd Regimental Combat Team, and at the time of their induction they had shared the same anxieties and hopes as other Nisei. Furthermore, the men were capable of competent performance. All of them had completed successfully their basic training, some in companies winning commendations, and at times Company K itself operated with efficiency that matched or even surpassed that of other Nisei companies. Had a statistical matching yielded any differences greater than chance, an effort would have been made to explain them in terms of "causal factors" presumably responsible for low morale. Explanations of this kind are speculative and have already proved inadequate. Thus, a different approach has been attempted.

A biologist investigating an unfamiliar organism begins by collecting specimens, which he examines with great care. With increasing familiarity he constructs typologies and natural histories, and eventually he formulates hypotheses that apply to broad classes of phenomena. Case studies are especially useful during the early stages of any inquiry, and sociology—still a pioneering discipline—can benefit from this procedure. This study differs from many sociological investigations in that it is more inductive than deductive in design. The chronicle of Company K will be taken as a specimen to be examined in great detail in order to formulate testable hypotheses. In this manner it is hoped that a contribution can be made to the development of a more reliable, empirically grounded theory (Glaser and Strauss, 1967).

The central problem in this inquiry is to ascertain the manner in which a group becomes demoralized. This study differs from most other investigations of primary groups in that attention has been focused on the feelings and beliefs of the participants. That human beings act in terms of their *definition of the situation* is a principle that has become axiomatic for most sociologists. Yet there has been a tendency for them in their empirical research to overlook or gloss over the intermediate processes between the occurrence of events and the overt reactions that follow. The formation of collective interpretations, though often described, is seldom taken systematically into account in the formulation of hypotheses. But it is not enough to know what actually happens. It is necessary to get a record of how things appear from the standpoint of the participants themselves, especially when their outlook differs from the official view. Emphasis has been placed on the manner in which each event is defined and evaluated in the spontaneous communication that takes place among those partaking in the transaction. (a) How do primary groups define the various tasks in which they are involved—whether in performing routine duties, in protesting, or in organizing violence? (b) How do the participants identify themselves—as members of a military unit, in terms of ethnic identity, or as an aggrieved party seeking vengeance? Who constitutes the audience for whom the performance is staged? (c) What regularities are there in the information processing that occurs on a person-to-person basis? What are the sources of information? How do such groups achieve consensus? What happens in situations in which consensus is not attained? In short, how do primary groups mobilize for action—whether in support of organizational goals or against them?

This inquiry also differs from most sociological case studies, which describe the structure of communities and organizations, in that emphasis has been placed on the *temporal* dimension. The formation of perspectives is a cumulative process. It is doubtful that the demoralization of any unit can be understood apart from its historical development. Disturbances cannot be explained adequately in terms of simple "causes." Most collective transactions are complex, and they are worked out in a group that has a unique history. People who are in sustained association usually develop a distinct point of view. Certain experiences acquire special desired topics. Many of the discussions in the barracks were recorded verbatim, but accounts of what happened elsewhere frequently had to be reconstructed later when opportunities for note taking were more favorable. No doubt something was lost in lapses of memory. Most of the enlisted men involved knew that a sociologist among them was taking notes preparatory to "writing a book." Apparently this explanation was satisfactory, for they cooperated in providing leads, vital information, and a protective shield against inquisitive officers. As one of the accepted members of the unit being studied, it was possible for the observer to get candid expressions normally kept from outsiders and to grasp nuances of meaning that were peculiar to the group. As the study proceeded, observation was facilitated by the cooperation of increasing numbers, and during the latter phases of the inquiry several volunteered to serve as regular informants.

Sociologists interested in disorders have been hampered by the fact that most of their data have come from ex post facto inquiries. After a disturbance has occurred, they interview the various parties involved—officials, participants, observers—and attempt to reconstruct the event on the basis of their testimony. Such depositions are of questionable value, however, not so much because of deliberate fabrication but because of the selective character of perception, memory, and reporting. Events are complicated, and no one can possibly remember in detail everything that happened, especially in emotionally charged situations. When questioned, one reports what enables him to "make sense" of what took place, even when the tendency toward self-justification is kept to a minimum. Other items that may be remembered are not mentioned; they seem irrelevant, for they do not fit into any plausible explanatory scheme. It is not surprising, therefore, that discrepant versions are given by different observers whose honesty is beyond question. One unusual feature of this study is that it does not rest upon such retrospective testimony. The disorders described here were not anticipated; they simply took place in a company that was already under observation by a sociologist collecting data for another inquiry—an analysis of rumor that has been published elsewhere (Shibutani, 1966). After each crisis erupted it was possible to check back over field notes to see just how the difficulty had developed from day to day. The basic data consist of records of direct observations.

Since the information available to enlisted men was limited, these materials were later supplemented by other sources. The collection of certain types of data customarily included in inquiries of this kind—such as the exact number and disposition of personnel—was proscribed by war-

2. «» the challenge to nisei loyalty

The attack on Pearl Harbor came like a thunderbolt from a clear and cloudless sky. Though all Americans were stunned, nowhere was the impact more jarring than among persons of Japanese ancestry living in the United States. For them the assault initiated a period of sustained insecurity, misunderstandings, persecution, and conflicting loyalties. Because of the despicable nature of the blow, anything that could be identified with Japan, even remotely, was denounced as having some link with the perfidy. Especially on the Pacific Coast "Jap baiting" became a safe and profitable pastime, and innocent groups were suddenly discovered to be spy rings or schools for saboteurs. Individuals whose personal integrity had never before been questioned found it necessary to demonstrate their worthiness and to defend themselves against fellow citizens with whom they had previously lived in harmony.

Their dismal fate was complicated by their already having been under some suspicion, much like so many minority groups in Europe and the Middle East. One of the consequences of modern nationalism has been the formation of "problem minorities." Ethnic groups, distinguished from the surrounding populace in culture and sometimes in appearance, have frequently become targets of popular distrust and agitation. Segregated into the least desirable areas and denied opportunities for full participation in community life, they have regarded themselves as objects of unfair treatment and have developed deep resentments. On occasion such groups have appealed to their "ethnic brothers" across national boundaries for protection. This tendency has led to accusations of nurturing sympathetic ties with a foreign power; in the event of war they would be unreliable, even dangerous. In the past, difficulties of this kind have resulted in diplomatic incidents, massacres, population transfers, and attempts to establish international tribunals for adjudicating differences. Those familiar with the problem anticipated the worst.

At the outbreak of the war most of the Japanese in the United States were concentrated on the Pacific Coast and in Hawaii. Although the two settings were quite different in many respects, in both their position was vulnerable, and they were a problem minority in the traditional sense.

A Problem Minority in Two Settings

Until 1898, when Hawaii was annexed by the United States, immigration was largely a matter of meeting the labor needs of sugar plantations. Agents went to remote corners of the world; and from 1852, when 293 Chinese were brought to Hawaii, until 1930 approximately 400,000 workers were imported. Each of the minority groups—Chinese, Filipinos, Japanese, Portuguese, as well as the native Hawaiians—started by working on plantations. The influx of Japanese laborers began before 1880, and at the time of the annexation they made up more than half the population of the islands. But plantation life was harsh. The most recent arrivals got the worst housing and lowest wages, and the *lunas*, or overseers, were severe. Such indignities soon led to departures to other pursuits. In 1902 Japanese made up 73.5 percent of the plantation employees; in 1922, 38.3 percent; and by 1932, only 18.8 percent. Most of the early arrivals saw themselves as transients; they tried to save money, and more than half the 180,000 immigrants did eventually return home. But the rest stayed on, either remaining on the plantations, drifting to Honolulu or Hilo, or moving on to the Pacific Coast (Lind, 1938:188–243; Lind, 1955:72–73; Fuchs, 1961:25–26, 113–21).

Although the Japanese had been received cordially in the beginning, the welcome cooled as their numbers multiplied; and eventually opposition to further immigration developed. Growers were displeased that the Japanese did not adjust quietly but responded to mistreatment with strikes and riots. A major strike in 1920, when the Filipinos and Japanese on six Oahu plantations joined forces to demand $1.25 for an eight-hour work day, was especially upsetting. Even after it was settled, the nagging, unpleasant feeling persisted that the Japanese had not responded correctly to paternalism, and they were viewed as secretive and sneaky. However, the tradition of hospitality to all ethnic groups kept these objections underground. The ancient Hawaiians had loved to travel about the islands, and strangers had traditionally been greeted with cordiality and respect. The spirit of *aloha* had survived even the viciousness of some of the newcomers, and those who violated this ideal were shamed, punished, and condemned as traitors (Adams, 1934; Kuykendall and Day, 1961:10). Thus, no overt discrimination in salutation, employment, sex relations, or use of public facilities ever developed. Although many of the *Haole*—Hawaiian for "stranger"—adopted the doctrine of white supremacy, it never received legislative recognition. Regardless of private feelings, the public code required an avowal of equality and good will. Thus, anti-Japanese sentiments were muted by Hawaiian ideology.

On the Pacific Coast the Japanese did not become an object of public concern until about 1900. They began entering in large numbers about 1882, when the exclusion of Chinese had created a temporary void in the supply of farm labor. By 1900 there were 24,326 in the continental United States. Regardless of their occupation in Japan, most immigrants started as common laborers—domestics or workers on railroads and in canneries, lumber mills, mines, and farms. Most of the early immigrants were young men under 35. They were especially welcomed as agricultural workers, for they were experienced in intensive cultivation. They were initially regarded as efficient, reliable, willing to accept low wages, and available in large numbers under gang bosses.

From 1901 to 1910, 54,929 immigrants were admitted from Japan; in addition an estimated 37,000 moved from Hawaii. The group thus became more visible. As long as the newcomers remained docile laborers, all was well. When they began to compete effectively, however, objections arose. Since they were organized in gangs, the foremen were able to bargain as spokesmen for an entire unit; by striking just at harvest time they were able to raise their wages above those commanded by white workers. In tenant farming they paid higher rents than other tenants; but their profits were higher, for they got a higher yield per acre and at harvest time could get ample Japanese workers. This practice was viewed as unfair competition. After nearly a half century of antipathy, abuse, and violence, the Chinese had either returned to China or had withdrawn into segregated enclaves. Now the vocal agitation against the "Yellow Peril"—the danger of being engulfed by coolies who reproduced at a phenomenal rate—was directed against the Japanese. In 1905 the Asiatic Exclusion League was formed, its purpose to preserve North America for Americans by preventing Asiatic immigration. It urged school segregation and economic boycott, and by 1908 the organization claimed a membership of 110,000 and had 238 affiliated bodies, most of them labor unions. Politicians, ever responsive to popular demands, joined in the fray. But opposition to the exclusionist movement developed among churchmen, educators, fruit growers who still needed labor, and businessmen who feared the loss of Japanese trade.

The Japanese were regarded as a separate "race," unalterably different from Caucasians in moral and intellectual qualities. The most common charges were unscrupulous competition, moral degradation, and subversive intent. They were polite and unoffending on the surface, but this only concealed their inner emotions; hence, one could never know what they were thinking. Their inscrutable Oriental cunning made them tricky, dishonest, and able to take unfair advantage of others. In trade they engaged in cut-throat competition. They were immoral; they corrupted young children and were a menace to white women. They were also clannish and unassimilable. It was an eternal law of nature that white people could not assimilate others without corrupting their own civilization. The children of Asians, though American citizens, could not be Americanized; they were by nature unassimilable. Because of their low standard of living, high birth rate, and vile habits, the Japanese were indigestible. After Japan's victory over Russia in 1904 fears arose that the nation had further territorial aspirations. The local Japanese were but the spearhead of an invasion; many were spies or soldiers in disguise. Their sole loyalty was to the emperor of Japan. The masses in Asia were waiting for the vanguard to do its work so that they could overrun the Pacific Coast. Such popular beliefs crystallized as they were reiterated in political charges, novels, newspaper articles, and motion pictures (U.S. House of Representatives, 1942:59–90; tenBroek et al., 1954:11–32; Daniels, 1962).

As agitation on the Pacific Coast intensified, relations between the United States and Japan became strained, and the demands of Californians attracted national attention. The earthquake and fire in San Francisco in April 1906 brought the controversy to a head. In the looting and violence that followed the breakdown of city administration, the Japanese suffered disproportionately from lack of police protection. The fire destroyed most of the schools. When planning their reconstruction the San Francisco Board of Education passed a resolution to segregate all school children of Asian descent—93 out of 29,000. The reason given was age difference. A few of the Japanese attending public schools were somewhat older, trying to get an American education while working as houseboys; the announced aim of the proposal was to protect American girls from Japanese adults. Politicians seeking union support encouraged the move. A formal protest was lodged by the Japanese ambassador, and President Theodore Roosevelt had to intervene. He summoned members of the school board to Washington, D.C. After considerable deliberation the board members agreed to rescind the order, and the federal government dropped its suit to test the constitutionality of the California school law. The government also agreed to prevent any more Japanese from entering the country by way of Mexico, Canada, or Hawaii and to restrict direct immigration from Japan. In March 1907 the indirect immigration of holders of Japanese passports was halted, and in 1908 the Gentlemen's Agreement was negotiated. The Japanese government agreed thereby to issue passports only to nonlaborers, to those already residents of the United States, to members of their immediate families, and to those intending to assume control of farm enterprises already acquired (tenBroek et al., 1954:11–67; Griswold, 1938:333–79). Thus, the influx of laborers came to a halt.

Since the Gentlemen's Agreement did not cut off all immigration, agitation continued. Immigrants who had left their families in Japan sent for them, and bachelors made hurried trips home in search of wives. Those who could not afford the voyage found an exchange of photographs more economical; their families in Japan located suitable mates, and proxy marriages were performed. Some indication of the number of Japanese women entering the country is given by the changing sex ratio in the group: in 1900 there were 2,370 males for every 100 females; by 1910 the figure had dropped to 694. Agitation focused on "picture brides"; this practice was further evidence of immorality—importing women for indecent purposes. Furthermore, the women worked in the fields, making competition harder for others; they bred children, reproducing at three times the rate for white people; they encouraged laborers to settle down so that they were no longer available as migratory workers.[1] During the 1910 campaign in California the platforms of the Republican, Democratic, and Socialist parties all contained exclusionist planks. In 1913 the California legislature passed an Alien Land Law. Land ownership by aliens ineligible for citizenship was prohibited; tenure of land leases was limited to three years; land already owned could be retained but could not be bequeathed to heirs. World War I brought

1. In 1920 the crude birth rate for the Japanese was 67.6 (per 1,000 population), four times that for Caucasians; but by 1940 this had declined to 15.3. Considering the peculiar age-sex distribution this was not unusual (Sabagh and Thomas, 1945).

a temporary abatement of hostility, and Japanese farmers prospered. Soon afterward, however, agitation was renewed. War workers released from factories were disturbed by Japanese control of land, and returning soldiers resented their economic security. In 1919 the Asiatic Exclusion League was reorganized into the California Joint Immigration Committee. During the following year an initiative to remove the right of Japanese to lease land, an amendment to the 1913 land law, passed by 688,438 to 22,086 (McGovney, 1947). The U.S. Supreme Court ruled in the Ozawa case in 1922 that Japanese could not be naturalized, and in 1924 a comprehensive law cut off virtually all immigration from Asia.

With the passage of restrictive legislation hostility abated, and a period of accommodation followed. The Japanese became a small minority group somewhere near the bottom of the social scale. Most of them were concentrated in the three coastal states; in 1940, 88.5 percent (93,717) lived in California. They were set apart from the remainder of the communities in which they lived by various institutional arrangements. Occupational choice was limited. Farming was the principal means of livelihood. According to the 1940 census 43 percent of the gainfully employed (22,027) were in farm work—7,000 as operators or managers and 13,000 as laborers. About 24 percent (11,472) were in wholesale or retail trade—mostly in the distribution of agricultural products. Approximately 17 percent (8,336) were in various service industries (U.S. House of Representatives, 1942:101–15). In the cities residential segregation was maintained by custom and by housing covenants. Most of the urban dwellers were concentrated in slum areas, although a few secondary areas of settlement developed in working class neighborhoods. In most western states intermarriage was prohibited by law. Such statutes were unnecessary, for intimate contacts of the sort likely to lead to marriage were uncommon.

After the Meiji restoration in 1868 the Japanese government had adopted a program of rapid industrialization. As the merit system replaced nepotism as the basis for advancement, great emphasis was placed on formal education. The immigrants also recognized the value of education, and they took advantage of the opportunities available to their children. Considerable pressure was placed on Nisei to attend school as long as possible and to take their studies seriously; those who did well were honored by the community. By 1940 the educational level attained by Nisei was considerably higher than the national average. Although a few small communities maintained segregated schools, in general Asian children on the mainland were assigned to public schools on the same basis as all others—by place of residence. While some of the schools near slum areas had a disproportionate number of Orientals, those who lived in outlying areas attended the schools nearest their homes. On the whole the Nisei fared well. Outstanding students received honors; some were elected to student-body offices; gifted athletes won acclaim. Many established close friendships with non-Japanese. English became the native tongue of most Nisei, and the younger generation thereby attained easy access to the mass media. They acquired a world view that their parents often found difficult to comprehend. Furthermore, most Nisei were trained during the height of the Americanization movement and were thoroughly indoctrinated with the ideology of the "melting pot," that Americans consisted of descendants of immigrants from all over the world. Even though there was some confusion concerning ethnic identity, most Nisei came to conceive of themselves as Americans.

Reared in the moral atmosphere of the Japanese family, most Nisei also learned Japanese values. Among other sources of influence were the Buddhist church, Japanese language schools, and Japanese motion pictures. Activities of the Buddhist church were conducted almost exclusively in Japanese. Most Nisei living in sizable settlements also enrolled in special language schools. Since these classes were held after regular school hours, attendance was marked by a singular lack of enthusiasm. Very few mastered the Japanese language, but they were exposed to values such as filial piety, thoughtfulness, self-reliance, obligation, and other virtues believed to be distinctively Japanese. They were taught the folklore of ancient Japan, and these values were further inculcated by repeated examples and anecdotes. Some exposure to the ideals of *bushido*, the code of honor of the samurai, was unavoidable. Moral precepts, tales of fortitude, plays, songs, novels, and movies frequently took their themes from stories of the medieval warriors; and ideals such as courage, benevolence, courtesy, honor, and self-control were constantly emphasized as the characteristics of *real* Japanese. The keystone of the philosophy—the concept of absolute loyalty to one's lord, whoever he might be—was backed by accounts of supreme sacrifices, including killing one's own parents, in living up to one's obligations (Nitobe, 1913:145–53). Parents subscribed to Japanese magazines, and many communities had their own newspapers as well as weekly showings of imported movies, many of them glorifying the samurai and his code. Though Americans, the Nisei were also conscious of being part of a proud "race"—one characterized by high ideals. But their understanding of Japanese values was both imperfect and stereotyped.

On the Pacific Coast many institutional barriers prevented Nisei from participating fully in community life. Those who were older entered the job market just when the nation was emerging from a severe depression, when competition for desirable positions was keen. Few college graduates were able to find employment in their field of specialization, and it was not uncommon for them to work as farm laborers, gardeners, or fruit-stand attendants. Although a few outstanding individuals managed to win jobs for which they were trained, they were exceptions. The extent to which opportunities were limited is revealed by 1940 census figures. In San Francisco, for example, 48 percent of the gainfully employed men and 69 percent of the women were in service industries; these percentages were somewhat lower in the Pacific Northwest. In California there were only 960 professional workers (2.38 percent of the gainfully employed) and 199 semiprofessional workers (.49 percent). The main source of white collar positions was the civil service. Although discrimination in the use of public facilities was illegal in most western states, Nisei had to learn to "pick their places." Few actually experienced rejection, for they tended to avoid situations in which some unpleasantness might occur. But there were many rumors of the fate of those who had dared to challenge the color line. Although little hostility was apparent in daily transactions, many West Coast Nisei felt themselves to be the objects of unfair treatment. As they approached adulthood in large numbers, they faced many of the same barriers that had confronted their parents.

Separated from other Americans and unable to speak Japanese well enough to communicate effectively with their parents, the Nisei in urban ghettoes formed a society of their own. Most of their meaningful contacts were confined to their own group; they lived in their own limited social world and pursued careers within it. Although most were bilingual, they relied primarily on English in addressing one another. But their speech was often interspersed with Japanese words, usually anglicized in pronunciation (Spencer, 1950). Their social institutions were more American than Japanese; in fact, their social world was a small-scale replica of American society. They accepted the standards presented over the mass media but used these behavior patterns primarily among themselves; few ventured forth into the surrounding world. Their organizations were patterned after similar groups in the larger community, which enabled them to practice their newly acquired roles without exposing themselves to humiliation or to direct competition with more experienced Americans (Broom and Kitsuse, 1955). They aspired to

American goals: charm, romance, success. They yearned for a life comparable to that of white people: a college education, romantic marriage, a lucrative occupation, and respect. In dress, language, demeanor, and life style the Nisei on the Pacific Coast were more American than Japanese. Though excluded from the mainstream of American life, they were rapidly becoming acculturated.

The status system in Hawaii was quite different. The Japanese were a large minority group in a situation resembling a European colony. The highest status, prestige, and power were enjoyed by a small Haole elite. This group consisted largely of people of northern European descent. They were the merchant and professional class; they spoke English that was free of any accent; and they married only within their own circle. Although they made up only about 5 percent of the total population, they controlled the government, land, labor, and business. Consciousness of ethnic identity was pronounced, and the people conceived of themselves primarily as members of one of six categories. Each ethnic group developed a distinctive way of life, and the fate of each individual depended to a considerable extent on his classification. The goal of the Haole was to maintain political control of the islands; the native Hawaiians aimed to recapture the idyllic existence of the past; the Portuguese endeavored to be included among the Haole; the Chinese struggled for economic independence; the Japanese strained to be accepted; and the Filipinos were primarily concerned with earning enough to return home. Thus, a small and powerful oligarchy of Haoles—believing they were destined to rule for the good of the islands—controlled large masses of colored minorities. They were the patrons, advisers, innovators. Although this oligarchy was more beneficent and charitable than most other elites, it ruled nonetheless (Fuchs, 1961:42–149, 152–259).

Honolulu was a western-style city, and most Haoles lived there in self-segregated areas. On the plantations the influence of other ethnic groups was more noticeable—in the games played, language spoken, gods worshipped, and other customs. Even there, however, there was a gap between the small number of salaried employees—clerical, administrative, and supervisory—and the large body of laborers paid on an hourly basis. The Haoles were for the most part college-bred, and their pastimes were middle class—cocktails before dinner, social dancing, conversations about current books and international affairs. Opportunities for upward mobility for the others were limited; the higher ranking jobs required education, and for some time very few were eligible. The Haoles on plantations received preferential treatment and higher wages even for the same work performed by others. If a Japanese and a Haole were equally qualified, the latter received the job. Honolulu firms recruited managerial personnel from the mainland rather than upgrade minority group assistants. In any case after a lifetime in the lower ranks Orientals, even if they had technical knowledge, lacked the social experience and poise necessary to deal with assurance with Haoles on an equal basis. The tradition of friendliness and hospitality was maintained, but Haoles discouraged intimacy with others. Those who habitually associated with outsiders jeopardized their own status (Norbeck, 1959:106; Fuchs, 1961:55).

After 1896 the Japanese constituted numerically the largest ethnic group; by 1940 they made up 37.2 percent of the total population of the islands. As increasing numbers left the plantations, many went into independent farming; and by 1930, 70 percent of the farms were Japanese. Increasing numbers also moved into cities. In 1900 they made up only 15.7 percent of the population of Honolulu; by 1910, 30 percent; and by 1940, about half. Another large settlement developed in Hilo, and they made up a disproportionate share in many smaller towns. Heavy concentrations formed on Hawaii (the "big island"), Kauai, and Maui. Upward mobility was marked. Although job ceilings existed and the more desirable positions were harder to get, opportunities were not closed. In 1896, 87.5 percent of the gainfully employed men were laborers; by 1920 this figure had dropped to 35.9 percent. Of the minority groups the Japanese held the highest percentage of clerical positions. On the plantations they made up the core of the skilled-labor class and were second only to the Portuguese as lunas (Lind, 1955:50–51, 74; Fuchs, 1961:122–25). Except during the depression, when competition intensified, they had been welcomed in skilled and semiskilled trades outside the plantations. By 1940 the Japanese made up more than half the owners of small retail stores and restaurants and over half the craftsmen—painters, carpenters, electricians, and plumbers. About 15 percent of the gainfully employed were in professional, proprietary, or management occupations (Lind, 1946:16–18; MacDonald, 1946:184–87). Thus, economically they fared far better than the Japanese on the Pacific Coast.

What is called "pidgin English" became the common language of all the minority groups in Hawaii. The term is used loosely to refer to the broken, somewhat mixed English dialect of old natives and immigrant coolies as well as the slightly dialectical variants of English spoken by the young. It ranges from early forms, used only on outlying islands, to semistandard English in Honolulu. Although exposure to the mass media made people conscious of standard English, pidgin remained the language of the laboring classes.[2]

The Haole oligarchy believed that wealth and power brought with them certain obligations. Faced with a large Asian population, the oligarchy decided to Americanize it. What was wanted was for the children to speak correct English, play baseball, eat hot dogs, and become Christians. Since missionary zeal for education was unbounded, public schools were soon established. Teachers were imported from the mainland. They taught patriotism, equality, and freedom. They were the principal transmitters of American culture, but they concentrated on values that in many ways were not suited to life on the islands. Although the Hawaiians, Portuguese, and Filipinos showed little concern for intellectual achievement, the Chinese and Japanese jumped at the opportunity. Most Nisei were educated in public schools. McKinley High School, which from 1910 to 1920 was the only public high school on Oahu, was sometimes referred to as "Tokyo High."

The public school became the meeting ground for all children but those of the Haole. One of the main concerns of Haole parents was the welfare and future of their offspring, who were surrounded by children with different, if not undesirable, ways. While subscribing to the ideal that everyone should be friendly, they feared their children's learning pidgin from their playmates and from some of their teachers. They were also afraid of miscegenation. Most Haole youngsters were sent to Punahou, to some other private school, or to the mainland. Punahou perpetuated the ways of the elite class. In 1896, when a Chinese applied for admission, the trustees, unwilling to adopt a racist policy, instituted a new rule: no pupil would be admitted who was "incapable of using the English language as a medium of instruction." Thus, inability to pass a language examination was cited as the reason for rejecting most Asian applicants (Fuchs, 1961:262–98). Beginning in 1924 the Department of Public Education established several "English Standard" schools. They were open to children of any ethnic group, provided they could pass an English proficiency test, and minorities were urged to attend them. Although the proportion of non-Haole accepted increased, resentment against the system continued; for such schools were seen as a symbol of Haole snobbishness. On the eve of Pearl Harbor, only 21 percent of the students in these schools were Oriental. Thus, in spite of the publicly

2. In linguistics "pidgin" designates a marginal and secondary language that is not the native tongue of any of the users. Hawaiian pidgin is actually a dialect of English (Reinecke and Tokimasa, 1934; Reinecke, 1938; Carr, 1960).

proclaimed philosophy of equality, the schools in Hawaii were in fact more segregated than those on the Pacific Coast.

Although they were taught standard English in the schools, youngsters from minority groups used it only during instruction. Once they were among themselves, they reverted to pidgin. Even those who became well educated continued to use it. Many professional men confessed that they were never at ease with standard English. It was like a foreign language; it did not seem natural, for they had to plan carefully what they were going to say. Thus, as the Nisei became acculturated to Hawaiian life, their native tongue became pidgin. Those who used standard English outside of school were scorned as snobbish. Pidgin symbolized equality among those who used it—the ethnic minorities.

Although social distance was maintained from the Haole, among the minorities ethnic affiliations were overtly disregarded. The atmosphere was one of mutual respect across ethnic lines, but intimacy was discouraged. The doctrine of equality minimized differential treatment in public, but much ill will existed beneath the amiable surface. As Nisei moved up the social scale faster than the other minorities, they were criticized for maintaining exclusiveness, much like the Haole. Although the Nisei acknowledged the superior status of Haoles and looked up to them, they regarded themselves as superior to all others. They developed a strong in-group feeling and remained more among themselves. Japanese language schools were singled out for condemnation. By 1919 there were 163 such schools with 20,000 students. The peak attendance came in 1934, when 85 percent of all those eligible were enrolled; by 1941 this figure had dropped to 74.5 percent (Lind, 1946:24; Kuykendall and Day, 1961:244). Although intermarriage and miscegenation had long been accepted in Hawaii, the Japanese intermarried less than any other immigrant group. Marriage to an outsider brought disgrace upon a Japanese family, and those who did so sometimes had to leave their community (Adams, 1937:160–73; Lind, 1946:20).

By 1940 the acculturation of the Nisei both on the Pacific Coast and in Hawaii had proceeded to the point where intergenerational conflicts were becoming commonplace. The generation gap had a demographic basis; the immigrant generation, called Issei, was more sharply distinguished from their children than in most populations. Almost half the men had migrated before 1907 and had married late; large numbers of women of child-bearing age had arrived after 1900, and their immigration had been cut off in 1924. By 1940 the median age for Issei men was 53; for women it was 45. But the median age for the Nisei was 15

(Thomas, 1952:132). Most Nisei could not speak Japanese well enough to communicate with their parents beyond a very superficial level. Furthermore, the two generations were regularly exposed to different sets of communication channels and thus developed quite different outlooks. Issei views were shaped largely by Japanese language newspapers, which relied heavily on releases of the Domei News Agency. Their children patronized the American press and radio. Many of the strictures of Japanese culture were seen by most Nisei as oppressive. The few who were obedient and accepted quietly what their parents demanded were praised as exemplary by their elders, but they were held in contempt by their fellow Nisei. Many Japanese institutions, especially those involving courtship and marriage, were rejected as "old fashioned." In Japan marriage was a family matter; the prime concern was to ensure the perpetuation and advancement of the family. The personal preferences of the individuals involved were a secondary consideration, and partners were chosen by collective decision. Most Nisei, embracing the American conception of marriage for romantic love, rebelled. Many Issei felt that their children behaved in ways that were scandalous, sometimes even sacrilegious. Some even sent their children to Japan to be educated in a more acceptable manner, and this practice created a third category of Japanese-Americans—the Kibei.

Japan's increasing military power alienated American public opinion, just as the treatment of its nationals in the United States was regarded by the Japanese government as insulting. International tension increased as the Japanese invaded Manchuria, withdrew from the League of Nations, and bombed the American gunboat *Panay*. When Japan joined the Rome-Berlin Axis in 1940, American citizens living there were ordered home. As war between the two nations appeared increasingly imminent, the question of Nisei loyalty became a matter of public concern. In 1940 there were 158,000 persons of Japanese ancestry in Hawaii and 127,000 on the mainland. Of this number 205,000 were American citizens, but were they reliable? Publicity given to the Sudeten Germans who had collaborated with the Nazis and to the "fifth column" in Norway made Americans increasingly sensitive to the perils of subversive activities by enemy agents in their midst. The House Un-American Activities Committee periodically publicized nefarious plots and spy rings, and this created further apprehension. In Hawaii the problem was especially acute, for three out of every ten voters were of Japanese ancestry. Did they see the islands as an American territory or as a Japanese province? Haole who had recently arrived from the mainland, called Malihini, were concerned more than the old-time Kamaaina; and military and naval officers stationed there were especially distrustful of the Japanese. Rumors developed of seemingly harmless Japanese being exposed as naval officers, of Japanese tourists photographing vital installations, of deliberate efforts to purchase homes near Pearl Harbor and Schofield Barracks, and of Nisei enlisting in the U.S. Army to spy for the emperor. Many Nisei were excluded from lucrative defense jobs for which they had technical qualifications. Kibei were under even greater suspicion. Officials in investigative agencies were genuinely troubled. How would these people act in the event of war? Would they engage in sabotage? Would they supply vital information to the enemy? Since the Japanese settlements were isolated, they seemed mysterious and impenetrable. Individual Nisei whom they knew might be trustworthy, but what of the others?

The Japanese were well aware of these suspicions and of the charges being made against them. But as the danger of war increased, they lacked a well-defined orientation. Many tried to ignore the question or were indecisive. Some Issei were definitely loyal to Japan. They attributed their status, which they regarded as more favorable than that of Chinese or Negroes, to their being protected by the Japanese government. They had supported the Japanese war effort in China by making donations, by collecting tinfoil, and by filling comfort bags for Japanese soldiers. By 1940, however, one-fifth of the Issei had been residents for over thirty years. Most of them had spent virtually all their adult lives carving out a place for themselves and their families in the United States and had developed a sense of belonging there. While some still looked forward to retirement in Japan, most had come to see America as their home. Their children were Americans, and they would encounter many difficulties living in Japan. Officials of the Japanese government were also inconsistent on the question. Some identified with the emigrants and protested their mistreatment. Others made no attempt to bid for the loyalty of the emigrants and noted explicitly that the Nisei were Americans who owed their loyalty to their country. They argued that Nisei should prove themselves as Americans to uphold the honor of the Japanese "race." Since most emigrants were of lowly origin in Japan, other officials were barely able to conceal their contempt for the *imin* and their offspring.

Most Nisei were less equivocal. They could see no future for themselves in Japan. What they had learned of life there from their parents, the Kibei, and from Japanese movies was not attractive. It was too harsh and restrictive. In spite of their resentments they could not see living in Japan as a feasible alternative or an escape from their problems. Thus, most Nisei were indifferent to Japanese military successes and rejected the notion of casting their lot with Japan. Organizations professing to represent all Nisei, though actually made up of successful businessmen and professionals, constantly reiterated their support for American ideals. The New American Conference in Hawaii and the Japanese American Citizens League on the Pacific Coast annually adopted high-sounding resolutions affirming their undying loyalty to America. Such expressions impressed neither the Nisei nor other Americans, but no one bothered to challenge them.

Manning the Defenses of Hawaii

Immediately after the attack on Pearl Harbor, life on the Hawaiian Islands was thrown into utter confusion. No one knew just what had happened or how to act. When commercial radio stations suspended operations, so that their signals could not be used to guide further attacks, residents were deprived of their last established source of reliable information. Groping desperately for news, they became easy victims of rumors: Enemy transports had been sighted; Japanese troops were landing on various beaches; all the other islands had been taken; Oahu was surrounded by enemy submarines; Japanese paratroopers had landed in the hills. Police and military authorities were besieged with frantic calls reporting signal lights, homing pigeons, and short-wave messages. Those who kept tuned to the police radio were exposed to reports that officers were instructed to check, which added to the pandemonium. The most widespread rumors were that a McKinley High School ring had been found on the body of an enemy pilot shot down during the attack; that Japanese plantation workers had cut large arrows in the cane fields to guide planes toward Pearl Harbor; that automobiles driven by traitors had blocked the narrow road from Honolulu to Pearl Harbor; that a message giving the precise time of the attack had been sent to the eagerly waiting local Japanese by a cryptic advertisement in a Honolulu newspaper on December 6; that many Japanese establishments had held open house that night, serving extra liquor to incapacitate military personnel; that Nisei in milk trucks had opened fire with machine guns on the defenders of Hickam Field; that the local Japanese had poisoned the water supply; that some Issei dressed in military uniform were standing by for orders to take over the government (Lind, 1946:38–47; Allen, 1950: 47–56; Lord, 1957). Thus, the catastrophe at Pearl Harbor was initially accounted for in terms of treason and sabotage by the local Japanese. This explanation was reinforced by an official announcement that two men on the tiny island of Mihau had in fact aided a Japanese pilot who had crash-landed there and by Secretary of the Navy Knox, who at a press conference following a brief visit ascribed the disaster to the "most effective fifth column work that has come out of this war except in Norway."

The Nisei and the resident Japanese stood accused of disloyalty. They were acutely conscious of their tenuous position. They realized that outsiders knew little about them, and they could well understand their fear and hostility. But there was little they could do to disprove the allegations. They underwent a sudden and drastic change of status; they were cast down to the lowest rung of the multiethnic ladder and suffered in addition the distasteful identification with a dangerous and hated enemy. Their first reaction was fear—fear of internment, fear of mob violence, fear of losing their livelihood. They felt like criminals awaiting execution. They withdrew and stayed home. Lost and bewildered, they lived from day to day, not knowing which of hundreds of people might turn on them. They felt that they were at the mercy of those who hated them. Individuals already resentful of Japanese gave vent to their feelings without fear of retribution. Haoles, Chinese, Hawaiians, Filipinos, and Koreans who had long harbored antipathies now turned against them. Even ministers, school teachers, and social workers expressed doubts about Nisei loyalty. Although many friendships survived the blitz, others wavered in their trust; sympathy and regard for personal friends contended with fear and uncertainty. But most of the statements vilifying the local Japanese were made privately, not in newspapers or over the radio (Lind, 1946:56–61). The general philosophy of ethnic tolerance, reinforced by police and military authority, was enough to maintain order. There were only sporadic cases of violence, mostly by Filipinos outraged by news of atrocities by the Japanese army in the Philippines.

The bombing changed overnight the way of life of the islands. Martial law was declared, and Hawaii's ruling oligarchy was replaced by the military. For the first six months the threat of possible invasion was too acute to permit much attention to anything except the pressing task of rebuilding defenses as quickly as possible. Every effort was expended to build up an island bastion. Damage was quickly repaired; industries were converted to war work; people flocked to volunteer for civil defense work, and blackout regulations were imposed; the territorial guard was activated. Travel restrictions were imposed; radio, telephone, and mail censorship began; contraband was confiscated; and curfew regulations went into effect. Enemy aliens—German and Italian as well as Japanese—were immediately placed under stringent control. The apprehension and imprisonment of those suspected of treason was selective. Even while the attack was still under way, intelligence officers and FBI agents were gathering the first group of Japanese suspects—representatives of the government, Shinto and Buddhist priests, language school teachers, some Issei leaders, and some fishermen. Other enemy aliens were forbidden to possess firearms, short-wave radios, cameras, or maps of military installations. They were also forbidden to change their residence or jobs without permission. The publication of twelve Japanese newspapers and three magazines was suspended; two newspapers continued to operate under the close supervision of the army. The Issei were told, however, that as long as they obeyed these regulations they would be given the same consideration as other residents, and all citizens were enjoined to treat them well (MacDonald, 1946:150–76; Fuchs, 1961: 299–300).

Protection of the nation was the first concern, and some things just had to be done. Rights of citizenship were stripped from some Nisei; 534 Kibei were arrested and interned. After the initial emergency many Nisei were removed from sensitive positions. Defense workers, after spending long hours repairing damage, were ordered off projects. Medical attendants were prevented from performing their duties. Those who responded to requests for guardsmen were told that "Japs are not wanted." Some employees, after years of loyal service, were summarily dismissed. Some Nisei resigned voluntarily. Candidates for public office, including those running for reelection, withdrew; other politicians declined nominations. In 1942 only one Nisei ran for public office. The Hawaiian Territorial Guard, made up of men from all ethnic groups, had helped protect various installations. But during the weeks after Pearl Harbor criticism mounted that sons of enemy aliens were guarding public utilities, communications, and vital waterfront areas. On 19 January 1942 on orders from Washington, D.C., the 317 members of Japanese ancestry were discharged without explanation. Neither the press nor the radio reported this action, but it soon became common knowledge among the Japanese.

Although checking rumors was a slow and tedious process, a serious effort was made to do so. As the falsity of each item was established, an announcement was made in at least one of the daily newspapers; in general the denials were accepted by the public. When intelligence officers were satisfied that no sabotage had occurred during the attack, despite many opportunities, official denials were issued by the chief of police of Honolulu, by the secretary of war, and by the director of the FBI. Demands that the Japanese be interned on the mainland were opposed by those who pointed to the economic impact of deporting over one-third of the population. Furthermore, shipping space was badly needed, and the military personnel necessary to manage a mass evacuation could not be spared. Although military and naval authorities considered the possibility of evacuation, they concluded that it was not feasible. In his first public statement Lieutenant General Delos Emmons, commander of the Hawaiian Department, declared:

> While we have been subjected to a serious attack by a ruthless and treacherous enemy, we must remember that this is America and we must do things the American way. We must distinguish between loyalty and disloyalty among our people. Sometimes this is difficult to do, especially under the stress of war. However, we must not knowingly and deliberately deny any loyal citizen the opportunity to exercise or demonstrate his loyalty in a concrete way.

Thus, the U.S. Army became committed publicly to a policy of fair play and individual justice. After such announcements public sentiment developed for equal treatment of the Japanese, and they were issued gas masks as well as rations of food, gasoline, and tires on the same basis as all other residents. The few who categorically condemned them were put on the defensive, for they were accused of violating the *aloha* spirit (Lind, 1946:42, 70–71; Allen, 1950:134; Murphy, 1954:54–60).

Rejected for being Japanese, many Nisei tried to conduct themselves in ways that were conspicuously American. Anxious to create a good impression, they rejected any Japanese institution that might give offense to the wider community. On orders from military authorities all twenty-two Japanese language schools were abandoned; eight Buddhist temples and Shinto shrines were dissolved, as were several cultural associations. Assets from their liquidation were donated to agencies such as the Red Cross and the YMCA. Some engaged in an indiscriminate destruction of everything of Japanese origin—photographs of the emperor, Japanese swords, dolls, books, clothing, and small family shrines. Some even stopped eating Japanese food. Issei were told not to bow or to speak Japanese in public, or to appear outdoors wearing a kimono or *geta*. They urged one another to be meek, to stay out of the way, and to avoid gatherings that might attract attention. Some Nisei felt the curse of a Japanese given name, and several thousand went through legal proceedings to change it. Some Nisei women even wore jade and long gowns so that they might "pass" for Chinese (Lind, 1946:143–44). Some even rejected use of the word "Nisei" to refer to themselves, on grounds that it was a Japanese word. Early in 1943 the expression "AJA" —Americans of Japanese ancestry—was popularized by the press and was quickly adopted in Hawaii.

The sudden internment soon after the outbreak of war of many recognized community leaders immobilized the Issei. They were reduced in status to enemy aliens, and leadership passed to the younger generation. Parent-child roles were interchanged. Parents became dependent on their children to translate the regulations with which they were anxious to abide. Many of the young blamed their elders for their plight, and the Japanese principle of filial piety came under attack. Many Issei found that they had lost respect in the eyes of their own children. Since they had been taught since childhood that Japan had not lost a war in its 2,600 year history, some Issei remained confident of an ultimate Japanese victory. But most of the immigrants were bewildered, and they accepted the definition of acceptable behavior laid down by AJA leaders (Kimura, 1943). When the evacuation of the Japanese on the Pacific Coast was announced, many became more apprehensive even though the army had announced that there would be no mass evacuation from Hawaii. Some Issei purchased suitcases and warm clothing—just to be on the safe side. Most Issei believed that AJA spies were employed by the FBI. Rumors pointed to various individuals, and the aliens became fearful of saying anything that might be misinterpreted. They strongly resented the *inu* (dogs), but there was little they could do about anonymous informers.

Although AJAs varied in the extent to which they actually supported the war effort, they were dissatisfied with their lot. Since they had not in fact collaborated with the enemy, they resented the wild accusations and distrust. They were particularly indignant over the stigma of being called "Japs." Concerned with their reputation in the larger community, many wanted a chance to demonstrate that they were loyal citizens fully prepared to meet their obligations. They encouraged each other to contribute actively; they gave generously to the blood bank and purchased war bonds. They volunteered in large numbers for civil defense organizations—manning first-aid stations and joining the Kiave Corps, an unpaid labor unit. The men who had been discharged from the Hawaiian Territorial Guard were especially anxious to dispel doubts; they wanted an opportunity to verify their verbal protestations. On 30 January 1942 a nucleus of former guardsmen, members of the University of

Hawaii ROTC, sent a letter to General Emmons offering themselves for "whatever service you may see fit to use us." The general accepted, and on February 23 many of the discharged guardsmen along with others who quit their jobs or dropped out of school formed the Varsity Victory Volunteers. It was attached to the 34th Combat Engineers Regiment and was officially designated the "Corps of Engineers Auxiliary." Actually it was a labor battalion. The men worked a 48-hour week building prefabricated houses, making repairs, stringing barbed wire, toiling in rock quarries, making articles in cabinet shops, and building and maintaining roads. They made three trips to the blood bank as a group and purchased $27,850 worth of war bonds (Nakahata and Toyota, 1943). Their efforts had a symbolic effect; they proved to the wider community that at least some of the AJAs were sincere. They also showed the local Japanese that it was possible to win respect and goodwill, and their spirit spread. When the American fliers shot down in the Doolittle raid over Tokyo were executed, over $10,000 was raised and presented to the army for "Bombs on Tokyo." Some outsiders were impressed; others saw it as a grandstand gesture.

At the outbreak of war approximately 2,000 Hawaiian AJAs were already serving in the U.S. Army—members of the 298th and 299th Infantry Regiments. On December 7 the 298th was on duty at Oahu, and Nisei soldiers participated in the defense of the island. Their showing was creditable, but they presented awkward problems. Officials feared that some of them might prove unreliable, and there was the additional danger of confusion with the enemy in the event of invasion. The difficulty was resolved by creating a segregated unit; the AJAs were taken out of their respective companies and reassigned to a newly formed Hawaiian Provisional Infantry Battalion. On 5 June 1942—in the midst of the Battle of Midway—the unit left for the mainland to receive further training at Camp McCoy, Wisconsin. While en route it was redesignated the 100th Infantry Battalion (separate). The soldiers realized that they were marked men; they were on trial, and many expressed their determination not to be found wanting.

Outwardly, the entire Japanese community supported the American war effort. As word came back from Wisconsin of the friendly reception given members of the 100th Battalion, many began giving parties for servicemen stationed in Hawaii; thousands of troops from the mainland receiving advanced training there before leaving for combat zones were entertained. There were so few Haole women on the islands that most of them dated only officers. Many enlisted men had to turn to other women, including Japanese. AJA girls walking arm-in-arm with Haole soldiers became commonplace. While fear of gossip or of being blacklisted by AJA men kept the number down, by the second year of the war one of every five Nisei brides married non-Japanese; one in ten married Caucasians (Lind, 1946:197).

On 28 January 1943 the Secretary of War announced the formation of a special combat team of Japanese-Americans and called for volunteers—1,500 from Hawaii and 3,500 from the mainland. Anticipating objections in principle to segregation, the War Department provided the following rationale: the important consideration for Nisei was that they be given the right to fight for their country. If troops of Japanese ancestry were diffused throughout the armed forces, they would count only as additional manpower, and there would be no way of taking special account of what the group had contributed. But the performance of a separate unit would be noticed and could serve as conclusive refutation of charges of disloyalty. In support of the proposal President Roosevelt declared, "The principle on which this country was founded and by which it has always been governed is that Americanism is a matter of the mind and heart. Americanism is not, and never was, a matter of race or ancestry." Nisei already in the army were urged to volunteer for the segregated unit—to form the training cadre of experienced NCOs. The 100th Infantry Battalion was transferred to Camp Shelby, Mississippi, for additional training. General Emmons took the occasion to praise the local Japanese for performing so admirably under adverse conditions.

The response among Hawaiian AJAs was enthusiastic. The proposal was seen as the long awaited opportunity to answer those who questioned their loyalty. Collective pressures mounted to make a good showing—to safeguard their future on the islands. On January 31 the Varsity Victory Volunteers disbanded after eleven months of service and enlisted. The quota of 1,500 men was exceeded in a few hours. Two weeks later the army disclosed that 9,507 men had volunteered—one-third of all AJA men between the ages of 18 and 38. Those who were rejected wept in disappointment, and officials were besieged with letters requesting special consideration for individuals not fully qualified. Not all AJAs were anxious to volunteer, but those who were indifferent or opposed found it difficult to speak out. Some community leaders, almost certain to be rejected because of age or number of dependents, were criticized for making a grandstand gesture in volunteering as well as for putting pressure on others to enlist. On March 28 the Japanese community gathered on the grounds of the Iolani Palace to bid *aloha* to the boys going to the mainland for training. The crowd was estimated at 20,000 (Lind, 1946:153; MacDonald, 1946:199). In April 1943 the 442nd Regimental Combat Team was activated and began training at Camp Shelby.

In August 1943 the 100th Infantry Battalion embarked for Europe, and shortly thereafter it went into action. Attached to the 34th Infantry Division, it was first committed at Salerno; it then participated in almost every major engagement in Italy from Volturno to Rome, including the Anzio beachhead. Its superb performance astounded all observers. The "outstanding record" of this battalion as well as the "excellent showing" of the troops training at Camp Shelby led the War Department to re-examine its policy concerning Nisei soldiers. Casualties were heavy and had to be replaced. In January 1944 selective service was once more re-instituted for all Nisei. This move was acclaimed, for in the eyes of Hawaiian AJAs it renewed in them the sense of really belonging in the United States. It was greeted as a victory, and soon thereafter a steady stream of draftees went into the army.

Not all Issei were sympathetic to the American war effort. When the war finally ended, a number revealed that they had not been for an American victory; many Issei had hoped for a negotiated peace. Some even felt personal humiliation over Japan's defeat and experienced difficulty in facing others. A few could not believe that Japan had lost, and they seriously entertained rumors that the war was over but that Japan had won! Nonetheless they were grateful for the way in which they had been treated, especially in receiving gas masks and rations on the same basis as others. They were surprised and relieved and more than pleased at the opportunity to participate. They provided more than their share of volunteers as block wardens, fire fighters, and first-aid workers; they even worked in labor battalions. They developed a sense of security from feeling trusted and being allowed to take part in a common cause; some Issei went out into the wider community for the first time since their arrival in the country. Although their formal allegiance was to the United States, the Issei often contributed with Japanese cultural tools. Those who bowed each morning did so toward the mainland. Families with sons in the army expressed pride; to have sons fighting for their country was an old Japanese tradition. As one mother put it, "I gave my sons to this country. I am very satisfied and proud." Another declared, "Of course, once he is in the army, he must not think of coming back safely. I won't have him return as a coward." On 23 November 1943 a plantation worker on Oahu announced that he was donating $400 to the Red Cross rather than distributing tea in memory of a son killed in action. Others followed suit, thus altering an old Japanese custom (Kimura, 1943; MacDonald, 1946:202). The chief effect of the war on the Japanese in Hawaii was to draw them from their somewhat isolated position toward more active participation in the larger community. Their collective effort eventually won them respect at home, pride in themselves, and admiration abroad.

From Pearl Harbor to Evacuation

Once the shock over the disaster at Pearl Harbor subsided, the populace on the mainland became calm. In various local communities civic leaders declared a "state of emergency" and appealed to citizens to avoid "hysteria." While there was excitement from a sense of being involved in something momentous, the danger did not seem immediate. In the San Francisco Bay Area, for example, a total blackout was ordered on the night of December 12. While everyone complied, they did not take it seriously; bands of young people romped merrily in the streets in a carnival atmosphere. But the catastrophe was explained in much the same manner as it was in Hawaii. That so gigantic a blow could be dealt against such a mighty fortress without the aid of enemies within seemed incredible. Many of the rumors that developed in Hawaii spread at once to the Pacific Coast, and there were additional accounts—that on December 7 American machine gunners had cleared the streets of Honolulu of all Japanese and that all over the city signs proclaimed: "Here a Jap traitor was killed." Suspicions that had been harbored for decades were now confirmed. But there was little overt hostility. Although curious onlookers besieged the segregated settlements and in some instances had to be dispersed by the police, relationships between persons of Japanese ancestry and others remained for the most part unchanged. Individuals who had Nisei friends went out of their way to be kind and reassuring.

Unlike Hawaii, the Pacific Coast remained under civilian control for more than two months, and the Department of Justice made a sharp distinction between enemy aliens and American citizens. The FBI immediately arrested "suspicious aliens," and by the end of the first week 1,370 Japanese, 1,002 Germans, and 169 Italians were in custody at Immigration Service Headquarters in San Francisco. In explaining these arrests Attorney General Biddle announced that there was "absolutely no evidence of fifth column activities, sabotage, or espionage" and that those seized were persons the government felt "it would be unwise not to apprehend"; in most cases they had been selected more than a year before. As in Hawaii, enemy aliens immediately became subject to travel restrictions, were required to turn in contraband, and had to observe curfew. However, these regulations did not apply to Nisei, who were treated as Americans. The Treasury Department seized all Japanese businesses, and their assets were placed under federal control. Issei business establishments—such as grocery stores, nurseries, pool halls, restaurants, and laundries—were closed temporarily, but after a week all of them were permitted to reopen. Enterprises owned by Nisei were not affected. Such differential treatment made the Nisei acutely conscious of the value of their citizenship; many appreciated their birthright for the first time.

There were many appeals for fair play, both by government officials and by others. In one of his first announcements the attorney general declared:

> There are in the United States many persons of Japanese extraction whose loyalty to this country, even in the present emergency, is unquestioned. It would therefore be a serious mistake to take any action against these people. State and local authorities are urged to take no direct action against Japanese in their communities but should consult with representatives of the FBI.

A few days after Pearl Harbor national church organizations issued a joint appeal to the American people, urging decent treatment of the Japanese in the country. During the first two months of the war the calm was shattered by only a handful of incidents, most of them involving Filipinos in California. On December 23, a 33-year-old Nisei, who had just been discharged from the U.S. Army, was stabbed to death in Los Angeles; two Filipinos were alleged to have been the last to be with him. The most violent outburst came in Stockton on Christmas day, when gangs of Filipinos smashed the windows of numerous stores, manhandled several Japanese, and killed a 55-year-old garage attendant. They then announced a boycott of all Japanese establishments. Police immediately brought skid row under control and closed all bars and clubs catering to Filipinos. When the open city of Manila was bombed, violence erupted again. On December 29 a 57-year-old Issei was shot in Sacramento, and on the same night a window was broken in a Japanese store in Oakland. On January 2 five shots were fired into the residence of a Costa Mesa couple, and on the following day two Japanese were seriously wounded in Gilroy when more than fifty shots were pumped into their home. That night an elderly couple was slain as they slept in El Centro. On January 6 five shots were fired into the home of a Mount Eden Nisei. On January 9 a young San Jose mother was shot in a night attack, and on January 15 another Issei was stabbed in Stockton. Local officials responded promptly; editorials denounced "hotheads"; and Filipino leaders appealed for reason. By the middle of January the violence had ended.

The rumors of treason at Pearl Harbor were denied publicly by high government officials, including the director of the FBI, who announced that not a single act of sabotage or espionage had been committed by anyone of Japanese ancestry—citizen or alien. Although these denials were accepted in Hawaii, they passed virtually unnoticed on the Pacific Coast. Indeed, some of the rumors were reaffirmed when they were dramatized in the popular movie *Air Force*. Widespread belief in the complicity of resident Japanese was further reinforced by sensational press coverage, especially in California, of FBI spot raids to uncover contraband. On 2 February 1942 four San Francisco Japanese were reported arrested, and four days later nine more were seized in Vallejo. On February 10 further raids took place in Monterey, Salinas, and Watsonville. A week later similar arrests were made in the Sacramento area, and on February 21 extensive raids began in southern California. Day after day people were jolted by huge, eye-catching headlines: FBI SEIZES 21 AXIS ALIENS IN VALLEJO SPY RING HUNT; TWO JAPANESE ARMY RESERVE OFFICERS SEIZED BY FBI IN SF RAID; MORE JAPS ARRESTED IN SACRAMENTO AREA; FBI ARRESTS ALIENS IN ROUNDUP HERE; ALL-OUT ALIEN ROUNDUP ON; MASS RAIDS TRAP SCORES IN COAST'S GREATEST SPY HUNT. The blazing headlines led to a crescendo of demands for some kind of action. Politicians and local officials, sensing the disquietude of the people, began demanding removal of all Japanese from their midst. The FBI raids, which had started with the outbreak of war and continued to the eve of evacuation, received extended publicity only during the month of February. Once government action was announced, interest waned, and news coverage ceased.

The Japanese were bewildered. That Issei were being arrested was not surprising. While published accounts of spy rings, huge stores of firearms and ammunition, and Japanese admirals in disguise frightened those outside the segregated settlements, to those within the charges were preposterous. Among the articles listed as contraband in the news articles were Japanese army uniforms (any man who had been in Japan after reaching 19 would have been conscripted), aerial bomb casings (empty and used for theatrical purposes), wrestling banners (for *sumo* and judo titles), ammunition (at a sporting goods store), detached headlights, maps, and technical books on radios. The grounds for detention given in the newspapers were patently absurd. What, then, were the *real* reasons for the arrests? Answers were provided in a flood of rumors. A variety of objects not listed in the regulations governing enemy aliens were stated to constitute contraband: anything printed in Japanese (even Bibles), Japanese flags, knives longer than eight inches, Japanese phonograph records, pictures of airplanes, field glasses, maps, portraits of the emperor. Why were some arrested while others were not? The general impression was that the FBI was picking up harmless old men, while those who might actually be dangerous were still free. All Buddhist priests and *kendo* (fencing) instructors as well as language school teachers were thought to be on the "wanted" list, even though many such persons had not even been questioned.

Issei were also frightened by rumors of the manner in which the accused were treated: the FBI refused to allow wives of prisoners to speak to their husbands in Japanese; suspects were not allowed to change their clothes or even to go to the toilet; in a raid in Concord old people were forced into their yard without bathrobes on a freezing cold morning; some FBI agents forced a girl with mumps to get out of bed in order to search beneath her. Issei busied themselves destroying anything in their homes that might possibly be forbidden, and those who thought they might qualify for arrest packed their suitcases and stoically prepared for jail. Considerable resentment developed against certain individuals suspected of being informers; they were accused of taking advantage of the situation to strike back at their personal enemies.

During the spring of 1942 Axis forces were advancing everywhere, and the general outlook was pessimistic. Japanese troops had overrun much of Southeast Asia, and Nazi Germany controlled virtually all of Europe. It was at this time that demands mounted for action against the local Japanese. Groups that had long been agitating against the Japanese, such as the California Joint Immigration Committee, were joined by many others in calling for military control. The stereotype of the "treacherous Jap" was reactivated. News commentators, editorial writers, and public officials began demanding vigorous steps. Congressman Dies of the House Un-American Activities Committee promised to disclose evidence of espionage and sabotage; the boards of supervisors of Los Angeles and of San Francisco passed resolutions urging the federal government to evacuate Japanese aliens from defense areas; the Speaker of the California Assembly urged removal of all Japanese—including American citizens. Public opinion was inflamed. It was emphasized that distinguishing a loyal Nisei from a disloyal one was impossible; another common charge was that no Japanese had ever "turned in" another to the FBI.

Some even cited humane grounds; it was necessary to intern the Japanese to protect them from mob violence. The evacuation of all Japanese was seen as necessary to protect the nation; humanitarian and economic considerations should not enter the decision (Grodzins, 1949; tenBroek et al., 1954:68–96).

Opponents of mass evacuation, led by clergymen and educators, pointed to the constitutional rights of citizens. They asked that each individual be treated separately, noting that those who were subversive could be handled by the FBI. They also cited the economic motives of those who were advocating evacuation most loudly. Religious groups were particularly active in urging fair play. Since they had their own communication channels, containing information that flatly contradicted the headlines, ministers were able to take a stand challenging the agitators. They urged their members to be understanding and set up special organizations to provide relief to Japanese in need because of work or travel restrictions. They arranged for storage in churches of the goods of those who had to move (Matsumoto, 1946:10–25). But defenders of the Nisei were at a disadvantage. When they urged tolerance, they were labeled "Jap-lovers," and their pleas were drowned in the mass hysteria. Furthermore, there was no way of proving conclusively that any Nisei was loyal. Indeed, no one could be certain of what the Japanese would do if an invasion force were to land. The American Civil Liberties Union was criticized for "yowling" about rights when the nation simply could not afford to take chances.

On 19 February 1942 President Roosevelt signed Executive Order 9066, which shifted control over enemy aliens from the Department of Justice to the War Department and authorized the designation of military areas from which all persons—citizen and alien alike—might be excluded. On the following day Lieutenant General John L. DeWitt, commander of the Western Defense Command, was appointed by the Secretary of War to carry out these duties. As soon as jurisdiction was transferred to military authorities, the distinction between citizen and alien was obliterated, and all subsequent action was taken on an ethnic basis. Regulations of the Western Defense Command applied to all Japanese, including American citizens, even though certain classes of Italian and German aliens were exempted. On March 2 Military Area Number 1 was designated; it included the western half of California, Oregon, Washington, and southern Arizona. Although this proclamation only defined the areas regarded as vital, it was announced that future proclamations would specify just who was to be excluded.

PLATE I *Evacuation, U.S.A., 1942*
(Courtesy of The Bancroft Library, University of California, Berkeley)

At first voluntary evacuation out of the military area was encouraged. The army requested the cooperation of those living in the interior in clearing the coast of Japanese, but reaction was generally negative. In Fresno and Tulare counties, both in California but outside the military area, Japanese who had been residents before the war were accepted; but the influx of new evacuees led to unpleasant incidents. Residents objected to their community becoming a "dumping ground for Japs." The governor of New Mexico advised landowners not to sell land to Japanese; he wanted to cooperate with the federal government but was opposed to Japanese immigration. The attorney general of Arizona announced that he was against the "colonization" of his state. In Nevada both the press and officials urged residents not to mistreat those traveling through, but they adamantly opposed settlement in the state. Governor Carr of Colorado was among the few who supported accepting Japanese as the state's "contribution to national defense." The state university indicated it would accept evacuee students. From 12 March to 31 October 1942, 1,963 voluntary evacuees settled in the state. But many newspapers and local officials were unhappy, and the governor's stand was widely assailed. In Idaho the press urged acceptance of Japanese as the "duty" of Americans. Sugar companies, badly in need of beet workers, supported the move; but the governor was opposed. Farm workers were also needed in Montana, but the unions objected. Thus, many Japanese tried to move on their own, but the program failed because of hostility in the receiving areas. The army then had to proceed with enforced mass evacuation.

On March 6 the army announced plans to construct temporary assembly centers for the processing of Japanese to be excluded from the military areas. On March 19 President Roosevelt announced the formation of the War Relocation Authority (WRA), a civilian agency that was to administer the relocation of persons excluded from vital areas. On March 24 the mass evacuation of all persons of Japanese ancestry from such areas was announced; the grounds were simple—*military necessity.* At the same time a curfew from 8 P.M. to 6 A.M. was announced for all Nisei as well as enemy aliens living in those areas. All persons in these categories were forbidden to travel more than five miles from their homes, and voluntary evacuation came to an abrupt halt. Contraband restrictions, previously applicable only to aliens, were now imposed on Nisei as well. Bainbridge Island, Washington, was the first area to be evacuated; then came Terminal Island, in Los Angeles. One community after another was cleared of Japanese; the evacuees were sent first to as-

PLATE 2 *A Relocation Center*
(Courtesy of The Bancroft Library, University of California, Berkeley)

sembly centers nearby and then were transferred to more permanent relocation centers in the interior. Subsequent proclamations expanded the areas from which "dangerous persons" were to be excluded. When the eastern half of California was defined as a military area, many voluntary evacuees who had just settled there were dislodged once more. By November 110,000 persons had been banished from their homes in Alaska, California, and much of Oregon, Washington, and Arizona and interned in ten relocation centers.

The evacuees had only a brief period in which to prepare for camp life. Each was permitted to bring only those possessions he himself could carry. Although governmental agencies tried to protect them from exploitation, losses were inevitable. Businessmen and farmers had to liquidate their holdings in a few weeks; they had to accept the best offer available, if they were to salvage anything at all (Thomas and Nishimoto, 1946:1–52). For some Nisei this was proof that there was no future for them in the United States; it confirmed the contention of those Issei who had insisted that they would never be treated as Americans, no matter how they themselves felt. Nisei who had always been suspicious of white people were not surprised. Those who were more assimilated were disillusioned, although many still did not give up hope.

Throughout this period there was constant interest in the fate of Nisei soldiers, for this served as an index of what might happen to the others. Nisei soldiers were ostensibly the most loyal, for they were pledged to defend their country. As long as some were still serving in the army, the citizenship of all others seemed reasonably safe. But the War Department made no announcements about this matter, and the personnel policies of military posts were shrouded in wartime secrecy. Those concerned with the fate of Nisei servicemen had to depend on rumors—that Nisei were being given the worst and dirtiest jobs in the army; that many were being released without explanation, some with dishonorable discharges; that Nisei at one post were locked up for twenty-one days after the war began; that Nisei who were sent to the West Coast from interior training camps were removed from their units and returned, while their comrades went on to the Pacific theater; that some Kibei in Washington had set fire to his barracks. There were other rumors of anti-Nisei riots at various posts, of suspected men being jailed, of transfers to labor battalions, and of plans to send Nisei into segregated units along with Negroes. Thus, the general impression created by rumors was one of discrimination, discharges, demotions, and violence. Some Nisei thought that those discharged were likely to be Kibei, but others reasoned that it would be inconsistent to send parents to concentration camps while their sons were in military service.

Since the Selective Service System left the matter to the discretion of local draft boards, much inconsistency and confusion developed. Like other Americans many Nisei volunteered their services immediately after the attack on Pearl Harbor. Some were accepted; the induction of others was delayed; still others were rejected, though qualified in all respects other than ancestry. During the spring of 1942 many boards were reluctant to draft Nisei, but others continued to call them for induction. Some men were drafted just a few weeks before the exclusion orders went into effect, just when they were most needed to pack and to dispose of family property. Thus, some draft boards were inducting Nisei at the same time that some military commanders were discharging them without cause. On 17 June 1942 the War Department advised the Selective Service System to discontinue the induction of Nisei until further notice. Soon afterward all Nisei were reclassified 4-C—"aliens ineligible for military service."

Since there was genuine fear of subversion in a war that was not going well, the program of mass evacuation enjoyed widespread public support. Although some individuals and organizations continued to protest the incarceration of American citizens without trial, by and large their voices were ignored. Signs such as "No Japs Wanted" and "We Do Not Solicit Jap Trade" were commonplace, and one barber achieved notoriety by putting up a sign: "Free Shave for Japs—Not Responsible for Razor Slips." An NORC poll taken in March 1942 showed that 93 percent of a national sample approved moving Japanese aliens from the Pacific Coast; 59 percent favored removing citizens as well; and 65 percent approved keeping the evacuees under strict guard. Only 25 percent objected to incarcerating citizens. By December the mass hysteria had waned. Even then, however, when the Gallup Poll asked people on the West Coast if they would be willing to hire Japanese, only 26 percent said they would; 69 percent said they would not. Asked if they would trade in Japanese stores, 58 percent answered that they would not (National Opinion Research Center, 1946; Cantril and Strunk, 1951:380–81).

The Formal Determination of Loyalty

Although tension on the Pacific Coast subsided with the evacuation, the question of Nisei loyalty remained a public issue. Those who had been agitating against the Japanese continued to press for further action. Various California groups urged the permanent exclusion of all Japanese; what they wanted was the denationalization of Nisei, internment for the duration of the war, and deportation afterward. Some objected to turning custody of evacuees over to the WRA, a civilian agency; they felt that control should remain in the hands of the military. Those favoring a firm stand were much encouraged by General DeWitt's testimony before the House Naval Affairs Sub-Committee: "A Jap's a Jap. It makes no difference whether he is an American citizen or not. You can't change him by giving him a piece of paper."[3] The House Un-American Activities Committee, the American Legion, and various local officials criticized the WRA for "coddling Japs." After visiting the Gila River Relocation Center in Arizona, Mrs. Eleanor Roosevelt countered that she saw no evidence of coddling; when she urged fair play and the release of loyal Nisei, she was roundly condemned. In the California legislature

3. Although the official transcript reads somewhat differently, the general was so quoted by reporters who were present. This wording, from the Associated Press release of 13 April 1943 is what received widespread publicity.

resolutions were introduced memorializing Congress to amend the Constitution to ban those of Japanese ancestry from citizenship. The Native Sons of the Golden West filed suit to remove Nisei names from voter registration lists. When the District Court of Appeals ruled that Nisei could not be denied the right to vote, the organization appealed its decision. In May 1943 the U.S. Supreme Court refused to hear the case, thus upholding this verdict. The Native Sons then launched a campaign for a constitutional amendment. Two bills were introduced in Congress—one to deprive Nisei of citizenship and the other to keep them interned for the duration (Thomas and Nishimoto, 1946:24–52).

Those who were convinced that most Nisei were loyal also continued to press their case. By the time the evacuation program had been completed, embarrassing questions began to arise. Only the Japanese—citizen and alien—had been removed, without trial or even the filing of charges; but 113,800 Italian and 97,000 German aliens living in the military areas had been left undisturbed. The reason given by the Western Defense Command was "military necessity," but this appeared more and more unconvincing as it became apparent that Hawaii would not be evacuated. One-third of the population there was of Japanese ancestry; they were free to carry on, and nothing had happened. In California a Pacific Coast Committee on American Principles and Fair Play was organized to campaign for Nisei rights. In New York forty-two prominent liberals sent a letter to the president condemning the House Un-American Activities Committee. The American Civil Liberties Union filed suits challenging the legality of the exclusion orders. Church groups continued to support the evacuees. They sent Christmas gifts to the centers, offered scholarships to those whose education had been interrupted, urged fair play in their local communities, pressured various industries to employ Nisei, and accused the government of using Nazi tactics. These efforts apparently did not pass unnoticed. In December 1942, when an AIPO poll asked whether evacuees should be allowed to return to their homes, 35 percent of a national sample approved, and only 17 percent disapproved. In the five western states, however, only 29 percent approved. When the rest were asked what should be done with those who had been interned, two-thirds stated that they should be deported (Cantril and Strunk, 1951:380–81).

High officials of the WRA were for the most part also convinced of the loyalty of the vast majority of Nisei, and they adopted a policy of releasing as rapidly as possible those who had been cleared by the FBI to resettle in areas not covered by military exclusion orders. The aim was to send out loyal citizens to help alleviate the acute labor shortage. Initially the resettlers were carefully screened, and the clearance procedure was long and cumbersome. Advised of the army's impending plan to form a combat team of Nisei, WRA officials saw the possibility of expediting clearance for everyone, and they decided on a routine registration of all evacuees. Since all volunteers for military service would have to be checked by security agencies to determine their eligibility, they seized on this occasion as an opportunity for setting up a simpler device to facilitate clearing everyone. When the registration of all adults over 17 was announced, however, suspicious evacuees misinterpreted WRA intentions. During the winter of 1942-1943 difficulties erupted in all ten centers.

Although a crucial decision was involved, the questionnaire was prepared carelessly. Most of the items on the registration form were routine, but two in particular precipitated the crisis:

Question 27

For men: Are you willing to serve in the armed forces of the United States on combat duty, wherever ordered?

For women: If the opportunity presents itself and you are found qualified, would you be willing to volunteer for the Army Nurses Corps or the WAAC?

Some saw this question as a stratagem for tricking Nisei into volunteering—after they had been classified as aliens ineligible for military service.

Question 28

For men: Will you swear unqualified allegiance to the United States of America and faithfully defend the United States from any or all attack by foreign or domestic forces, and foreswear any form of allegiance or obedience to the Japanese emperor or any other foreign government, power, or organization?

For women: Will you swear unqualified allegiance to the United States of America and foreswear any form of allegiance or obedience to the Japanese emperor, or any other foreign government, power, or organization?

This constituted no problem for Nisei, even though it was ridiculous for them to foreswear an allegiance most of them never had. But many of the aliens balked. Since they were ineligible for American citizenship by law, giving up their Japanese citizenship would leave them without any legal status. Furthermore, many reacted bitterly to still another test of loyalty after they had accepted evacuation so quietly. Pro-Japanese agitators seized on this blunder. In each center those willing to register clashed with protesters. Demonstrations were mounted, and in some camps advocates of cooperation with the WRA became the objects of gang violence (Leighton, 1946:162–210; Thomas and Nishimoto, 1946:53–83).

Since each citizen had to declare himself either loyal or disloyal to the United States, all Nisei were forced to make a conscious choice on a matter on which most people never exercise an option. The diverse orientations among them were highlighted in the deliberations on this issue. Nisei differed considerably among themselves in sophistication, degree of acculturation, and political interest; they also held varying conceptions of ethnic identity. Some were thoroughly assimilated; they conceived of themselves as American and could not fail to serve their country without losing self-respect. Some conceived of themselves as Japanese by "blood" but American by nationality; they regarded the obligations of citizenship as paramount and wanted to devote to their country the characteristics they believed they had inherited genetically. Many of them confessed that they disliked fighting against Japan, but they saw it as their duty. Some were Japanese through and through and could see no future for themselves in America. A great many were utterly confused about their identity and did not know what to do. Bitter arguments broke out. Radical intellectuals and the conservative Nisei bourgeoisie temporarily forgot their differences and sided with the Christians to make up the nucleus of those favoring registration. Some Kibei extremists vowed to kill all enemies of the emperor wherever they may be, and several harsh confrontations occurred. They terrorized the parents and siblings of those who were willing to cooperate with the WRA, and it was during this period that some Nisei developed a deep hatred of Japanese nationalism. When the registration period ended in February 1943, 28 percent of the male and 18 percent of the female citizens had refused to affirm their loyalty to the United States. Among aliens the percentage of refusals was smaller (Thomas and Nishimoto, 1946:53–112).

The Tule Lake Relocation Center, in northern California, had the largest number refusing to cooperate, and it was converted into a segregation camp for the "disloyal." By May 1944, 18,422 evacuees—including 12,489 citizens—were interned there. All other evacuees were distributed to the nine other centers, and most of them were eventually cleared for resettlement. Although the people detained at Tule Lake were regarded as dangerous, the division was not strictly along lines of national sympathy. Some loved Japan dearly and were eagerly looking forward to a Japanese victory, but for most the decision rested on more practical and immediate grounds. Some feared being forced to go "outside" into a hostile world after being deprived of their means of livelihood; they saw the center as a haven of refuge and believed that refusing to register was the safest way to stay there. Others were protesting what they regarded as unfair treatment. A number declared themselves disloyal to avoid separation from other members of their family. Still others feared being drafted and saw this as the most certain way of escaping military service. Some were merely pragmatic; they were convinced that Japan would win the war and did not wish to risk punishment by the victors (Grodzins, 1956:105–31). On 1 July 1944 Congress passed a law making the denationalization of Japanese-Americans possible; by the time the war ended 5,371 Nisei had signed applications renouncing their citizenship, and some of them were subsequently deported.

It was in the midst of the registration crisis that the War Department announced the formation of the special combat team and called for volunteers—3,500 from the mainland. In sharp contrast to Hawaii, the response in the relocation centers was reserved and phlegmatic. Many objections were voiced against serving in a segregated unit, even among those who favored registration. Many called attention to the possibility of its being sent on "suicide missions" and charged that Negro troops were being used recklessly. Many parents did not wish their sons to fight, especially against Japan, and urged them to declare themselves disloyal, regardless of their actual feelings. Rumors of the manner in which Nisei soldiers had been treated certainly did not make army life seem attractive. Nisei were alleged to be "broomstick soldiers" who were not permitted to handle weapons. Who wanted to volunteer to become gardeners or cook's assistants for $50 a month? Many rumors developed of the treatment of Nisei troops attached to various service commands during President Roosevelt's tour of military installations. One widespread version was that all Nisei had been removed from their respective companies and placed in custody while the president reviewed the other troops. Furthermore, at that time Nisei soldiers were still excluded from the military areas on the Pacific Coast, for that order was not rescinded until 18 April 1943. Other questions arose. Would the sacrifices that Nisei might make on the battlefield really make any difference? Some agitators pointed to the performance of Negro troops in World War I and charged that their status had not improved as a consequence of it.[4] They also

4. During World War I members of the 369th Infantry Regiment—the "Harlem Hell Fighters"—fought so gallantly that they were awarded the *Croix de Guerre* by the grateful French government. In 191 days of combat the unit took every offensive objective but one and never lost a foot of ground on defense. It is

noted that Issei who had won American citizenship through service in the AEF during World War I had been evacuated along with those who were still aliens. Whenever Nisei objected that they were still citizens, they were asked, "If you are an American, what are you doing in here?" But many Nisei felt that their future in America was at stake. They were acutely aware of the charges being made against them and of the efforts being made to take away their citizenship, and they felt that it was essential to make a good impression. Since national loyalty was a subjective matter that could be concealed so easily, military service appeared to be the only conclusive proof that could be offered. In some centers the more rabid pro-Japanese elements threatened the families of men contemplating volunteering; some were frightened into relenting, but others so resented the terrorism that they were spurred on to enlist. Eventually about 1,200 volunteered from the relocation centers and soon joined those from Hawaii at Camp Shelby.

As a result of the registration crisis the overwhelming majority of Nisei became publicly committed to allegiance to the United States and to willingness to serve in the armed forces. A great many were still confused about their identity, but the registration erased all doubts of their official status. Once the formal commitment had been made, ambiguity and indecision were gone, and there was no turning back. The majority of aliens also agreed to abide by American laws and not to hamper the war effort—a statement substituted by the WRA for aliens answering Question 28. They had given their word, and they considered this a binding obligation, regardless of how they felt. Once the matter had been settled, President Roosevelt reported to Congress that most Nisei were loyal and that the rest had been segregated. The WRA then escalated its program to encourage all remaining Nisei to leave the centers.

Among the first to leave were college students, who were aided by religious and philanthropic organizations. The National Japanese-American Student Relocation Council was first organized in Berkeley in March 1942 and was initially financed by the YMCA and YWCA. After the evacuation its headquarters were moved to Philadelphia, and its staff worked diligently to get Nisei accepted at accredited schools, to have tuition fees waived, to get a generous evaluation of credits already earned in West Coast schools, and to help find part-time jobs and housing. Since much advance work had been done to assure their acceptance, Nisei were

---

doubtful, however, that many evacuees actually knew of this record (Little, 1936).

welcomed. In the autumn of 1944 the council reported that more than 3,000 Nisei had been placed in 550 colleges in 46 states (Matsumoto, 1946:111–16).

Farm labor was also recruited to meet a critical shortage, especially in the beet fields in the Rocky Mountain states. Although employers were delighted at the large reservoir of experienced workers, others were displeased. Since many Japanese had already settled in this area during the period of voluntary evacuation, many took short-term leaves to join friends in seasonal work. But the new migration of Japanese from the centers not only led to objections among other residents of "too many Japs" in their midst; the Japanese already there, who by now were established, felt their own positions threatened. The presence of so many Orientals made them feel conspicuous. One firm that employed large numbers of resettlers was the Seabrook Farms in southern New Jersey, a frozen food firm that raised much of its own produce. The company provided jobs, housing, and paid going wages. Migrations to that area began in 1943, and the resettlers were well received by the surrounding communities. By May 1943, 11,198 evacuees were out on various types of leaves.

At first only the adventurous few departed. Although there were isolated incidents, on the whole the resettlers were received courteously. Except where they were known, most people apparently regarded them as "cute little Chinese" and often commented approvingly on Madame Chiang Kai-shek, who had just made a goodwill tour of the country. Those in the vanguard were astonished to discover that there were no institutional barriers against Asians as there had been on the Pacific Coast. Although housing was scarce, they were not forced to congregate in ghettoes; indeed, resettlers experienced no more difficulty than others in finding suitable places to live. They were pleasantly surprised to find most Americans friendly and receptive. Although most of their coworkers knew little or nothing of the evacuation, they were fair. Many became indignant when told of what had happened, and some went out of their way to make the newcomers comfortable. Unfavorable articles appearing in the press had little effect on daily contacts. As news of their reception reached the relocation centers, young people began leaving in increasing numbers, especially to cities in the East and Midwest. Letters described the freedom and excitement of city life, which appealed to the young, who were becoming bored with their camp existence. As the proportion of young people was reduced, the daily routine became even more dull and futile. Going out somewhere became the "thing to do," and parents found it increasingly difficult to object, even to the departure of their daughters. Thus, when the anticipated resistance did not materialize, resettlement became a mass movement. Only the old and infirm, those with ailing parents or too many children to support, and those still distrustful of white people remained.

Once the mass migration got under way, Nisei were aided by powerful organizations and individuals. Liberal groups and religious bodies established hostels for resettlers and helped find housing and jobs. Field offices set up by the American Friends Service Committee were especially helpful. Much of the personal contact work in hundreds of local communities was handled by various Protestant churches that worked closely with federal officials. By December 1944, when the army rescinded most of its exclusion orders, approximately one-third of the evacuees had already left the centers.

The resettlement program turned out to be an unexpected success. The Nisei entered the job market during the peak of a critical manpower shortage. Their superior West Coast education made them better qualified than many of their coworkers. Furthermore, the 4-C draft classification of the men reassured employers who had to put them through long and expensive training programs. Many college graduates found themselves working for the first time in jobs for which they had been trained, in fields from which they had been excluded on the Pacific Coast. Once contacts were actually established, many of the Nisei's cultural characteristics apparently facilitated their acceptance. Most of them spoke fluent English and shared many of the interests promoted in the mass media. In addition, several observers have noted the remarkable similarity between the values of the Nisei social world and those of the American middle class. Japanese culture, to the extent that it had been transmitted to the Nisei, emphasized such traits as cleanliness, politeness, and respect for authority. As with those imbued with the Protestant ethic, individual achievement through frugality and diligence was stressed. Once the resettlement got under way, many employers openly expressed their preference for Nisei. In their characterization of the newcomers, both employers and fellow workers cited attributes prized in the American middle class: neatness, intelligence, and dependability. The emerging stereotype of the Nisei worker included traits that the managers themselves possessed and prized, and Nisei were often addressed with a respect not shown to white employees from working class backgrounds (Caudill,

1952; Jacobson and Rainwater, 1953). Some employers were so delighted that they advertised openly in newspapers for more "Japanese workers."

Although the resettlers were a heterogeneous lot, all were acutely conscious of the possibilities of being misunderstood. In spite of other differences they shared a common definition of the situation. Even though they had now been classified as Americans by the government, other Americans might still refuse to acknowledge their claims. The prevailing view was that they were still vulnerable and on trial. They had to demonstrate the correctness of their claims. Each felt obligated to refrain from doing anything that might invite adverse comment; the rash act of a single individual might well jeopardize the safety of everyone. Officials advising those departing from the centers frequently cautioned them to be careful, but such advice was not necessary; the Nisei themselves fully appreciated the situation. Those who were not sufficiently acculturated and were likely to make an unfavorable impression were sometimes discouraged from leaving. Resettlers generally agreed that they should dress neatly, leave large tips, and avoid boisterous conduct. They tried to avoid large gatherings, for they did not want to be conspicuous. They frowned on dating Caucasian women. They avoided using the word "Japanese" in public; since the term "Jap" was offensive to them, for purposes of ethnic references they used substitute expressions that were meaningless to outsiders—such as "Boochie," "Buddhahead," "Nihonjin," and "Yabbo." Violators of these norms became subject to group sanctions; those who had been careless sometimes found themselves admonished or even threatened by Nisei who were complete strangers. Those accosted in this manner generally acquiesced rather than objecting to such meddling. Persons who were responsible for unfavorable notice, such as those who had touched off riots in the centers or who had become involved in a stabbing incident in Chicago, were highly resented. Publicity given to further riots at the Tule Lake Segregation Center evoked considerable hostility against pro-Japanese agitators. In spite of differences in their self-conceptions Nisei resettlers developed a common interest in winning acceptance. The performance of those who had elected to remain American was a collective effort. They recognized explicitly that in the public eye the entire group could be held responsible for the acts of a few, and most Nisei felt honor bound not to compromise the chances of their fellows.

The reinstitution of selective service for Nisei in January 1944 was not welcomed by those on the mainland. Many resettlers were just getting established in their new careers. They had located satisfactory places to live, had finally found work they enjoyed, and had made new friends. Some had just sent for their families. What was especially frightening, however, were the reports of casualties sustained by Nisei units. The *Pacific Citizen*, the only Nisei periodical with a national circulation, featured each week in boldface type the names of servicemen killed, wounded, or missing in action. (See Table 1.) Although the early casualties were virtually all Hawaiian, after July 1944 names of mainland

Table 1
Nisei Casualties Listed in the *Pacific Citizen*

	Official			*Unofficial*			*Total Casualties Listed*
Month	*Dead*	*Wounded*	*Missing**	*Dead*	*Wounded*	*Missing**	
1943							
November		18					18
December	26	159					185
1944							
January	59	189	17				265
February	25	36					61
March	21	42	6				69
April	9	41	1				51
May	2						2
June	2	7					9
July	13	24	2				39
August	23	112	1	22	37		195
September	69	455	23	2	4		553
October	68	80	1				149
November	9	9		41	140	3	202
December	19	164		25	96	5	309
1945							
January	54	596	1	2	6	2	661
February	75	61	2	2	8	3	151
March	9	21	5	2	6	1	44
April	10	36	2	8	49	1	106
May	94	415		50	203		762
June	1	3		2	2		8
July							0
August	1						1
Totals	589	2,468	61	156	551	15	3,840

Source: *Pacific Citizen*, November 13, 1943, to August 4, 1945.
*Includes prisoners of war.

Nisei began to appear. Each week the paper was studied carefully for the names of relatives, friends, and acquaintances. The listing was not entirely accurate. Most of the names were taken from official War Department releases. Since such announcements were delayed, however, only Hawaiian names continued to appear even after several mainland Nisei were known to have been killed. Since the newspaper's distribution was primarily on the mainland, reporters in relocation centers and resettler communities began collecting the names of those whose next-of-kin had been notified, and these were also included. Although the lists were clearly labeled "official" and "unofficial," the names were printed in the same heavy type. Most of the names reported informally appeared again within a month or so. Furthermore, some men were wounded more than once, and those who died from their wounds were listed a second time. Thus, the names of some individuals appeared two or three times.[5] Given the small number of Nisei in action, the casualties seemed excessive. In some issues the rows of names took up a substantial portion of the eight-page tabloid. Furthermore, seeing so many names of individuals one knew personally brought the whole matter even closer to home. Both on the mainland and in Hawaii rumors that Nisei were being used as "cannon fodder" were revived. Since draftees would serve as replacements for casualties, they would be expected to perform like the volunteers. This was not an appealing prospect.

In some relocation centers considerable resistance developed against the draft. In March 1944 the Spanish consul, who represented the Japanese government during the war, announced that the Nisei had no obligation to serve in the U.S. Army. American officials disagreed. A month later in the Poston Relocation Center, Arizona, one man was charged with sedition for inciting Nisei to resist the draft; nine refused induction and were sentenced to three years in prison; but 221 others appeared for their physical examinations. Organized resistance was strongest in the Rocky Mountain area. In early 1944 the *Rocky Shimpo*, published in Denver, started a campaign against the draft, and in the Heart Mountain Relocation Center, Wyoming, a Fair Play Committee was organized to lead the resistance. In April the committee was labeled subversive, and twelve men were jailed. The *Heart Mountain Sentinel*, the camp paper, attacked the *Rocky Shimpo*, accusing the editor of ruining the lives of

5. Despite the duplications, final figures subsequently released by the War Department revealed that the actual casualty rate was even higher than what was reported in the *Pacific Citizen*.

forty-one men by distorting news. In June fourteen men from the Granada Relocation Center, Colorado, who had failed to report for physical examinations, went on trial before the District Court in Denver; two were found guilty of draft evasion. During that month sixty-three Heart Mountain men also refused to report, although fifty-one of them indicated their willingness to serve if their citizenship status were clarified. The trial of the sixty-three began on June 12 in a Cheyenne federal court; all were found guilty, and on June 26 each was sentenced to three years in prison. In July the editor of the *Rocky Shimpo* and seven others were arrested on draft conspiracy charges; in November a Cheyenne jury acquitted the editor but convicted the others. In March 1945 the Tenth Court of Appeals upheld the conviction of the sixty-three Heart Mountain men.

Most Nisei, however, accepted their fate and reported for induction. With their reclassification from 4-C to 1-A a disproportionate number were called, for the median age of Nisei at that time was 19. Since most draft boards had been straining to meet their quotas, they had no alternative to tapping at once their reservoir of Nisei. Thus, all eligible men who would normally have been called from 1942 onward were inducted at once, and hundreds poured into the army. On 31 May 1944 the War Department announced that since the reinstitution of selective service for Nisei, 3,312 had been called; of this number 669 had been rejected, and only 139 had declined to report. Most of the new recruits were assigned to segregated training units at Camp Shelby; at Camp Blanding, Florida; or at Fort McClellan, Alabama. By the winter of 1944 almost everyone had at least one close friend or relative in the service, overseas or in training. As increasing numbers were inducted, those who had not yet been called became restless. Most of their friends were gone, and they were bored. Though they were treated well and received high pay, many began to feel that they did not really enjoy their jobs. Many were lonely; they longed for life in the segregated settlements on the Pacific Coast. These men did not resist the draft; some even volunteered to end the uncertainty. Once in uniform, most Nisei discovered a sudden sense of security; their loyalty could no longer be questioned.

With so many of their men in the army the war finally came home, and mainland Nisei began to feel that they were doing their share in the collective effort. Resettlers were increasingly caught up in the general war fever. They developed an interest in things military—uniforms, ranks, insignia, and decorations. The Red Bull shoulder patch of the 34th Division and the Statue of Liberty emblem of the 442nd were noted with pride. There were many tales of the agonies of basic training—the short haircuts, the infiltration course, the 20-mile hike—and prospective recruits listened avidly as 17-week veterans recounted their ordeals. Women volunteered for USO work and began to favor soldiers; special consideration was given to men on their last furlough before going overseas. Those who could not qualify began to feel left out, and young men found it increasingly embarrassing to explain their civilian clothing. Japanese names began to appear in large numbers on neighborhood honor rolls. Gold and silver stars appeared on windows—even in the relocation centers. Going-away parties were given; letters were written; women shared the problems of camp followers and of meeting husbands or sweethearts on furlough. Friendly inquiries from other Americans about Nisei soldiers in Italy made the resettlers feel that they were at last part of the war effort. On 14 July 1945 the War Department announced that 20,529 Nisei—20,289 enlisted men, 153 officers, and 78 WACs—were serving in the armed forces. Eventually all those outside the Tule Lake Segregation Center were drawn into the war work of the community in which they lived. Resettlers in particular came to feel that they were accepted as part of the mainstream of American life.

Transformation of the Stereotype

War correspondents quickly saw both the human interest and political significance of Nisei soldiers. Racism was a prominent feature of Axis ideologies, and the Japanese in particular had accused Americans of preaching democracy while persecuting ethnic minorities at home. What more dramatic demonstration of the "melting pot" philosophy than the active participation in the struggle of sons of Japanese immigrants! Although Nisei troops made up only a tiny segment of Americans fighting in Europe, they received a disproportionate share of publicity, as newsmen sent back a succession of dispatches on the exploits of these colorful men. Federal officials also assisted in giving widespread notice to Nisei contributions to the war effort through a flood of press releases.

Given the opportunity, Nisei soldiers established a battle record that astonished even their most optimistic supporters. The achievements of the 100th Infantry Battalion became legendary among American combat men even before the war ended; it was rated as one of the "crack outfits" of the 5th Army. The unit sustained so many casualties—reported to exceed 1,000 by the end of its first year in action—that it was designated the "Purple Heart Battalion." When the 442nd Regimental Combat Team reached Italy in the early summer of 1944, the veteran 100th was incorporated into it as its first battalion, and the entire unit went on to fight at the same tempo—with what some observers called a "disregard for personal safety." During 225 days in the line in Italy and France only a few were taken prisoner, and not one man deserted under fire. The adjutant of the regiment subsequently declared, "We never had a real disciplinary problem in the states and haven't had one AWOL overseas." When the fighting ended in Europe, the regiment had suffered 650 dead, 3,536 wounded, and 67 missing. With 10 group citations and 3,915 individual medals—including 1 Congressional Medal of Honor, 47 Distinguished Service Crosses, 342 Silver Stars, 810 Bronze Stars, and 2,022 Purple Hearts—the regiment became one of the most highly decorated units in the U.S. Army (Shirey, 1946). Such accounts of Nisei heroism were difficult to counter. Those who were still determined to distrust all Japanese dismissed the releases as "WRA propaganda," but for others it was irrefutable evidence that most Nisei were loyal to the United States.

Since the activities of the military intelligence service were cloaked in secrecy, the accomplishments of Nisei troops in the Pacific could not be disclosed until late in the war. In August 1944 the awarding of Bronze Stars to six Nisei who had participated in the conquest of Saipan was announced. Later that year and during the spring of 1945 various newspapers in the areas where their families resided published articles on individual Nisei who had been killed or decorated in the Pacific theater. In April 1945 Joe Rosenthal, who took the memorable picture of marines raising the American flag on Mount Suribachi, revealed that many Nisei serving in the Pacific had volunteered for dangerous missions and that they had coaxed countless enemy soldiers to surrender—thus saving American lives. Such articles were frequently accompanied by editorial comment condemning bigots and their unfair charges. Gradually, increasing numbers of men in prominent positions began to refer to Nisei contributions to the war effort and joined in criticizing those who continued to demand punitive measures.

Human interest stories appeared in the mass media presenting the plight of the group in sentimental terms. In February 1944 *Life* magazine ran a full-page photograph of a bandaged Nisei veteran, blinded in combat in Italy. *Time* magazine featured a two-column article on "Ben Kuroki, American," the story of a Nisei from Nebraska who had volunteered soon after Pearl Harbor and had flown his full quota of combat missions as an aerial gunner. There were numerous reports from combat zones: photographs of Nisei soldiers in action, reports of the astonishment of German prisoners on being captured by Japanese, and accounts of how Hawaiian AJAs used pidgin on their radios to confuse German eavesdroppers. Newspapers and magazines throughout the nation printed photographs of a forlorn Japanese mother in a relocation center sitting under a four-star service flag, of still interned Japanese parents being awarded posthumous medals won by sons killed in action, of a baby born to the widow of a Nisei recently killed in Italy. These touching accounts brought the story of the injustices of evacuation to the American people in human terms, and the impact was apparently more jolting than abstract arguments concerning infringement of constitutional rights. Public opinion polls reflected the change of attitude. An NORC poll in September 1944 revealed that 56 percent of a national sample felt that after the war Nisei should have the same opportunities for jobs as white people.

In intellectual circles, where the evacuation had been accepted with little enthusiasm to begin with, there were second thoughts. Some were guilt stricken. The whole program was reminiscent of Nazi concentration camps, for the Western Defense Command had carried it out on a strictly ethnic basis. News commentators began to question the "Pacific Coast attitude" on evacuation. Columnists and editorial writers in increasing numbers began criticizing the mass expulsion as a stupid blunder—an expensive one both financially and in terms of propaganda. Groups of congressmen condemned inflammatory statements still being made against Nisei. Since the Roosevelt administration was responsive to the intellectual community, protests and suggestions reached some of the highest offices in government. Even when they were preoccupied with conducting a war for survival, key decision makers had to concern themselves with fair play for a tiny minority group. In April 1944 a few months after jurisdiction over the WRA had been transferred to his department, Secretary of the Interior Ickes denounced the "hate-mongers" on the Pacific Coast and accused them of not believing in the Constitution; he warned that the government would not be stampeded into undemocratic action. Restrictions were progressively relaxed. In September 1944 the army permitted the return of a few evacuees to the Pacific Coast, largely to test sentiment there. A Nisei girl accepted at Pasadena Junior College was granted permission to return.

As exclusion measures were relaxed, a flood of protests arose on the Pacific Coast. Especially in California various groups reiterated their demands for the continued exclusion of all Japanese, including Nisei war veterans. Stickers began to appear on automobiles: "No Japs." In July

1944 the state of California began law suits to recover land it regarded as illegally held by Japanese aliens; it won several of the early escheat cases. This continued opposition to the return of evacuees attracted nationwide attention, and various organizations, officials, and commentators decried the "California attitude"—charging that selfish economic interests lay behind it. Labor leaders, churchmen, and educators on the Pacific Coast mobilized to urge welcoming back the evacuees, and residents in the western states became divided. In July 1944 an initiative petition in California to bar Nisei from land ownership failed to get enough signatures to be placed on the ballot. But opposition continued. In December the California State Senate passed a resolution against the return of evacuees. As late as May 1945 California congressmen were denouncing "WRA propaganda"; three of them asked the House Appropriations Committee to cut the agency's publicity funds. Several leaders of fair-play groups received anonymous telephone calls threatening them with violence.

Three events in particular attracted widespread attention. On 11 November 1944 a barber in Parker, Arizona, refused to serve a crippled Nisei veteran and ejected him forcibly from his shop. At the time of the incident the soldier was on crutches, wearing a uniform bearing seven service ribbons—including the Purple Heart. When questioned by reporters, the owner denied shoving the disabled man, but he admitted refusing service and vowed that he would continue to do so even if "they close me up." A few days later this occurrence was dramatized on the program "Five Star Final" over radio station WMCA, New York. At the same time a moving letter from a Nisei soldier to his father was read, describing how his brother had died in action. It was then disclosed that the letter writer had also been killed. At a news conference on November 21 President Roosevelt paid tribute to the courage of Nisei soldiers and added that he failed to see how anyone could become upset over the release of a few Japanese from relocation centers.

Another incident to attract national attention was an act by the American Legion post in Hood River, Oregon. On 2 December 1944 it erased the names of sixteen Nisei servicemen from the county war memorial. At a press conference on December 14, Secretary of War Stimson condemned this action as undemocratic and disclosed that three of the men whose names had been removed had been wounded in action. Protests arose all over the country, and several American Legion posts joined in the condemnation. When the *Stars and Stripes* published an account of what had happened, war correspondents reported that soldiers overseas were outraged. Several from Oregon sent indignant letters to their local newspapers. But the legionnaires were undaunted and announced that they had found a seventeenth name to erase, one that had been overlooked. When it was disclosed that the seventeen had all been inducted before evacuation and that by this time more than sixty Nisei from the area were in service, the response was that Nisei were dual citizens and could not be trusted. On 2 January 1945 one of the Hood River men was awarded a Bronze Star, and the national commander of the American Legion described the removal of names as "ill considered and ill advised"; he urged the post not only to put back the names but to add thirty-nine others. But the post remained adamant. When confronted with the Nisei combat record, it cited the argument commonly used by those still opposing the return of evacuees—that Nisei achievements in Europe were to be discounted because they were fighting against Germans and that Japanese enjoyed fighting against white men. It also reiterated the charge that WRA releases were merely propaganda.

On 16 February 1945 the War Department announced that one of the sixteen Nisei had been killed in action in the Philippines. The circumstances of his death were widely publicized, and the hero was subsequently awarded a posthumous Silver Star. These disclosures evoked such a violent national reaction that even local newspapers joined in condemning the legionnaires. National officers of the American Legion rebuked the Hood River post, saying it had "dishonored itself," and acknowledged the Nisei war record. Faced with censure and possible expulsion from the national organization, the Hood River post finally capitulated. In March 1945 it agreed to restore the names, although even at that time a spokesman repeated the charge that all Nisei owed their allegiance to the emperor of Japan.

On 17 December 1944 exclusion orders covering most of the Pacific Coast were rescinded. Slowly a trickle of evacuees began to return to their homes. During the first six months of 1945, 10,288 Japanese requested leaves from the centers, and 32 percent gave their destination as the Pacific Coast. Although some were welcomed by their friends and neighbors, many encountered a campaign of terror—night riders, arson, bombings, vandalism, and boycotts. In March 1945, when night riders fired into the home of a Nisei veteran, the FBI immediately entered the investigation; and the Secretary of War denounced the attack on the ex-soldier as an "inexcusable and dastardly outrage." A month later anti-evacuee leaders met in Sacramento to organize statewide resistance. In May a girl who had returned to Orange County was threatened; she had four brothers in the army, one of whom had been killed in action. In June a farmer near Fresno—the only one arrested in twenty-one shootings since January—pleaded guilty to firing four shotgun blasts into a Nisei home on May 22. When he received a suspended sentence of six months, the American Civil Liberties Union asked the attorney general of California to file a felony charge, arguing that an unpunished misdemeanor was not enough. Secretary of the Interior Ickes also attacked the light verdict.

Another incident that shocked the nation occurred in Placer County, California. In 1940 there had been 1,637 Japanese in the county; all but 66 were farmers. On 18 January 1945 the packing shed of a Nisei who had returned to Auburn was set afire. On the following night several carloads of night riders fired shots into his home. When the Nisei announced his determination to stay, the sheriff placed a 24-hour guard around his home. In February a campaign was initiated to intimidate returnees by violence and boycott. Six stores put up signs on their windows: "No Jap Trade Solicited." Arguing that this would bring discredit to the community, a citizens' group was organized to help the evacuees. Four men were then arrested for the dynamiting in Auburn and charged with arson; in April three of them were brought to trial. After the prosecution had presented evidence against them, the defense attorney did not call a single witness, nor did he dispute the state's evidence. "This is a white man's country," he pleaded, and he urged the jury to keep it that way. He described the trial as "a battle between the WRA and the people of Placer County." The jury acquitted the three men.

During the first six months of 1945, there had been thirty-three attacks on former evacuees. In only two cases had anyone been arrested and brought to trial, and no one had been punished. Liberal organizations began demanding federal intervention, with troops if necessary, to protect the returnees. On 12 May 1945 Secretary Ickes denounced the intimidation. Pointing out that several of the victims were relatives of soldiers whose homes displayed service flags, he accused the terrorists of using Nazi Storm Troop tactics. National press comment on his speech was favorable, even in newspapers that usually disagreed with him. The night riders were condemned as cowards. As news of the Auburn case evoked cries of outrage, state officials were forced to take a firm stand. The attorney general joined in condemning the violence and ordered sheriffs throughout the state to maintain order.

By the time Japan surrendered public opinion on the national level had changed. In April 1945, when NORC asked how many Japanese would try to do something against the United States if they had the opportunity, 53 percent of a national sample replied "none" or "only a few." Another survey in 1946 revealed that 50 percent of a national sample believed resident Japanese to be loyal to the United States; only 25 percent regarded them as disloyal; and the rest were uncertain. When asked if Japanese immigrants should be allowed to become American citizens, 46 percent expressed approval. Further analysis of these data revealed a high correlation between education and favorable judgments. When respondents were asked to amplify their views, many mentioned the combat record of Nisei troops, and others spoke of their reputation as reliable workers (National Opinion Research Center, 1946; Butler, 1946). Thus, the reputation won by Nisei infantrymen among American troops in Europe followed them home, and the group had been redefined. Politicians attacked the injustices of evacuation, and on the Pacific Coast some of the very officials who had taken credit for "getting rid of the Japs" participated in ceremonies welcoming home the war heroes.

During the latter part of 1945 all wartime regulations concerning persons of Japanese ancestry were rescinded. Effective 1 June 1945 Nisei inducted into the armed forces were no longer segregated but were processed and assigned on the same basis as other troops. On September 4 the army revoked all remaining exclusion orders. The closing of all relocation centers on December 1 was announced, and the government paid the transportation of those still left in the camps to their point of origin. During the spring and summer of 1946 approximately 55,000 Japanese returned to the western states. Many problems arose, for those remaining in the centers to the end were the people least able to care for themselves. Former farmers found that they could not start over, for they had no money. Many businessmen found their stores now located in areas occupied by other ethnic groups. Many found that their homes had been sold or rented to war workers, and they had to live in hostels, trailer camps, or double up with friends. Property that had been stored was damaged; machinery and household goods were ruined. The WRA appealed for local relief, and for the first time large numbers of Japanese were placed on welfare rolls (Bogardus, 1947). Local politicians, some of whom had been avid supporters of evacuation, now made demands on the federal government to make reparations.

The evacuation program came to be recognized as a blunder. Legal experts attacked the treatment of Nisei as unjust, a move that had set a dangerous precedent—incarcerating citizens without due process. They criticized the U.S. Supreme Court for decisions supporting the military

(Rostow, 1945; tenBroek et al., 1954). Motion pictures, such as *A Bad Day at Black Rock* and *Go for Broke*, further popularized the unfairness of the program and pointed to Nisei contributions under duress. In 1948 Congress enacted legislation to compensate evacuees for their losses; and 22,000 claims, totaling over $133,000,000, were subsequently filed. Since the procedure initially established was so complicated, the claims department was able to handle only fifty-five cases a month; in 1951 the act was amended for faster service. The last claim against the government was settled in November 1965, after two of the original three plaintiffs had died (Fisher, 1952; Petersen, 1966). Japan was granted a small immigration quota in the controversial McCarren-Walter Act, and in 1952 Congress voted naturalization rights for the Issei.

In Hawaii questions of Nisei loyalty could no longer be raised, for of all casualties suffered by Hawaiian servicemen 80 percent of the dead and 88 percent of the wounded were men of Japanese ancestry. Opposition to statehood on grounds of possible treason could no longer be taken seriously. Thus, in 1959 Hawaii joined the union, and one of the first men elected to Congress from the new state was a veteran of the 100th Battalion.

Analytic Summary and Discussion

Thus, Japanese-Americans were redefined as desirable objects under the most adverse conditions imaginable—in the midst of a war against Japan. The manner in which members of any ethnic minority are treated depends on popular beliefs concerning the group, which are usually stereotyped. To account for the reassessment of the Nisei we must look at the manner in which objects are constructed in a mass society. One possible explanation is suggested by Klapp (1948) in his discussion of the creation of popular heroes. A hero-making situation is one that is dramatic, full of human interest, and at the focus of public concern. It is one marked by suspense; an important issue is unresolved; and the outcome is in doubt. In such a context anyone who contributes conspicuously to a favorable solution becomes a hero. One common type of hero is the "Cinderella"—an unpromising candidate who has been ridiculed but rises to succeed over his favored opponents. Should a Cinderella prevail at a time of dire need, he also becomes a delivering hero. During World War II representatives of the Nisei performed magnificently just at the decisive moment when national attention was focused on them. The situation had "color"; it was one that excited the imagination. Furthermore, the Nisei were definitely underdogs, and Americans traditionally have been sympathetic to those struggling against great odds. Spontaneous popular acclaim is often followed by formal recognition; once it is formed, the new definition is then supported by popular legends, publication, and motion pictures. Activities based on the new definition then became institutionalized.

The new stereotype appears to have persisted, and in the decades that have followed many persons of Japanese ancestry appear well on their way to realizing the "American dream." The history of the United States has been a saga of a succession of immigrant groups, each starting at the bottom of the social scale, taking advantage of free education, and climbing in a generation or two into positions of respectability. Since this pattern has not held as clearly for colored minorities, it has been assumed that high visibility was a decisive factor. Since the Nisei are readily distinguishable from the general population, they appear to be an exception. No provisions were made for them in the antipoverty program under the Economic Opportunity Act of 1964. Although private antagonisms undoubtedly persist in some circles, organized agitation against them has disappeared, even on the Pacific Coast. Nisei have access to all public accommodations. Residential segregation is rare; except in Los Angeles it is difficult to find any large concentrations of Japanese. Upward mobility has been marked. Although many Nisei are doing menial work, many others occupy positions of responsibility. In 1960, 56 percent of Japanese males were in white-collar occupations, compared with 42.1 percent of the white population; 26.1 percent were in professional or technical work, compared with 12.5 percent for Caucasians (Schmid and Nobbe, 1965). Some Japanese newspapers are still being published, and old friends manage to keep in touch; but in many areas ethnic ties are becoming increasingly more difficult to maintain. The third generation is becoming more fully assimilated; many of them are unable to communicate with their grandparents and become impatient with parental talk about the trials of evacuation. Especially in areas east of the Rocky Mountains intermarriage has become commonplace. Although it is premature to speak of the integration of Nisei into American life, there can be little doubt that the group is no longer a problem minority.

3. «» the social world of nisei servicemen

On a lovely evening in the spring of 1701 Lord Asano committed *seppuku* (suicide) in a dignified ceremony befitting his high rank. In a moment of rage he had drawn his sword in the shogun's palace, a capital offense. His holdings were confiscated, leaving his samurai unemployed. Since Lord Kira, whose greed had goaded the young nobleman into the transgression, had escaped unpunished, Asano's retainers swore vengeance. More than 120 signed a blood oath, but most of them defected or drifted away. The others endured privation and humiliation to deceive Kira's ubiquitous spies. Finally, on December 14, 1702, the remaining forty-six stormed Lord Kira's mansion and avenged the martyrdom of their master, knowing full well that they would have to die for this attack, whether or not it succeeded. *Chushingura*—the tale of the forty-six (sometimes forty-seven) *ronin* (unemployed samurai)—has been so perennially popular that it might be regarded as the national epic of Japan. Although it is about a vendetta, it stresses loyalty, honor, and justice as keystones of the samurai code. Although the story has a historical basis, it has been embellished over centuries of retelling in puppet plays, Kabuki performances, motion pictures, and now television serials. Since swashbuckling samurai movies were as popular among the young as American westerns, it would be difficult to find a Nisei who had not seen or heard at least one version of *Chushingura.* Thus, selected features of Japanese culture became incorporated into the outlook of most Nisei.

Ethnic Stereotypes among the Nisei

Culture consists of those common understandings that characterize a particular group. These models of appropriate conduct are not always carried out, but they do provide expectations of which all parties are aware. Although American in language, custom, and dress, most Nisei learned many Japanese values, including the code of honor of the samurai. These standards were not only stressed at home but were constantly reiterated within the segregated communities. Athletic contests were especially important. In major American sports—especially baseball and basketball—and in Japanese sports—judo, *kendo*, and *sumo*—intercity leagues were formed. In the spirited competition for championships Nisei athletes were tested for their endurance and courage. Whenever someone persevered under duress until victory, the Nisei joked about his *Yamato damashi*—the warrior spirit presumably inherited from Japanese ancestors. Even in jest, however, it was understood that this was a desirable trait. Such models were constantly reinforced not only in repeated praise of exemplary performances but also in ridicule of those who fell below acceptable standards. As with any ideal code, violations were commonplace; yet these were some of the criteria in terms of which Nisei evaluated themselves and each other.

The unusually broad generation gap facilitated the formation of a highly stereotyped conception of *real* Japanese—a people characterized by traits that placed them apart from others. To a remarkable extent Issei parents were able to conceal their own misconduct and that of other adult Japanese from their children. With reference to sex, for example, the Japanese in Japan were not puritanical; indeed, they considered pleasures of the flesh as worthy of cultivating as a fine art. They accepted geishas, prostitutes, and extramarital affairs as a natural part of life (Benedict, 1946:177-94). But the Issei were especially reticent about erotic matters. Control over sexual relations was very strict, and most Nisei grew up believing that Japanese lived up to austere, almost ascetic, standards. Thus, certain features of Japanese culture developed among the Nisei in highly exaggerated and idealized forms. Unlike people in Japan, the Nisei had only limited opportunities for reality-testing. They were too far away to see, and they could not temper their ideals through a recognition of life as it actually went on in Japan.

A strong sense of being Japanese was instilled in the children. They were often told, "Never forget that you are a *Nihonjin*." The ideals of Japan were frequently contrasted with the slovenliness of the *Hakujin* (white people). When parents were asked why Nisei had to comply with a code of conduct to which others were not held, the explanation often given was: "Because you are Japanese." Many Issei felt that this was sufficient reason for doing anything. Although some Nisei were sophisticated enough to know that most behavior patterns are learned, many accepted the naive popular conception of the genetic inheritance of such traits.

Before World War II Japanese were taught from childhood to conceive of themselves as representatives of groups, most important of all the family. Each person was a trustee of a spiritual estate. He was bound by a network of obligations and was expected to subordinate his personal desires to collective requirements. Filial piety (*on*) was stressed, and each individual took his proper station within his family by generation, age, and sex. *On* was a debt, a burden, that one carried as best he could. It included his debt to his parents. *On* always overrode personal preferences. Thus filial piety became more than mere deference or obedience to parents; one repaid his debts to his forebears by passing on to his children the care he himself had received. While the prerogatives of rank were great, they were not unconditional. Those who exercised authority also acted as trustees. Father and elder brother were responsible for the household—living, dead, and the yet unborn. They acted to uphold the honor of the house. Elders who in their day had had to submit to decisions of the family council demanded the same of their juniors. Love, while desirable, was not necessary; even if one were resentful, he still had to meet his obligations. Individualism on this score was discouraged from childhood, and rebels were punished at home as well as scorned by neighbors. Thus, righteousness consisted of recognizing one's place in a network of mutual indebtedness that embraced forebears, contemporaries, and future generations. Most Japanese were bound by a deep sense of obligation to family, community, and country (Benedict, 1946:98–99).

Great stress was placed on family honor. Self-restraint to prevent disgracing the family was expected. The individual was a representative of the larger unit, and children were cautioned, "Don't do that; people will laugh at our family." They were also challenged to be worthy: "When you grow up, you must bring glory to the family." The reputation of the family had to be kept shining, whether one was right or wrong. Nisei could not help but notice their parents' condemning another family for the misdeed of a single member. If the name of a Japanese caught in a vice raid were published in a local newspaper, for example, one heard scorn heaped on the culprit's family and expressions of sympathy for his mother and sisters. How could they show their faces in public again? If one violated a norm, he was punished by his family. If outsiders did not know of it, it was hidden from them; but within the family the offender was punished nonetheless. Many Nisei were ambivalent about such responsibility. The notion of self-sacrifice for the family did not appeal to those exposed to American ideals of individual freedom, but most of them complied.

The primary task of life was to fulfill one's obligations. The pursuit of happiness was not seen as a serious goal. To give up pleasure requires strength, and it was the most admired virtue. Most esteemed in folktales were self-sacrificing heroes and heroines. Giving in to personal desires when they conflicted with one's obligations was a sign of weakness. Moral obligation always took precedence over compassion. One had a *giri* to his community—the fulfillment of contractual obligations. *Giri* included loyalty to one's lord and to fellows of his own class; on occasion it might even override his personal sense of justice—when one felt his hands were tied. He had to live up to the requirements of the etiquette of his station, for the most dreaded condemnation was that one did not know his *giri*. Many stories were told to children of the virtuous —those who had earned the right to self-respect by proving themselves strong enough to accept personal frustration. A man who suffered in fulfilling his duty at great sacrifice often imputed to his neighbors admiration for himself. One also had a *giri* to his name. One had a duty to clear his reputation of insult or charges of failure. This might even justify a vendetta. He had to keep his reputation intact at all cost. Thus, inner controls were expected of each individual. Stoicism and self-control were required, if necessary, for self-respect.

Most Japanese were very responsive to public scorn and ridicule, even when it was unjustified. There was constant apprehension concerning the opinions of others regarding one's ability to fulfill his obligations. There was acute sensitivity to gossip. Japanese tended to fear failure— bringing shame and dishonor on his family or community. They felt as strongly about insults as they did of benefits received. There was extreme sensitivity to slights, even those that others might regard as trivial. Any questioning of competence became a criticism of the person and of the unit to which he belonged. Mockery was a drastic sanction, and reprimand before others was regarded as a disgrace. Thus, great significance was placed on responsibility. Each fearfully assumed that others sat in judgment of every act. If one failed, he was tortured for having shirked his duty. Failure to meet one's responsibilities elicited shame which, unlike guilt, is a reaction to the criticisms of others (Benedict, 1946:133–76; Iga, 1957; Nishi, 1963). The sense of shame was well developed in most Nisei.

Achievement through individual effort was applauded. Status was seen as something that could be earned or improved through self-discipline, frugality, and hard work. The ideal was the attainment of success through perseverance, even in the face of adversity. Abraham Lincoln was much admired, and many Japanese heroes were men who through diligence had risen from humble beginnings to greatness. One was expected to complete any task he undertook and to do it well. As in the American middle class, deferred gratification was encouraged.

Japanese society was clearly stratified, and one common feature of Japanese culture was the preoccupation with status gradations. Each person was located in several complex hierarchies in which neither those of higher nor lower rank could without penalty overstep their prerogatives. The differences were not merely those of class, but of age and sex as well. Such inequality had persisted for centuries. Although the various castes were legally abolished during the Meiji restoration, patterns of hierarchical discourse had been retained. Different words were used in addressing those of equal, higher, or lower rank. Thus, the Japanese had a "respect language" and two others. Righteousness consisted of recognizing one's place in the network of mutual indebtedness, respecting the authority of those above, and showing consideration for those below. Respect rules were learned and observed within each family. Persons meeting for the first time began with inquiries concerning one another's work and family position; it was necessary to establish the pedigree of each participant in any transaction in order to know which language to use and what degree of deference to show.

The Japanese relied on habits of deference set up in the past and formalized in their etiquette. In interpersonal contacts formality was stressed rather than spontaneity. Competence in relations with people of different rank was expected. Politeness and respect for those of higher status was emphasized; each person was expected to give deference to all those who outranked him—on whatever grounds. This façade was maintained in public, even if it was not always observed in private. In spite of linguistic differences Nisei could not help but observe the stance with which their parents addressed outsiders, especially those of high status.

Cleanliness has frequently been noted as a mark of Japanese culture—neatness and ritualistic tidiness as well as hygienic cleanliness. Scholars were often pedantic, preoccupied with minutiae, and artists also focused meticulously on details. Cleanliness of one's body and possessions was so emphasized that everyone, even a poor laborer, was expected to take a daily bath. Nisei observed the manner in which their parents struggled to keep themselves clean, despite the dirty work that many had to do. Furthermore, those who failed to bathe with sufficient regularity were ridiculed.

A high value was also placed on honesty and on obeying the law, and many American officials have noted that most Japanese were orderly and law abiding. But respect for authority had a special meaning to the Issei. In Japan laws were promulgated in the name of the emperor, and civil administrators and the police were regarded as his representatives. Thus, obeying the law was a repayment of indebtedness—*chu*, a form of fealty to the emperor. Since a man's self-respect depended on his ability to meet his obligations, violation of the law had a significance to Japanese that was difficult for outsiders to comprehend (Benedict, 1946: 129–30). A widespread belief existed among Nisei that Japanese did not steal or commit other such offenses, and governmental statistics tend to support this view (Kitano, 1969:117–19). It should be added, however, that strong community pressure developed in the segregated areas to maintain appearances. While others committed crimes or went on relief, such conduct was regarded as beneath the dignity of Japanese. Minor violations were hidden from authorities, and offenders were often viewed as having done something that was unnatural. They became objects of dismay and were regarded as aberrations, even in instances in which extenuating circumstances might have led others to sympathize with the offender. It is doubtful that most Nisei understood the concept of *chu*, but they could not fail to notice the abhorrence shown by their parents toward violating even petty regulations.

With so many striving for excellence, competition was inevitable. Yet various procedures had evolved to avoid direct confrontation. In contexts in which one party might fall short, intermediaries were used. It was a matter of etiquette to avoid shame-producing situations. Each person, furthermore, tried to avoid circumstances in which failure might occur, for this would lead to unbearable shame (Benedict, 1946:154–58). Among Nisei a strong tendency developed to stress equality of status. Those who moved ahead too conspicuously were condemned, as were those who lagged behind. On many basketball teams, for example, the captain was merely a formal leader who conferred with the referee. Decisions about the team were made collectively, and substitutes on the sidelines decided on their own when to replace a player who was losing his effectiveness.

Despite claims of being members of a proud "race," Nisei were well aware that they lived in communities dominated by Caucasians. Furthermore, in school and through the mass media they acquired many American values. Most Nisei on the Pacific Coast acknowledged feeling inferior to the *Hakujin*. They formed a stereotyped conception of the white man toward whom they were ambivalent. Many openly admired Caucasians, envying and emulating their way of life. White people were often idealized and overvalued. White women in particular were regarded as something special. A friendly waitress, for example, was viewed with astonishment—a white woman who did not mind serving Japanese! A Caucasian fellow student who was considerate and solicitous was regarded as an exception, an unusual person who really "understands us." Some Nisei scorned the more esoteric features of Japanese culture, not from lack of affection for their parents but to escape feelings of inferiority in relation to Caucasians. Individual Nisei who succeeded in the larger community were also viewed with ambivalence. On the one hand, they were praised for bringing honor to the group; at the same time they were criticized as trying to imitate white people. As is true with so many other ethnic minorities (Frazier, 1957:217–28), the exceptional Nisei who achieved high status was often accused of being "snooty" and of thinking "he is too good for us."

In Hawaii the idealization of the Haole was even more exaggerated, as was the ambivalence. Most members of minority groups had difficulty in seeing Haoles as human beings; they were viewed in stereotyped terms, as haughty and bigoted. Most Haole children went to Punahou, private schools, or English Standard schools; and the subtle class distinctions among them, keenly felt among themselves, were not perceptible to the uninitiated. They were all seen as part of a privileged group (Fuchs, 1961:61, 122). When a Haole tried to be friendly, he was required to prove over and over that he was not like the others. Even after a person had passed many tests a lingering suspicion remained that, despite outward appearances of friendliness, his *real* nature would be revealed in a crisis. There were many tales of Haoles who had thus betrayed their trust. Yet, they also looked up to the Haole, and most AJAs acknowledged feeling inferior to them. Men successful in business and the professions confessed being uncomfortable in the presence of Caucasians; they felt socially inadequate. AJA students at the University of Hawaii indicated some of the sources of their feelings of inferiority; they were too short, inappropriately dressed, awkward, lacked knowledge of etiquette, spoke English poorly, and had parents who were ignorant of American customs (Masuoka, 1936; Smith, 1945). In brief, they fell short of Haole standards. Thus, white people were both hated and emulated. Their way of life was desired, but most AJAs did not want to mingle socially with them; open competition would expose them to shame.

Mastery of standard English became the index for measuring one's degree of "Haolefication." Since Caucasians were the only people on the islands for whom English was the native tongue, Haole and standard English became associated in the popular mind. Pidgin was used by all the workers and by none of the masters; it thus became a badge of inferiority. It symbolized equality among those who used it. It was also the language used at home, the language of warmth and intimacy. Speakers of standard English somehow seemed distant and aloof (Reinecke, 1938; Hormann, 1960; Morimoto, 1966). To qualify for preferred positions one had to approximate standard English, but members of minority groups who did develop such linguistic skills were often accused of imitating Haoles. Since those who had not mastered the language had difficulties at the University of Hawaii, increasing numbers of Orientals struggled to qualify for the English Standard high schools. But those who succeeded were often accused of being sissies. To belong to an adolescent gang one had to speak pidgin; failure to do so was taken as an indication of putting on airs. Those who used standard English outside of school were ridiculed. Sometimes they were warned: "No talk like Haole!" [Don't talk like a white man!] While the Haole way of life was envied, those who got too far ahead of the others in emulating it were condemned; in some cases they were even assaulted.

To outsiders most Nisei soldiers looked alike. They were short, stocky, dark skinned, and uncomfortably like the enemy in the Pacific. They wore American uniforms; yet they seemed out of place, incongruous. Since they had been treated alike on the basis of ethnic identity, they were forced to act together as a unit. But the men were actually drawn from diverse backgrounds, and differential acculturation resulted in considerable heterogeneity. This was recognized among the Nisei themselves, who distinguished those who were more "Americanized" from those who were more "Japanesy." Least acculturated were the Kibei. Although some Kibei spoke fluent English and graduated from American universities, most were more at home in Japanese. Nisei who grew up in rural communities, where they attended segregated schools, were often mistaken for Kibei because of their attitudes, their general lack of knowledge of American life, and their heavy Japanese accent. Hawaiians from plantations relied primarily on pidgin, and in many cases their English was difficult to understand. Their mastery of the Japanese language varied considerably, but most mainland Nisei regarded them as "Japanesy" in their orientation. Most Nisei from mainland ghettoes, where there was some de facto segregation of schools, spoke English with a slight accent. In general Christians were more acculturated than Buddhists, for they had friendly contacts with other Americans of their denomination. Since they had shared pleasant experiences with Caucasians, most of them had no particular fear or distrust of white people. Nisei from the Pacific Northwest were in general more urbanized in their outlook and spoke more fluent English than Californians. Most acculturated, of course, were the isolated individuals who had grown up from infancy among Caucasians. They spoke little Japanese; most of their friends were Caucasian, and their outlook differed little from that of other Americans.

Assimilation, as distinguished from acculturation, is more a matter of the way in which a person conceives of himself and of the audience for whom he performs (Shibutani and Kwan, 1965:504–15). Many Nisei were confused about their personal identity. They knew they were American citizens and were sensitive to the reactions of other Americans; yet most of them felt left out of the mainstream of American life. Both in Hawaii and on the mainland many conceived of themselves as Japanese in "race." Since most of their daily contacts were with people like themselves, they were most concerned with the opinions of those with whom they identified on an ethnic basis. Most Nisei lived for an audience consisting largely of fellow Nisei and their parents. Those who were more assimilated were less responsive to the demands of this audience and more concerned with the expectations they imputed to other Americans. Nisei varied considerably in the extent to which they had become Americanized, but it should be noted that even the most assimilated pursued some Japanese values. Although they rejected the more readily visible features of Japanese culture—such as bowing, dress, language, gestures, walking gait—most of them were not aware of the extent to which they had incorporated basic Japanese values into their personal outlook. Some even disavowed the very values by which they lived. They were more Japanese than they themselves realized.

Mobilization for a Special Mission

At the outbreak of war Nisei were serving in the army on the same basis as other Americans, although they were excluded from the navy and seldom got into the air force. In the absence of a clear national policy the fate of these men after Pearl Harbor depended on their respective commanding officers. On some posts Nisei were immediately disarmed and reassigned to menial tasks; many were transferred to service commands. Some continued to perform the same duties that they had before; their officers stood by them until ordered to make changes. A few apparently escaped notice and remained with their old assignments. Some men were discharged without explanation—most of them honorably but others for the "good of the service." Others were transferred to the Enlisted Reserve Corps (ERC) and sent home. On September 26, 1942, a directive from the adjutant general ruled that Nisei in the ERC were not to be recalled to active duty but to be discharged. Some of these men were drafted again in 1944, but others were never reactivated.

Within a few months after Pearl Harbor the only Nisei soldiers bearing arms were infantrymen in Hawaii; because of the troop shortage Nisei in the 298th and 299th Infantry Regiments were retained. Since their status as Americans was in question, they were eyed with mingled curiosity, wonder, and resentment. It was widely recognized that the Nisei soldier was in a unique position to counteract charges of disloyalty. The intense pressure of being the only hope of exonerating the suspected group fell on them, and they realized that they had to perform more effectively than other soldiers. It is not surprising, therefore, that these men were imbued with a sense of mission. They were fighting two enemies—those abroad and those at home. Their cause was the vindication of the Japanese in America.

In June 1942 the 1,250 enlisted men and 30 officers who made up the newly formed 100th Infantry Battalion were transferred to Camp McCoy, Wisconsin. Even in training they made a deliberate effort to create a favorable impression. Their conduct in public was exemplary. They were careful to be in proper uniform at all times and were meticulous about observing military courtesy. The use of the Japanese language was forbidden. Whenever occasions arose for demonstrations of loyalty, the men participated with a show of enthusiasm. By common agreement they made repeated donations to blood banks; 90 percent of them purchased war bonds; they conserved sugar, and from one monthly allotment they returned 2,400 pounds to the commissary. They were always ready to cooperate in various patriotic projects in nearby communities, and the officers made several public appearances. They put in extra effort to impress a succession of skeptical inspecting officers. Such conduct did not arise from orders issued from above but from social pressures among the men themselves. The few who got into drunken brawls and especially those who became involved in difficulties with Caucasian women were condemned by their comrades far more severely than they were reprimanded by their officers. Those who were careless were warned by older and more responsible men. The few who failed to cooperate were tormented and sometimes even threatened with violence. The entire battalion was on trial, and the men would not permit the heedless few to ruin its reputation (Murphy, 1954:81–96).

The 442nd Regimental Combat Team—consisting of the 442nd Infantry Regiment, the 522nd Artillery Battalion, the 232nd Combat Engineering Company, and the 206th Army Ground Forces Band—was activated on 1 February 1943. Most of the NCOs in charge of maintaining camp and of training were drawn from Nisei already in the army, who had been transferred from various service commands. Early in February the cadre arrived at Camp Shelby, Mississippi, and began repairing the barracks. On April 13, 2,686 volunteers arrived from Hawaii; mainlanders came in smaller contingents from the various relocation centers. When the full complement of volunteers was achieved, the unit started its basic training. Although some of the officers were Nisei, most of them were Caucasian. The majority of NCOs were mainland Nisei. The 100th Battalion was already at Camp Shelby when the volunteers arrived. Many Hawaiians had hoped that the combat team could be built around the already trained 100th, but the two units remained separated.

It was at Camp Shelby that large numbers of Hawaiian AJAs and mainland Nisei first came into close contact. Difficulties arose almost at once. Linguistic differences were an initial source of misunderstandings. Most mainlanders spoke standard English; their manners seemed more refined; their speech was less blunt. Hawaiians complained that the mainlanders could not understand what they said, that they were snobbish, and that they were too "Haolefied." They were designated as "Kotonks"; the most common explanation for this term was that tapping the head of a mainland Nisei would produce a hollow sound, much like that of an empty coconut. When some Kotonks learned pidgin in an attempt to establish cordial relations, many Hawaiians were resentful; it seemed to imply that the latter could not understand English. In the beginning, then, Kotonks were disliked because they approximated Haole standards too closely. Since both groups had faced similar disadvantages because of their minority status, some Hawaiians resented Kotonks as a source of shame—a reminder of their own inadequacies.

The stereotype of the Kotonk developed quickly. One common complaint was that the mainlanders were cliquish; they did not associate with Hawaiians, keeping to themselves. Furthermore, they were not sufficiently candid in their personal relations. They were never open; they were friendly, but hesitant; they were quick to complain and critical behind others' backs. They were petty, holding grudges over trivial matters. Kotonks lacked warmth; they were too much like the Haole. They were much too ambitious for personal advancement. The high proportion of mainlanders with rank exacerbated the difficulties. They were called "overnight sergeants," lacking in courage and conviction. It was charged that they were constantly "brown nosing" the Haole for promotions. In contrast, Hawaiians saw themselves as more open, happy-go-lucky, frank, friendly, and never trying to get ahead at the expense of others.

This suggests that another source of friction was the inability of mainlanders to understand the *aloha* tradition—the spirit of working and living in friendly cooperation, sharing whatever one has with needy friends. *Aloha* is given freely, not in expectation of return. Hawaiians gave themselves more completely than mainlanders, and they were impatient with what appeared to be a negative, calculating attitude. People who are themselves hospitable are often hurt when others do not reciprocate; they feel snubbed, left out (Fuchs, 1961:6–9, 73–74). Thus, the Hawaiians complained that mainland Nisei were *manini*—self-centered.

Hawaiians were also critical of what they regarded as the mainland Nisei's negative orientation toward the war. Since the conduct of Kotonks could affect the status of AJAs in Hawaii, many were apprehensive. The mainlanders appeared to be handling the relocation problem in a weak, passive way. It meant so much to AJAs to have pride in their kind, but the Nisei did not seem to be fighting hard enough to become an integral part of American life. Since all Japanese-Americans were involved in a "now-or-never" opportunity to prove their worth, everyone should make an all-out effort. Hawaiians who visited relocation centers reported being disgusted. There were so many young men who refused to volunteer; some would not even go outside the camp to work. They seemed drenched in self-pity and in a "they-done-me-wrong" attitude. In general, the feeling was that mainland Nisei were culturally more American but that they lacked the spirit of Americanism. They complained of white racism but looked down on others—especially Negroes and even AJAs. They seemed to lack the courage to fight for their rights (*Social Process in Hawaii*, 1945).

The initial reaction of mainlanders to Hawaiians was equally stereotyped. They were uncouth, not quite civilized. They were uneducated, amusing, and picturesque; their speech was crude, though "catchy." Many were "roughnecks." What was most disturbing, however, was the apparent readiness of Hawaiians to fight over matters that seemed trivial. Mainlanders disapproved in particular of their style of fighting—gang beatings in which all available Hawaiians would participate, regardless of the merit of the case. This pattern of violence had long been common in Honolulu, though less so on the plantations. Gangs of all ethnic minorities had engaged in collective beatings—sometimes just for "kicks." Personal quarrels were settled in man-to-man fights, but gang beatings were administered on isolated, vulnerable individuals. The victim was beaten with fists by everyone in the gang until he fell; then, when he was down, he was kicked. Typical victims of such violence were hitchhikers, stool pigeons, overconformists (someone whose superior performance made others appear inadequate in comparison), and symbolic victims (a member of some resented category who had himself done nothing offensive). The code of mutual aid in such beatings was essentially an extension of the *aloha* spirit of helping one's friends. Those who failed to participate were condemned as *manini*. It was not long before rumors spread among Nisei civilians of the unfair manner in which Hawaiians fought. Five or six men would attack a single victim; if they still did not win, they would draw switchblades.

Although misunderstandings and antagonisms persisted for some time, the volunteers had to adjust to one another. The intense pressure to vindicate themselves in the eyes of other Americans as well as their appreciation of the common dangers they would face in combat forced them to cooperate effectively.

On 11 August 1943, while the 442nd was still in training, the 100th Battalion left Camp Shelby. It landed in Oran on September 2 and in Naples on September 22. It was attached to the 34th (Red Bull) Infantry Division, and the battalion moved into the line for the first time on September 26 at Salerno. War correspondents and combat veterans alike were astonished at the eagerness of these men. Unlike most troops facing their baptism of fire, they were smiling; at last the opportunity had come to prove themselves. On October 21 the War Department announced that the 100th Battalion had come through its first test "with flying colors." By Christmas the unit was involved in the fighting in the hills around Cassino, and in January 1944 it participated in the attack on the citadel itself. It entered Italy with almost 1,300 men; by January 1 there were only 832 left; and just before the final assault on Cassino only 521 remained who were capable of effective combat. On 26 March 1944 the battalion landed on the Anzio beachhead, and it was here that replacements arrived from Camp Shelby—from the 1st Battalion of the 442nd Infantry Regiment. The troop complement was brought up to 1,095 men.

On 22 April 1944 the 2nd and 3rd Battalions of the 442nd Infantry Regiment left Camp Shelby. The combat team marched into Rome on June 9 and on the following day was attached to the 34th Infantry Division. The 100th Battalion, which had been fighting as a separate unit, was incorporated into the 442nd as its 1st Battalion. Before dawn on June 26 the entire combat team moved into the line and made its initial contact with the enemy. Until July 21 it participated in the bloody fighting that marked the 5th Army advance from Rome to the Arno River. Men from the 442nd were among the first to enter Livorno, and war correspondents wrote of their high morale and of the dismay of captured Germans. On September 26 the combat team was detached from the 5th Army and sent to France. Attached to the 7th Army, it participated in the Battle of Bruyères in the Vosges Mountains. It was there that it took part in the saving of the "lost battalion"—the trapped 1st Battalion of the 141st Infantry Division. When others were unable to reach it, the 442nd was assigned the task. On October 26 it succeeded in breaking through, losing more men in the assault than were left alive to save. On November 9 most of the combat team was relieved, and on November 17 it was detached from the 36th Division and ordered to Nice. The unit remained in the Maritime Alps, largely on patrol duty, until March 1945. The final action of the combat team was again in Italy—in the Po Valley campaign, the last 5th Army offensive of the war. On 5 April 1945 the 442nd was ordered to launch a diversionary attack on the Ligurian Coast, while the main body of the 5th Army moved against Bologna. In four days the combat team destroyed positions that had withstood attacks for five months and turned the diversion into a full-scale offensive on the west coast—action for which it received a special citation. The fighting ended on May 2, when the Germans surrendered to Italian partisans (Shirey, 1946).

Battle experience broke down the barriers of distrust and resentment that had separated Hawaiian AJAs and mainland Nisei. As they lived and worked together, they learned to get along and to appreciate one another. Since most of the volunteers were from Hawaii, the language and spirit of the AJAs prevailed over the entire combat team. As differences were resolved, pidgin was accepted by everyone in the camaraderie. Hawaiians were amused at the amateurish attempts of mainlanders to speak pidgin. While most Kotonks had learned the more common words, they could not master the proper inflections; rhythm and intonation were also difficult. As the fighting progressed and as replacements filled the ranks, the proportion of mainlanders increased; by the end of the war almost half the personnel consisted of mainlanders. But the use of pidgin continued (Murphy, 1954:113–16; Morimoto, 1966). Along with the Statue of Liberty shoulder patch, pidgin had become one of the symbols of group identity.

Volunteers for the Intelligence Service

The procurement and training of Japanese language specialists began before the war, and the program developed quickly thereafter. The Military Intelligence Service Language School (MISLS) was activated on 1 November 1941 under the G-2 section of the 4th Army. It opened at the Presidio in San Francisco with 4 instructors and 60 students. With the outbreak of hostilities the importance of this work was recognized, and the operation was quickly expanded. Since the 4th Army's ban on persons of Japanese ancestry from military areas applied to soldiers as well as to civilians, the school was moved on 25 May 1942 to Camp Savage, Minnesota.[1] On June 1 MISLS was placed under the direct jurisdiction of the War Department. Its first graduates were assigned to units in the South Pacific and in Alaska. Commanders were initially skeptical, but the linguists proved so indispensable in battle that requisitions soon arrived for more. The first major campaign in which the Nisei proved their utility was at Guadalcanal. By October 1943 there were 41 academic sections. In a training program of six months the men took courses in reading, writing, and speaking Japanese; translation, interpretation, and interrogation; analysis of captured documents; military and technical terms; order of battle of the Japanese army; geography and map reading; radio monitoring; social, political, economic, and cultural history of Japan; and some *sosho* (cursive writing). A section for officer candidates

1. After the war it was revealed that a handful of Nisei and Kibei in the Foreign Broadcast Intelligence Service had been permitted to remain in Portland, Oregon, to monitor Japanese broadcasts (Greene, 1946).

was established—with 35 Caucasians who had previously lived in Japan or who had completed intensive training at the University of Michigan. By January 1944 the school had grown to 52 academic sections with 27 civilian and 65 enlisted men as instructors; the training period was lengthened to nine months. By this time there were 1,100 students, including 107 in Officer Candidate School (OCS). Aside from turning out interpreters and translators for combat intelligence, special instruction was provided for officers of the Canadian army, the Marine Corps, and the Office of Strategic Services. By June 8, 1946, MISLS had graduated some 6,000 men (U.S. Military Intelligence Service Language School, 1946).

With the suspension of selective service in June 1942, the only avenue remaining for Nisei who aspired for military duty was MISLS. Although isolated individuals slipped by draft boards and were scattered in various branches of the service, officially G-2 was the only branch that was open. In the autumn of 1942 recruiters were sent to the relocation centers to urge qualified evacuees to enter the service, and Colonel Andersen, commandant of the school, made frequent visits to the camps to interview prospective volunteers. At that time 160 enlisted from the relocation centers. In May 1943 an appeal for volunteers was made in Hawaii; 773 applied, and 250 were accepted. A second call was made in November 1943; 300 applied, and 200 were accepted.

Most mainlanders who enlisted did so in the face of parental opposition. In the relocation centers pleas that Nisei could win back their status as Americans by proving their loyalty were met with cynicism; many objected that the volunteers would be exploited by the very parties that denied them their rights. Further objections arose from misconceptions concerning the school. Because it was known to be part of the intelligence service, many Issei feared that their sons would be trained to become spies. Graduates would obviously serve in the Pacific, and parents who were convinced that Japan was invincible did not wish their sons exposed to such danger. They objected both on sentimental and pragmatic grounds; they believed Japan to be a more formidable foe than Germany. Many assimilated Nisei failed to appreciate the position of their parents, who had to face pressure from other Issei. By volunteering they placed the reputation of their family in jeopardy.

The men who trained at Camp Savage were all volunteers. Ironically, those who wanted most to enlist were the very ones who were least likely to qualify. The recruiters were seeking men with a fluent command of Japanese, and the men who wanted to volunteer were those who were most assimilated to American life. In desperation many persuaded Kibei friends, parents, and even former Japanese language school teachers to help fill out their examination papers in order to make a favorable showing. Once admitted, they found the work in school extremely difficult. Since they were given the equivalent of a Japanese high-school education in six to nine months, anyone with an inadequate command of Japanese was confronted with an almost impossible task. Limited to seven hours of classroom work a day and two hours of enforced study each night, they simply did not have enough time. Duty officers frequently found it necessary to stop illicit studying after 23:00 hours. Many crammed for examinations after barracks lights had been turned off by using flashlights under their blankets. Some claimed they had diarrhea and studied in latrines far into the night. Others spent their weekends in hotels, pouring over their books. But these men did not complain. They saw themselves as test cases. They were among the few in the army who were to be entrusted with responsible tasks; most other Nisei soldiers were doing menial work. On them rested the future of all Nisei in America. Like those in the 100th Battalion, they went out of their way to make a good impression, doing nothing that would lead others to question their loyalty. It was at Camp Savage that the school grew, and it was the men trained there who were largely responsible for the reputation made by the unit.

As in the infantry the volunteers were marked by a sense of mission. In spite of their ostentatious display of enthusiasm, however, most men in the language school were privately miserable. There was widespread agreement that Camp Savage was a "hellhole." Physical facilities were far below the standards generally maintained by the U.S. Army. The barracks consisted of buildings formerly used for a state home for indigent old men. Some classes and services were held in a makeshift barn. Construction was continually under way, but it always lagged behind the rapid growth of the school. Many were bothered by the severe cold, especially because they were not adequately housed for it. Many rumors developed in Nisei communities of the difficulties these men encountered in trying to study under such adverse conditions. Even going out on pass presented difficulties, for the camp was isolated—fifteen miles southwest of the Twin Cities area.

By the autumn of 1944 the school had become accepted as an essential part of the army. It had graduated 1,600 enlisted men, 142 officer candidates, and 53 officers. These men had been assigned to approximately 130 different army and navy units; some had been loaned to allies as well. Teams of at least ten had been attached to the headquarters of marine and infantry divisions, where they worked as interrogators, interpreters, and cave flushers. By translating Japanese plans for attack and defense, they gave the Allied forces a definite advantage. They had served in several theaters of operation: China-Burma-India, Alaska, Pacific, and Southwest Pacific. Some had been assigned to the Office of War Information and had proved their worth in psychological warfare as propaganda writers and radio announcers. A few had served with the Office of Strategic Services.

By 1944 the demand for graduates was so great that the school had to be expanded. On August 15 MISLS was moved to Fort Snelling, Minnesota. The fort, which had been established in 1820 for the Indian wars, had expansive lawns, giant elm trees, a chapel, a library, and permanent brick buildings to protect troops from the inclement winters. It was like a park except for one remote sector—contemptuously called the "Turkey Farm"—where men were billeted in tar-papered barracks similar to those in the relocation centers. The school took over most of these facilities, for only a small detachment of military police, some engineers, the service crew, and the personnel for an induction and separation center remained to share the resources. By this time the curriculum was well organized. A highly secret radio-monitoring unit—the Military Research and Liaison Section—was also maintained at the fort. In February 1945 a Chinese division was activated; in June another division was opened for Nisei WACs; and in October instruction in Korean was introduced. By 1945, when the number of personnel hit its peak, there were 162 instructors and over 3,000 students. It was at Fort Snelling that Caucasian OCS men were first segregated. Officer candidates had hitherto taken their training with Nisei enlisted men, assigned to sections according to linguistic ability. They were now billeted in Company A and took their instruction in Division E.

Officers of the MISLS administration were well informed about the evacuation, and they were sympathetic. Convinced of the loyalty and worth of Nisei, they had struggled within the army for their acceptance. Because of the unusual difficulties encountered by soldiers whose parents were interned, these officers granted them special considerations. Getting a leave of absence to settle problems in relocation centers was relatively easy. The organization adopted a pass and furlough policy that was very lenient compared with the infantry. Food was especially prepared by Nisei cooks, and few could complain about it. The regular rations provided by the army were supplemented by rice purchased from PX profits. But only a handful of Nisei—officers in administrative positions and headquarters clerks—came into direct contact with these officials.

The general evaluation of MISLS by Nisei—civilians and soldiers alike—was negative. Anyone who talked to the Nisei personnel—instructor, cadre, or student—could not help but get the impression that all was not well. An air of distrust pervaded the entire organization. No one seemed to be happy at Fort Snelling. Those on pass or furlough complained constantly, and there seemed to be marked hostility against Caucasians—especially Caucasian officers. The administrative officials of MISLS were often accused of using the Nisei to further their personal ambitions. Another widespread complaint was that the presence of so many Kibei, whose loyalty was questionable, made things difficult for Nisei who lacked comparable linguistic skills. Apparently the administrative officers were unaware of the depth and extent of the disaffection and resentment against them. They pointed to the overseas record of their graduates as evidence of devotion to the organization. Actually, most of the men hated the organization but performed well in spite of it.

One major source of resentment was the discriminatory policy in assigning ratings to graduates. A Nisei graduate, no matter how competent a linguist in both English and Japanese, was sent overseas with the rank of T/4 (a sergeant) or T/5 (a corporal), usually the latter. Occasionally, an outstanding student left as a staff sergeant. On the other hand, Caucasian students, even if they were virtually incompetent in Japanese, were commissioned on graduation and went overseas as second lieutenants. Many rumors developed of how resentful Nisei had gotten even, when overseas, by exposing the ineptitude of Caucasian language officers to their superiors. Since most Nisei assumed that no non-Japanese could possibly master the difficult language, they believed that all Caucasian graduates were incompetent. While this practice was accepted quietly during the early part of the war, as the position of the Nisei in the United States became better established, it was condemned as completely unjustified.

Part of the tradition developed by the early volunteers was strict discipline and military courtesy. They had conducted themselves as model soldiers, and MISLS officials became accustomed to it. Those who followed the early volunteers complained repeatedly of the unusual emphasis placed on petty regulations, citing ridiculous cases of severe penalties for minor infractions. In the infantry, trivia were stressed in basic training but tended to be overlooked later, especially overseas. Thus, among servicemen MISLS developed a reputation of being a "chicken shit outfit."

Even after infantry service was reopened to Nisei, many continued to volunteer for MISLS. This did not signify approval of the organization; in most cases those who volunteered did so as a matter of expediency. Publication of the long casualty lists made a deep impression on the Japanese communities. The death of close friends and relatives as well as letters from men overseas describing the horrors of ground combat forced prospective recruits to reconsider their rejection of intelligence work. In the infantry they would be treated with more respect, but infantry life was hard, dangerous, and could turn out to be very brief. There was no doubt that work in the intelligence service would not only be easier but safer. There had been some casualties in the Pacific, but linguists were too valuable to risk in the front lines; the chances of being killed were minimized. Furthermore, ratings were higher and the pay was better. In addition, the school provided an excellent opportunity to learn enough Japanese to be able to communicate effectively for the first time with their parents. Many parents who had formerly opposed G-2 work now pleaded with their sons to volunteer rather than wait to be drafted for infantry duty. Thus, in spite of its reputation many practical considerations led Nisei to enlist, and by the summer of 1944 a somewhat different type of soldier began to enter MISLS.

Military Regulations and Nisei Norms

As with most American soldiers in World War II, among Nisei troops the highest prestige was enjoyed by the tough combat veterans of the ground forces. Emphasis was placed on virility; the code was to be a *man*. Values that were stressed were courage, endurance, lack of squeamishness, avoiding displays of weakness, sexual competence, and reticence concerning ideals. The infantry was highly regarded, but it was the one branch most men wanted to avoid (Stouffer et al., 1949: I, 296; II, 131–35, 302–12). Most soldiers developed a stereotype of the "combat man": he was tough, sloppy, straightforward, gentle, and able to "take it"—the "dogfaces" depicted in Bill Mauldin's cartoons (Mauldin, 1944). These were men who were dependable under fire. When not fighting, however, they would relax and be informal. They would dress comfortably and practically rather than wear the uniform of the day; they did not bother to salute officers, nor did they address them as "sir" unless necessary. Like most stereotypes this one was not entirely accurate, but it was widely entertained. The symbol of membership in this honored fraternity was the Combat Infantryman's Badge; it was awarded only to those who had actually been under fire. It was one of the few coveted awards, although most men were ambivalent about the price that had to be paid to get one.

This stereotype was sharply contrasted with another—the "rear-echelon" man. They were clean, fastidious about petty matters, effeminate, always observed military courtesy, and were careful to obey all regulations. They were often dismissed as sissies—"chair-borne" troops, "male WACs," and "USO commandoes." Although support troops were targets of contemptuous remarks, most soldiers were ambivalent about them. They resented the comparatively safe and easy life that such men enjoyed, but they also envied them (Stouffer et al., 1949: II, 290–316).

The informality and spontaneity of the Hawaiians coincided closely with the stereotype of the combat man. Mainlanders generally agreed that the Hawaiians exemplified the fighting spirit of American riflemen, and many Hawaiian ways were adopted as status symbols. Nisei soldiers believed the men of the 442nd Infantry to be extremely rugged on the battlefield, but otherwise informal. Their motto—"Go for Broke"—was heard with increasing frequency, as mainlanders began playing dice for large sums of money, sometimes risking their entire monthly pay on one roll. It was also believed that Hawaiians were less hesitant about approaching white women, and many lurid tales of conquests developed. Thus, the Hawaiians enjoyed high status among all Nisei soldiers, and they were emulated even by those who spoke contemptuously of them.

The identification of the style of the AJA soldier with the stereotype of the combat man illustrates one way in which the social control of Nisei society reached into the military establishment. Whether they came from Hawaii or the mainland, Nisei servicemen were bound by a multitude of personal ties. Since most of the men came from a limited number of relocation centers or from segregated communities on the Pacific Coast or Hawaii, almost everyone had relatives, friends, or acquaintances in his company. Many who had never met before discovered that they had friends in common. Since most soldiers kept in touch with Nisei civilians by mail, any violation of Nisei norms would be reported to those who knew the offender. A number of special codes of conduct developed, and these placed Nisei officers—commissioned and noncommissioned—in an especially difficult position. Although their formal roles were clearly defined by military regulations, their commands had to be temporized in terms of Nisei expectation. Where military regulations clashed with

Nisei norms, the latter usually prevailed, unless a Caucasian officer happened to be present.

Commissioned officers were in an unenviable position. Most Nisei considered one another as equal, regardless of military rank, and those in positions of authority had to earn respect by demonstrating their special competence. Nisei enlisted men deferred to Caucasian officers, even those who were incompetent; their spontaneous and widespread resentment against Nisei officers reflected their tacit acceptance of inferior minority status. They were accustomed to taking orders from Caucasians, but they resented having to pay deference to those whom they regarded as no better than themselves. Although army regulations required all enlisted men to salute all officers, many Nisei officers did not expect to be saluted by other Nisei. As one enlisted man put it, "I saw a Boochie lieutenant in town yesterday, and I didn't salute him. He looked at me like he thought I might salute, but the hell with him. Who does he think he is? He may be an officer, but he's still a Boochie." In many situations officers had to *request* the cooperation of the men in their charge. Any violation of such understandings by an officer led to his being labeled as "chicken shit," and malicious gossip spread both among troops and civilians that he was "drunk with power" or "bucking for promotion" at the expense of others. The personal reputation of some Nisei officers suffered permanent damage as a result of minor altercations in the army.

Nisei with high rank enjoyed certain advantages both in camp and among civilians. High rank was admired by outsiders, especially women. It also meant special privileges in the army. But they had to pay dearly for that prestige, for they were bound by obligations that were almost entirely in favor of enlisted men. They were held to absolute responsibility, but they did not enjoy absolute authority. Under some circumstances their men could with relative impunity flatly refuse to obey a direct order. Thus, officers were in a marginal position; they performed for two audiences—the Nisei social world and the military.

In critical situations the highest ranking soldier was expected to take charge. He was expected to distribute assignments but not to push his men too hard. He was also expected to participate directly in disagreeable undertakings. A Nisei in charge of a work detail to scrub floors, for example, had to get down to do some of the scrubbing himself; his rank did not give him the privilege of standing by and ordering others to do the work. Leaders were expected to avoid being "GI," that is, too fastidious about enforcing petty regulations. They were also expected to let their charges "take it easy" as much as possible. That a task had to be done and done reasonably well was taken for granted; however, the men were to do only what was necessary and no more. Nisei NCOs were expected to let their men "goof off" and to cover for them. On the other hand, it was understood that anyone caught by a Caucasian officer would never disclose that he had received permission. Anyone who "squealed" on his benefactor would be in serious trouble. Thus, a Nisei officer who "played ball" was also protected by his men, who sometimes made a special effort in his behalf when his superior officers were watching.

Most Nisei troops placed a high evaluation on themselves. They were convinced that Nisei on the whole were more intelligent than other soldiers and that officers who "know us" would take this for granted. In speaking of the "dumb *Keto*" they invariably pointed to someone from the rural South they had met in the army or near the two relocation centers in Arkansas. Those who had served in integrated units recounted over and over examples of incredible ineptitude they had witnessed. Of course, not all Nisei were alert, and those who could not keep up with the others were tormented; stupid errors were not tolerated. They also took pride in their resourcefulness. If the "army way" of doing something was clumsy, they got together among themselves to work out a more effective procedure. They resented being given detailed instructions on simple tasks; they considered it insulting, even though they realized that there were soldiers who might encounter difficulties without such guidance.

Reliability and perseverance were especially prized. It was assumed that any Nisei confronted by an especially challenging situation would go on either until he succeeded or dropped from exhaustion. In important undertakings Nisei could be counted on to do their best. They might shirk minor duties, but they would be ready when really needed. Such standards of conduct were enforced by the men among themselves. Anyone who dropped out of a road march would be tormented mercilessly for months. Those who lagged behind were reminded that they were Japanese and that they had no right to disgrace the others. Those who were more assimilated felt that some of their comrades went too far in this regard; privately they scoffed at the "Boochie hard head," but in difficult situations they too performed as expected.

Officers who had observed Nisei troops in training, on maneuvers, or in combat were frequently astonished at their performance, and several made public statements commending various units. Since every shred of relevant news was reported in the *Pacific Citizen* and the Japanese language press, sometimes with considerable exaggeration, these remarks were given wide publicity in the Nisei social world. Servicemen on furlough often confirmed the published reports, and their claims indicated that Nisei troops were indeed a group apart. In these ways the group norms crystallized, and even those who had not volunteered for military duty felt an obligation to their friends, some of whom had been killed in action, to keep up the high level of performance. There was to be no compromising publicity to buttress the arguments of racists. By the autumn of 1944 the expected pattern of conduct was being defined with increasing clarity, and there developed a set image of the Nisei soldier. Although no one spoke seriously of *Yamato damashii*, group expectations were in fact tantamount to conformity with much of the samurai code.

A premium was placed on excellence, and even those who were not entirely in sympathy with the war effort met their obligations. Coupled with the sense of mission was bitter resentment on the part of some Nisei against white men. Since few understood what was involved in such a complex phenomenon, there was no consensus among Nisei in fixing responsibility for the injustices of evacuation. In general, it was simply blamed on *Hakujin*—referred to vaguely as "they." Although their hostility was suppressed in public, among themselves they made no secret of their attitudes. Whenever an opportunity arose to compete with white men, some Nisei soldiers made special efforts to surpass them. Even in basic training they worked desperately to establish new camp records for endurance and speed. They took secret delight in deriding the inferiority of Caucasian troops, who were not even aware of being involved in any kind of contest. Thus, for some Nisei, resentment was channeled into efforts to exceed white men in various military feats. These performances added to the stature of Nisei troops in the eyes of Caucasian officers, most of whom were apparently unaware of this negative orientation.

Thus, the Nisei soldier saw himself in terms of a role that was circumscribed by military regulations, informal unit codes, and by Nisei mores. The relative importance of regulations and the manner in which individuals were expected to obey them was determined by the network of understandings that developed among the members of each platoon. In general, the fact that one was a Nisei was more important than his being a soldier. Military duties were performed in accordance with Nisei norms.

The Ambivalence of Nisei Draftees

The reinstitution of selective service in January 1944 brought Nisei soldiers with a somewhat different orientation into the army. The move had been anticipated in rumors long before it was announced. Casualties had to be replaced, and as long as Nisei were limited to segregated units, those replacements would have to be Nisei. It seemed unlikely that there would be a sufficient flow of volunteers. The move was not popular on the mainland. For the government to ask for volunteers was one thing, but the conscription of men whose parents were still incarcerated was another. Who would look after the dislocated families if all able bodied men were drafted? When the reclassification of Nisei to 1-A was announced, many tried to find work in war industries in order to get exemptions. Many discovered, however, that they could not get the necessary clearance from security agencies. This aroused considerable indignation. Why should Nisei be considered sufficiently loyal to die in infantry combat but not loyal enough for lucrative, draft-exempt jobs in war industries? Some Nisei became even more embittered after their status in America had been secured.

The most aggravating feature of the draft arose from segregation. Although infantry replacements in World War II were trained and assigned on an individual basis (U.S. Department of the Army, 1954), the standard procedure could not be used for Nisei because they were serving only in special units. There were not enough Nisei to justify separate induction, as for Negroes. Thus, Nisei were called on the same basis as others; they took their physical examination and were sworn in with other draftees. After that, however, it was necessary for them to wait until enough Nisei had been inducted to fill a training unit—usually about 250. Following induction, most Nisei were placed in the Enlisted Reserve Corps and sent home. Everyone understood why this procedure was necessary, but it worked many hardships—especially on resettlers. Since they were subject to recall to active duty on 24-hours' notice, they could not make definite plans. Men in ERC had difficulty in getting a new job, and often they could not get back the position they had given up prior to their induction. Those who were in school did not know whether they would be able to complete another semester. Many visited their friends and relatives while waiting to be called and soon found themselves without funds. Many complained that they felt silly seeing their friends again after a final preinduction party. Some had given away all their civilian clothes and were embarrassed to have to get them back. Although the practice was well known, one could not count on being placed in ERC. Inductees had to settle their affairs, for there was always the possibility of being in the last group needed to make up a required complement. Furthermore, there was no way of finding out how long one would be in ERC. Some were called up in a week; others were not called for three months. By the autumn of 1944 most Nisei simply resigned themselves to fate. If they were physically fit and did not have too many dependents, sooner or later they would be called into active duty. But the induction procedure led to so much inconvenience that it added to complaints of unfair treatment.

Most draftees lacked the zeal that characterized the volunteers. The latter had fought with reckless disregard for personal safety to accomplish their mission—the vindication of the Nisei. Most draftees felt that the battle for Nisei rights had largely been won. The mere fact that they were serving in the armed forces while their parents were still incarcerated was sufficient proof of their loyalty. Once in uniform, these men felt they had an unquestioned right to be treated as Americans. Unlike other Americans, it was relatively easy for a Nisei to escape military service; all he had to do was to declare himself disloyal. Yet, each was risking being killed in combat. Since the major mission had been accomplished, attention was increasingly focused on survival. By the winter of 1944 many Nisei felt that they had been subjected to several unreasonable tests and exonerated. Therefore, they objected to further discrimination of any kind.

Once inducted, especially into the infantry, Nisei were treated in the same manner as other soldiers, and they began to realize that their uniforms gave them immunity from suspicion. The extent to which the draftees felt established is shown in the manner in which they were able to joke about anomalies of ethnic identity. Some began to refer to themselves as "Japs" in a manner that they had previously not dared to do. By early 1945 they were calling friends who had been assigned as "enemies" in training problems "Saipan Joes"; occasionally they exuberantly shouted "*Banzai*!" as they charged toward their final objective. Their loyalty was sufficiently established so that they could identify openly with things Japanese.

Although the draftees felt more secure, they were still sensitive to unfavorable publicity. In March 1944, twenty-eight Kibei soldiers at Fort McClellan refused to bear arms. They had been in the army from before Pearl Harbor and had been performing menial tasks in various service commands. When they were transferred to the infantry and ordered to take basic training, they refused. In their court martial nineteen were found guilty and sentenced to five to thirty years in prison; two were acquitted, and charges against the others were withdrawn. Although most Nisei could understand their position, there was little sympathy for them. The Kibei were roundly cursed and condemned.

Sooner or later draftees found themselves in Infantry Replacement Training Command (IRTC) camps for basic training. Within a few weeks they became acutely conscious of how easy it was to become a casualty in modern infantry combat. They also saw how much pressure Nisei placed on one another. Although proud of the magnificent battle record and cognizant of what it had done for the future of the Nisei, men of military age and their parents became increasingly preoccupied with casualties. Articles in the *Pacific Citizen* were frequently written in a manner that exaggerated the contribution Nisei were making to the war effort, and some of them gave the impression that Nisei troops were being given exceptionally difficult assignments. The issue of 19 February 1944, for example, described a "suicide mission" undertaken by a platoon of the 100th Battalion from which only eleven men had returned. The 23 September 1944 issue reported that the 442nd was "spearheading" the 34th Division drive north of Livorno. Such articles reinforced rumors that the Nisei were getting the most dangerous assignments in Italy, and the heavy casualty rates made these charges plausible. It was also rumored that 90 percent of the 100th Battalion had been "wiped out" and that the hospital at the Gila River Relocation Center was being converted into a convalescent home for wounded veterans. Unlike civilians, however, most Nisei soldiers tended to blame the casualties on Nisei zeal rather than military policy. Combat duty in any infantry unit was dangerous enough, but many became especially wary of being in a segregated Nisei regiment. It appeared that they were increasing their chances of being killed. Thus, considerable ambivalence developed concerning service in the 442nd.

Many draftees privately deplored what they regarded as the fanaticism of the volunteers. There was unmistakable pride, but they felt that some Nisei were going too far in demanding selfless sacrifices. Emphasis continued to be placed on endurance and physical stamina, alertness, daring, and courage. A Nisei soldier was never to break under fire or run away. Once in the infantry, the men appreciated more fully how difficult it would be for them to live up to such expectations. Among close friends they condemned the strong pressures for compliance that characterized all Nisei groups. Since everyone else was conforming, however, trainees who were exhausted did not dare drop out of formation. A common explanation of such obstinacy and tenacity was that "Boochies don't like to get behind." Some expressed fear that Nisei units might volunteer for impossible missions merely to outdo Caucasians who had sense enough to remain behind. Many regarded this as stupid, but they could not be certain that most of the others shared their views.

As increasing numbers of mainlanders went into the army, issues of the *Pacific Citizen* devoted more and more space to their activities. There were accounts of winning athletic teams made up of Nisei trainees, of promotions, of decorations, of unusual assignments, and of individual heroism. By 1945 many resettlers had been approached by friendly Caucasians with inquiries about "your boys in Italy," and finding employment and housing became noticeably easier. Few had anticipated such a quick and generous response from the American public; and when it came, they acknowledged that the efforts of the volunteers had not been in vain. Most Nisei pointed with unconcealed pride to their men in uniform. In spite of reservations about discrimination and dangerous assignments, Nisei civilians stood solidly behind their servicemen. The draftees realized, therefore, that they could do nothing to spoil this record.

During the spring and summer of 1944 Nisei draftees were trained in segregated units—most of them at Camp Blanding, Florida, and at Fort McClellan, Alabama. Those who were hostile toward Caucasians tried to outdo them in training. They urged each other to do well in classification tests; a rumor developed that the average I.Q. of Nisei soldiers was 118 and that personnel officers were dumfounded. Those who were especially bitter about the evacuation expressed a desire to "get even with them." They seemed to feel that Caucasian troops would be humiliated on being surpassed and that this would constitute vengeance. When praised for cleanliness, some of the men became so fastidious about sanitation that others found them irritating. In each company it became apparent that there were a few who were determined to eclipse the hated *Keto* in every respect. They gloated over the enjoyment they would derive from surpassing them, and this led others to scorn them as fanatics.

In the autumn of 1944 some IRTC centers—Camp Wheeler, Georgia, and Camp Wolters and Camp Hood, both in Texas—adopted a policy of desegregating Nisei. The recruits had expected to train in segregated units, and their reaction was mixed. Everyone agreed that the new policy was more democratic, but most Nisei were disappointed. Some protested bitterly that it would now be necessary to purchase locks; *Keto* were dishonest. Others pointed out that extra precautions would have to be taken because *Keto* were unsanitary and frequently infected with venereal disease. Still others anticipated having to do unnecessary work because of the ineptitude of some *Keto* in their platoon. Some declared simply that they preferred a segregated unit because they would feel more relaxed and could borrow things more readily from one another. But the disappointment most commonly voiced was that they had been robbed of an opportunity to compete against Caucasians. Some had already been looking forward to the first hike in which they had hoped to establish a new camp record of no one dropping out.

But the more assimilated men were delighted at the new policy. Their joy did not arise primarily from their conviction that segregation was undemocratic and insulting; they were also indifferent to the prospect of training with Caucasians. Rather, they were relieved to be freed from the special obligations of the Nisei world. They could foresee the possibility of having to drop out of some formation and wanted to be free to do so without paying the price exacted in all-Nisei units. In addition, there were other characteristics of Nisei groups that they disliked. Some pointed to excessive pettiness; men stopped speaking to one another over a minor slight. Others disliked the solemnity that characterized so many Nisei undertakings; they pointed out that Caucasians on the whole were more relaxed and informal. Even among those who were wary of Caucasians there were many who were not keen on setting records. They had no objections to the records being set by someone else; their primary concern was survival in combat.

The majority of the men were disappointed with desegregation. Although the more assimilated men thought nothing of sleeping and showering with Caucasians, those who had never before been so close to outsiders were uncomfortable. They stayed among themselves. Even when Caucasian soldiers made friendly overtures, they maintained a polite distance. Most of them were uneasy, for they did not know how to act in their presence. Although they realized that segregation meant harder work and considerably more risk, most Nisei still preferred to be with their "own kind."

Analytic Summary and Discussion

The distinguished battle record made by Japanese-American troops in World War II was forged by men consumed by a sense of mission. But only a handful—the politically sophisticated who preferred second class citizenship in America to living in a world dominated by Nazi Germany and militarist Japan—had an ideological commitment to safeguarding democracy. The concern of most of the warriors was more immediate. However disillusioned they may have been about American morality, most Nisei could see no future for themselves in Japan. They were concerned with winning a place for themselves in American life and realized that conspicuous gallantry in military service was the only way to counteract conclusively the ridiculous charges that had been made against them. High standards of performance were set by the early volunteers, ironically with some cultural tools derived from the enemy. Once the norms were established, however, draftee replacements had little choice but to comply. Many of the men felt abused and unjustly treated, but their personal resentments did not matter. All Nisei were expected to do their part.

Within the Nisei social world the soldier was initially an object of pity—one who was powerless to assist his family during evacuation. During relocation center days the soldier was viewed by some as a "sucker" fighting for a lost cause; to others he was the last ray of hope for a suspected group. In the end, for those who had cast their lot with the United States, the soldier became an object of great pride. The unit to be described in the remainder of this book was the first to deviate from the pattern of exemplary conduct that brought recognition and honor to a maligned ethnic minority.

4. «» the initial break in discipline

The war in Europe was going well for the Allies in the early spring of 1945, but infantrymen were still much in demand. Although a bridgehead across the Rhine had been secured on March 7, the *Wehrmacht* was still intact and a formidable fighting force. As nine Allied armies closed a ring around Nazi Germany, resistance was fierce, especially as they entered German soil. Even in April, when the German High Command lost effective control over several of its armies, isolated units continued to resist. Such pockets had to be cleared out, and riflemen were needed for mopping up operations behind the speeding armored thrusts. The main replacement depot of the Army Ground Forces at Fort Meade, Maryland, was a very busy place during this period. After their last furloughs at home IRTC graduates reported there for final processing before being sent to a port of embarkation.

Several contingents of Nisei replacements had passed through Fort Meade, but from the middle of March they began to accumulate there in large numbers. They could be assigned only to the 442nd Regimental Combat Team, which at that time was incurring few casualties. From November 1944 it had been manning defensive positions in the Maritime Alps, and the men not on patrol duty were enjoying rest furloughs on the French Riviera. The unit was not committed to action again until April 5, and by that time a substantial number of replacements were already on hand in Italy. Thus, just when field commanders in Europe were calling for all available infantrymen, hundreds of Nisei replacements—most of them riflemen—were stranded at Fort Meade. By V-E Day they numbered over 1,200.

It was during this long sojourn at the replacement depot that difficulties erupted in one of the Nisei companies. The trouble started in A Company, 9th Battalion, 3rd Regiment. It was made up largely of mainland recruits who had trained at Camp Wheeler and Camp Blanding, along with some Hawaiians who had completed their basic training in Texas. Before long, A-9-3 was designated by other Nisei as a "fuck-up company," and the initial question to be addressed is how discipline broke down in this particular group.

The Accumulation of Nisei Replacements

Replacements did not expect to remain for long at Fort Meade. It was their last stop until shipping space became available for the type of personnel requisitioned by overseas commands. It was here that records were checked by the Classification and Assignment Section. Each man's pay account was brought up to date, and adjustments could be made concerning allotments to dependents, insurance, and purchasing war bonds. Advisers were available to go over personal problems and to help prepare one's will. Medical records were checked, and required immunization shots were given. If any prescribed training was not on a man's record, he was required to take it. Each replacement was issued new clothing and equipment, which were tested for serviceability in combat. Instructions were given on censorship and security, and the rights and responsibilities of a prisoner of war were explained. Drills were conducted on techniques for abandoning ship (U.S. Department of the Army, 1954:375–85). Everything that was done contributed to the general atmosphere: this was the final respite in the United States before going overseas—perhaps never to return.

Since it was their last stop before facing hazardous duty, replacements were given every consideration. Regulations were lax, and work was light. There were days on which they were permitted just to sleep or to write letters. The food was excellent, and most men agreed that it was matched only by the meals they had had at their induction centers. Furthermore, the depot was conveniently located, and the pass policy was liberal. Buses were readily available, and the fort was only forty-five minutes from Baltimore, fifteen minutes from Laurel, ninety minutes from Washington, D.C., and four hours by train to New York. Anyone who was not on a temporary work detail or not needed for processing was issued an overnight pass; weekend passes were granted on application. Those who were assigned to KP or some other duty on Sunday were issued passes on Monday. Participants in regimental boxing matches were issued three-day passes. Those who returned from leave a few hours late, though technically AWOL, were signed in without charges; as long as they were back in time to catch the bus for the port of embarkation (POE), nothing was said. This was such a sharp contrast to the strictness of IRTC that the men were astonished; most of them agreed in the beginning that Fort Meade was a "pretty nice place."

Because of serious morale problems that had arisen in the past, every effort was made to reduce to a minimum the period of transit. The issuing and checking of clothing and equipment required only a single day, and the final medical examination for most men consisted of little more than a token check that took but a few minutes. The testing of new gas masks required an hour, and the lectures took a few more. At most, the processing took about three days; when men were needed in a hurry, it was done in one. Those not called to POE soon after the required processing had been completed were assigned to a program of training, conditioning, and orientation. The rationale given was that it was necessary to keep replacements from forgetting what they had learned in IRTC and going stale; the men assumed that the exercises were intended solely to take up time—to keep them busy until sufficient shipping space became available. Since they believed that they would not remain for more than a few weeks, most men initially took these exercises in good humor.

With few exceptions Nisei replacements—designated in company morning reports as "Race: Jap"—were assigned to the 9th and 10th Battalions of the 3rd Regiment. Caucasians were also assigned to these companies, but most of them were there for just a few days; only the few with unusual MOS (military occupation specialty) numbers or temporarily detained on medical grounds remained. Since the men were assigned to bunks in the order in which they arrived, Nisei and Caucasian troops were at first intermingled in the same platoons. As the Caucasians left, however, the Nisei became segregated. By the end of March the entire 10th Battalion was predominantly Nisei, as were two companies of the 9th Battalion. When it became apparent that Nisei replacements would not be requisitioned for some time, they were regrouped into separate platoons within each company; this was a matter of administrative convenience, since the others had to be processed in a hurry, while the Nisei had to be kept busy doing other things. Thus, the eventual segregation of Nisei was not the result of deliberate depot policy.

The first Nisei to reach A-9-3 arrived on March 17. After that, throughout the months of March and April, others reported from furloughs in small numbers—from 3 to 23 a day. The largest contingent came from Camp Blanding, but sizable numbers reported from Fort McClellan, Camp Crofts, Fort Knox, Camp Wheeler, Camp Hood, Camp Wolters, and Camp Robinson. Almost daily, joyous reunions took place. Friends who had been inducted together and then sent to different IRTC centers were rejoined. Old schoolmates who had been separated by evacuation saw each other for the first time in three years; those who had been in the same relocation centers before resettling in different cities were reunited. In various parts of the fort—the post exchange, the theater, the recreation halls, the telephone exchange, and even on the streets—men unexpectedly encountered old friends. Some went from platoon to platoon, seeking out someone they might know. On March 30, 107 men were transferred from D-10-3, which was being closed; with this large order A-9-3 became predominantly Nisei. Those who had completed their processing were then reassigned to the 2nd, 3rd, and 4th Platoons, leaving the 1st Platoon for new replacements.

As Nisei mingled with Caucasian replacements on the company grounds, the difference between those who had trained in segregated IRTC units and those from desegregated units became readily discernible. Nisei from Camp Wheeler, Fort Knox, and the centers in Texas assumed that they would be treated like all other soldiers; and they mixed freely with the Caucasians. They struck up conversations, did not hesitate to ask questions, and sometimes even joined in the general camaraderie. Most of the men from Camp Blanding, however, were initially hesitant; they were uncertain of where they stood. Many confessed to feeling insecure, still fearing that Caucasians stood ready to condemn them at the slightest provocation. One man even had a nightmare of being a victim of mob violence. They rarely referred to themselves as "Japanese" or even as "Japanese-Americans" in the presence of outsiders. They seemed convinced that they had been treated well in basic training only because the cadre had been told of the special plight of the group so that "they understood us." Now that they were with soldiers who knew nothing about them, they were wary. They regarded themselves as a group apart and felt that their fate depended on the reputation built by those who had preceded them. As one man commented, "Christ, I hope the Buddhaheads who came before us were on the ball. If they did good, then we don't have to worry; but if they weren't on the ball, we would have a bad reputation. It's tough enough to be a Boochie without having some sergeant chewing your ass all of the time. Yeah, if those guys were on the ball, we have no *shimpai* [worry]." In a few weeks, however, as it became apparent that most other soldiers knew what they were and obviously did not care, the initial anxieties broke down.

Many continental Nisei had their first actual encounter with Hawaiian AJAs, about whom they had heard so much. Conspicuous cultural differences between the two groups led almost at once to misunderstandings, although men on both sides strained to avoid unpleasantness. In spite of the unfavorable stereotype, Hawaiian soldiers enjoyed high prestige; and it was not long before many continental Nisei were imitating them, perhaps thinking that they were thereby following the tradition of the 100th Battalion. Men who had dressed neatly in IRTC now became slovenly and began to walk about without tucking in their shirt tails. Several even began imitating Hawaiian pidgin, an affectation that annoyed some of their friends as well as the Hawaiians.

Many replacements encountered for the first time fellow Nisei who held ranks higher than private. In A-9-3 NCO replacements were given no special privileges. They were billeted with the privates and performed the same duties. Although a Nisei corporal or sergeant was occasionally put in charge of a work detail or given temporary command of a small unit in a field problem or in close-order drill, such responsibilities were usually assumed by the depot cadre. At first, many of the privates refused to take orders from another Nisei, and some even appeared to resent their presence. As one man noted, "That fuckin' staff sergeant thinks his shit don't stink. He told me to help him clean up the can. Just because he's got a couple of stripes he thinks he's good. In camp [relocation center] we used to beat up guys like him. He thinks he's as good as a *Keto*." It was not long, however, before such animosity was replaced by some measure of admiration. Most of the NCOs, aware of the resentment of other Nisei, did not exercise the prerogatives of their rank. Instead of ordering men about they suggested that certain things be done or requested cooperation; only a few of them were so bold as to scold another Nisei. Furthermore, many of the privates recognized the awkward position in which these men had been placed and did what they could to ease tensions. What won over many of the privates in time, however, was their growing realization that most Nisei NCOs were far more efficient and reliable than most of the Caucasian depot cadre. As their competence became increasingly apparent, many Nisei began to express an open preference for their own "zebras." The position of the NCOs was also eased by a rumor that the Nisei bearing stripes had served on the cadre at Camp Shelby, that they had remained behind after the departure of the main body of the 442nd to close the installation. To some extent, however, ambivalence remained; and even those who preferred sergeants who were "on the ball" resented being directed by individuals they regarded as their equals.

Nisei communication channels were quickly established, reaching other companies in the fort as well as resettler communities in New York and Washington, D.C. Those with friends in these cities visited frequently, bringing back the latest news. Since so many had close friends or relatives in other companies, information spread rapidly. During the first week of April, for example, a Nisei soldier was seriously injured when the bus on which he was riding was sideswiped by another bus. News of the accident spread quickly: the victim's arm had been dangling out the window just as the other bus struck. A few days later it was rumored that his arm had been amputated. Although virtually every Nisei in the fort had heard of the accident, some of the A-9-3 cadremen knew nothing of it until told by the Nisei. There were also rumors about the number of Nisei replacements at the depot, whence they had come, and how long they would remain. When difficulties broke out in A-9-3, there were rumors of punishment; one was that thirty Nisei had been confined in the stockade. By the middle of April, when most Nisei had become accustomed to their new routine, they identified themselves primarily on an ethnic basis. Whenever a Nisei spoke of "we" or "us," he was not referring to his platoon or his company, but to all GIs of Japanese ancestry. For the most part Caucasian replacements were excluded; they were not mistreated or insulted—just politely ignored.

As consciousness of kind became more pronounced, pressures to conform to Nisei norms became more noticeable. This was a new experience for some, especially for those who had not lived in segregated communities before the war. One practice that soon became fashionable among mainland Nisei was the blousing of trousers over highly polished combat boots. Immediately after receiving their boots—issued to infantrymen only for overseas duty—they spent hours dyeing and polishing them to make them resemble paratrooper boots. Those who had not trained in segregated units also had to learn the special language of Nisei soldiers. It was a mixture of the argot of Americans troops, certain Hawaiian terms, and common Japanese words. Other widely used expressions included "lose fight" to designate the sense of being deflated when conscientious effort turns out to be in vain because of an error by someone else; "waste time" to refer to anything undesirable or ridiculous and therefore not worth the required effort; "one time" to exhort others to make one last concerted effort; and "no lie," used synonymously with "Is that so?" [You're not lying, are you?]. It was not long before all training at Fort Meade was regarded as "waste time," and everything that the depot cadre did made them "lose fight."

Although it became clear that all Nisei were on a separate roster, most of them still did not expect to remain at the depot much longer. On April 9, sixty-nine Hawaiians—part of a large contingent of riflemen—departed for POE. Watching them make their final preparations reinforced the transitory orientation. It was assumed that other large units would be similarly alerted from time to time. As each man tried to allocate his remaining funds by anticipating when he was likely to be called, much speculation developed over the manner in which replacements would be assigned to shipping orders. Several principles were considered—order of arrival at Fort Meade, alphabetical listing, IRTC camps from which they had come. Since such orders were given without prior warning and since anyone on a shipping order was restricted to the fort, each day could conceivably be the last before going on to an unknown fate.

Under the circumstances passes took on special significance. Although the Germans were retreating, the war was not yet over; riflemen in particular were preoccupied with the possibility of not returning. This might be their last opportunity to see American life, American women, and to eat American food of their own choice. Many had friends and relatives in nearby cities, especially in New York. Although the practice was discouraged by the army, a number of wives had accompanied their husbands to the depot and were living temporarily in Baltimore or Washington, D.C. Since each night could be their last, virtually all men became preoccupied with passes. Even those who had exhausted their funds and had nowhere to go remained concerned with passes, and it was generally agreed that no one should do anything that might jeopardize any replacement's pass privilege. This was the one privilege treasured above all else.

Initial Reactions to Gross Incompetence

Most of the Nisei in A-9-3 completed their processing during the latter part of March, and on April 2 a program of military training was instituted. At first most men were quite pleased. They had become sluggish from their long inactivity, and the training itself was not nearly as strenuous as that at IRTC. Before long, however, several irritations developed. In IRTC everything was done efficiently and on schedule, and most of the cadre were clearly competent. Since Fort Meade was a replacement depot, however, neither the facilities nor the personnel necessary for effective infantry training were available. Difficulties arose from the inability of the cadre to adjust to the unusual circumstances. They were unable to coordinate their work, and the trainees found themselves bearing the burden of the ensuing disorganization. The ineptitude at first elicited amusement—as well as some embarrassment—but the men soon began expressing openly their contempt for their keepers. In time they became bitter and sullen.

The major source of exasperation during the first week of training was the chaotic manner in which troops were assigned to their daily tasks. Instead of having men fall into formation by squads and platoons, they were required to line up in whatever order in which they appeared on the scene. Since everyone was supposed to be present, roll call was held at each formation. None of the cadre ever mastered the pronunciation of Japanese names, and this alone took up considerable time. Following roll call the replacements were grouped and regrouped to make up units of the size required for the various duties to which they were to be dispatched. Thus, the first seventy-five would go to one exercise, and the next twenty would be sent elsewhere. Some of the exercises were pleasant—playing softball or sleeping through a training film—but some of them were irksome—running through an obstacle course, bayonet drill, or a work detail sorting salvage in the warehouse. Since the cadre refused to divulge which group would have the less desirable assignments, some privates decided that the safest way to avoid onerous duties would be to be absent altogether. At first, individuals who were slow in getting up or expected to be a few hours late in returning from pass arranged to have their friends answer for them in roll call. As soon as it became apparent that no particular soldier would be missed, the practice of "disappearing" became more widespread. Men began to hide in latrines, to linger in the mess hall after breakfast, or to walk casually over to the recreation center. Friends answered for anyone who was absent. As the discrepancy between the number answering roll call and the number of bodies actually present in the courtyard became apparent, the cadremen began to count over and to call roll again. No matter how often this was done, the discrepancy remained—a puzzle that never ceased to perplex some of the cadre. The Nisei were astonished at the continuation of such an ineffective procedure. In good weather they took the repeated counting in good humor. When they were forced to stand in formation in the rain, however, tempers flared, and some began to howl and hoot from ranks.

Another source of irritation was the apparent indifference of the officers to the physical discomforts of their subordinates. This, too, was in sharp contrast to IRTC. The incessant roll calls were always held outdoors, although the names were read more quickly when it was raining heavily. On one very cold morning most of the troops fell into reveille formation wearing gloves. Since the "uniform of the day" prescribed by battalion headquarters did not include gloves, Lieutenant Delaney, one of the company officers, ordered that they be removed. Then, when some of the men put their hands into their pockets, he became infuriated and castigated them severely. As he was doing this, even the other cadremen could not help but notice that the lieutenant himself was wearing gloves. On another occasion the company was aroused at 04:30 and rushed to a grenade range. It was still closed, and the troops had to stand in the rain for over an hour to wait for the range officers to arrive. That afternoon, when it became sunny and hot, they were required to hike without removing their wet field jackets. That night, even though a substantial portion of the company had caught colds, orders were issued to remove the shelter halves (half of a pup tent) that the men had erected voluntarily between their beds to minimize the chances of contagion. A common complaint was that the orders did not make sense.

On the evening of April 4 an especially frustrating incident occurred that contributed to the crystallization of judgment concerning the depot cadre. As the exhausted troops marched into the company area after a full day of strenuous training, they were informed by Lieutenant Williams, the executive officer, that a night problem had been scheduled for that evening. The lieutenant explained apologetically that in order for replacements to be excused on Saturday afternoons it was necessary for them to participate in one night problem each week. He promised that the exercise would require at most only an hour and a half; to be sure that the tedious requirement could be accomplished with dispatch he requested that all men remain in the company area after supper. He suggested further that they be ready to leave at 18:45 instead of 19:45; then they could finish an hour earlier. Most men agreed that this was a reasonable request, and they refrained from going to the PX, the barber shop, or the laundry. Some even surrendered their day's ration of cigarettes in order to comply. Since the lieutenant had been reasonable, the men agreed to cooperate.

When the whistle was blown at 18:45, every Nisei in the company was present, in prescribed uniform and equipment and ready to march. They discovered, however, that Lieutenant Williams had been replaced by a strange officer, who apparently had not been briefed on the details of the night problem. For the next hour and forty-five minutes they were grouped and regrouped, counted and recounted. From time to time they were given a ten-minute break; then they would be called to attention and the counting resumed. Finally, when it was almost dark, the company departed after being regrouped for the last time—in regular platoon formation! After a long approach march the platoons deployed at 22:30 to attack a hill. Most of the men were so disgusted by this time that they gave vent to their frustrations by storming vigorously up the hill. They encountered no opposition. As they set up a defensive perimeter to secure the position, it became apparent that the "enemy" was nowhere in the vicinity. The lieutenant had lost his way in the dark and had led them up the wrong hill. The men were crestfallen. In an effort to expedite matters each had tolerated the numerous roll calls, the long hike, and had even tried hard in the strenuous part of the exercise. In spite of their best efforts, however, the night problem had become a ludicrous flop because an officer did not know how to read a compass. It was well past midnight when the troops returned to the company area. No one said much. There was little that needed saying. They had acted in good faith, and from their standpoint the cadre appeared incompetent, dishonest, and callous.

A second incident that occurred on the following day, April 5, further convinced the replacements that their officers were both incompetent and indifferent to their welfare. The training schedule called for marching to a range about seven miles from the barracks, and the prescribed uniform was "Class C with arms." Although it had been raining all morning, raincoats were not mentioned. Finally, when the men were in formation, one of the company sergeants asked Lieutenant Delaney if it would not be advisable to telephone battalion headquarters to ask permission to carry raincoats over cartridge belts. After some discussion the company commander was consulted; the call was made, and the request was granted. There was no rain for the first four miles of the march; then, it began to sprinkle. A heavy downpour was readily visible a short distance away, and it was apparent that the gloomy rainclouds were moving in the direction of the company. The column was halted; four lieutenants and a captain held a discussion on whether or not the troops should put on their raincoats. The men could overhear easily what the officers were saying. One lieutenant protested vigorously that the schedule did not call for raincoats, and it was decided to adhere to the original schedule. As the storm swept over the area, everyone was drenched. After the downpour gave way to light sprinkling, a major riding on a jeep stopped the column. He stared aghast at the condition of the troops and demanded to know why the men were marching with their raincoats hanging over their belts "when any fool can see that it's raining." Indignantly, the major issued orders to put on raincoats. But it was too late; everyone was already so thoroughly soaked that the airtight raincoats prevented the fatigue suits from drying. Nonetheless, most of the men appreciated the major's gesture.

Since the men believed they would be joining the Hawaiians shortly at POE, they wanted to make the most of their stay at Fort Meade. Hence, they resented anything that cut into their "free time." Indeed, some of the most impassioned complaints against the callousness and inefficiency of the cadre stemmed from frequent delays in issuing passes. The type of incident that enraged even the most patient occurred on the evening of April 5. About half the company had applied for passes, and most of them were anxious to leave as soon as they were off duty. Many had dates and did not wish to keep the women waiting. However, an unannounced rifle inspection was held during the retreat formation, and several were "gigged." Since the rifle racks had been locked immediately after the troops returned from the field, no one had had an opportunity to clean his rifle. Some men, whose rifles had not been fired that day, were "gigged" for not having sufficiently short haircuts. Well over half the company was restricted. After being reminded that soldiers were subject to inspections at any time, the replacements were told that they could have their passes after supper if they cleaned their rifles and presented themselves with adequate haircuts to an inspecting officer. Many missed supper in a scramble to clean their weapons. When they ran to the orderly room, however, they discovered that the lieutenant had disappeared and that no one knew where he had gone. They waited and waited. At 19:00 all men needing haircuts were ordered into formation, and everyone else—even those whose haircuts and rifles had passed inspection—was restricted to the barracks. At 19:15 the entire company was called into formation. Fifteen men had not been paid for the month of March; after reading their names the first sergeant announced that everyone was restricted until these men had been paid. After the payrolls had been signed and the haircuts completed, a token rifle inspection was conducted by the company NCOs. The officer in charge had still not returned; at 20:15 the CQ (Charge of Quarters) began issuing passes on his own authority. What appeared to be callous disregard on the part of the cadre for their pass privileges embittered even those who were not interested in passes for themselves.

By the end of the first week of April discipline in A-9-3 began to break down. At first, recalcitrance took the form of yelling from ranks after being called to attention, of falling into formations sluggishly, and of deliberately marching out of step. In virtually every formation, as they were being counted and grouped, shouts erupted: "Let's go!" "Let's get on the ball!" "What's the holdup?" "One time! On the ball! One time!" "What are we waiting for?" The shouting often continued even when some officer ordered, "At ease!" Whenever they were required to stand in the rain without raincoats, someone would yell: "Anybody but a moron can tell it's raining!" Although the men only glowered sullenly when scolded by officers, they did not hesitate to talk back to NCOs. Occasionally, they muttered threats of gang beatings under their breaths, and some of the cadremen appeared genuinely concerned over the possibility.

The training facilities available at Fort Meade were limited—enough to keep replacements occupied for a few weeks. Before long the Nisei were forced to repeat them, and repetition added to the general sense of futility. Those who had completed basic training in communications or in armored units indicated that they found the program beneficial, but most of the replacements were riflemen. They were already familiar with most of the weapons; only the heavy machine gun, the heavy mortar, and the portable flamethrower were new. Nor did they find the tactical problems particularly challenging; the whole program was a poor review of what they had already been taught in more detail and with far greater skill. Furthermore, all the battle ranges were several miles from the barracks, making it necessary to march long distances just to run through a simple problem in which they felt they learned nothing new. Even the officers assigned as instructors at the various ranges were embarrassed; by the end of April they realized that all the Nisei had heard their lectures and jokes at least two or three times. Since inspecting officers were ubiquitous, however, there was little choice but to continue the boring routine.

Two exercises in particular continually reinforced the growing sense of uselessness. One called for removing boots and stockings and holding apart one's toes as an inspecting officer passed by, frequently without looking. The second required marching a few miles to a creek at which each man was required to wash his feet. This was followed by a lecture on the importance of foot hygiene—for riflemen largely a repetition of what had been stressed repeatedly in IRTC. At first no one objected, since neither task was strenuous, but in time the routine became more and more irksome. On one occasion, when three foot inspections were held on a single day, even the cadre agreed that the comedy was being carried too far. What proved most disgusting, however, was the stark contrast between what was said in the lectures and the total disregard shown in other situations. For example, on May 3 the troops were marched to the mortar range in a rainstorm. Because of the weather the officer in charge refused to take the weapons outdoors. Instead of allowing the men to go indoors, Lieutenant Delaney ordered a road march; and by the time the troops returned to the barracks they were drenched with rain and sweat. Some were especially annoyed by the squishing of water inside their soggy boots. When they requested permission to change their shoes, the lieutenant replied, "Oh, don't bother. Wet feet won't hurt you. In a couple of months you'll be at a place where your feet will always be wet." The men realized that this was very likely, but it was shocking to hear it from the very officer who had lectured so solemnly on foot hygiene: "Goddam! Talk about chicken shit! That fuckin' Delaney made us take off our shoes three times in one day because he was so fuckin' worried about our feet. Now that the schedule don't say nothin', he don't give a shit. We could catch trench foot for all he cares. He probably hopes we get it."

Some of the Nisei were now seething with resentment. They felt insulted. They regarded themselves as sufficiently intelligent to perform routine tasks without supervision. If they were wanted at some formation, they needed only to be told, and each man would assume the responsibility for getting there on time with the proper equipment. But they were being treated like imbeciles. Mutual hostility became even more aggravated as increasing numbers began falling out late for all formations. The practice began at reveille. Many argued that it was pointless to go out on time, since they did nothing but stand about shivering in the cold while the first sergeant walked back and forth counting them. Why hurry? But one morning the captain appeared unexpectedly and caught several men still in bed as reveille was being sounded. The cadremen were reprimanded severely for not having completed roll call before the bugle, and thereafter they began to call out the troops even earlier. Since reveille time was posted daily on the bulletin board, everyone knew when he *had* to be outside, and they continued to take their time. Before long a similar sluggishness characterized their response to all calls to formation. The IRTC practice of virtually knocking one another down in the rush to get out was now only a memory of another world.

From time to time the men got together to devise specific stratagems. One procedure for embarrassing march officers they disliked was deliberately to walk out of step in what came to be called a "staggered formation." Difficulties arose initially from inconsistent army policy. Those who had trained in integrated units had become accustomed to marching in the regulation 30-inch steps required of Caucasian troops, but those who had trained in segregated units had become accustomed to 25-inch steps permitted at Camp Blanding. From the first day replacements from different IRTC centers found themselves stepping on one another's heels or having to trot in order to keep up. When one Camp Wheeler man complained, "I can't get used to this Boochie 6-inch step," a Camp Blanding man retorted, "What's the use of running when there's plenty of time?" To retaliate against selected officers the men made no serious effort to adjust to each other but continued to march in irregular patterns. At times the platoons only vaguely resembled a military formation. On one occasion when an inspecting colonel witnessed the disorder and reprimanded the march officer, the men congratulated one another in glee.

Events following the death of President Roosevelt reinforced even further beliefs of cadre incompetence and further intensified the mounting resentment. News of the tragedy aroused much excitement and speculation. Because of the belief in some Nisei circles that the president had personally ordered the evacuation, some animosity against him was apparent. Nonetheless, the general atmosphere was one of gloom, and the prevailing view was that the world had suffered a great loss. Discussions were soon cut short, however, for a number of strange officers appeared in the company area to supervise groups assigned to a variety of tasks. For some reason all the Nisei were kept busy, although it soon became apparent that there was actually very little that needed to be done. But the sudden break in the routine aroused expectations that something unusual was about to happen, and before long a rumor developed that A-9-3 would participate in a gigantic funeral parade. This was discussed avidly, and on the following morning details were elaborated in additional rumors. One was that everyone would have to march for five miles with rifles at "right shoulder arms." This seemed plausible, for movies of important parades showed troops with rifles on their shoulders. But this was the most tiring way to hold a rifle! Another rumor was that a group of Nisei would be selected on the basis of height; most men agreed that this would be desirable, for it would assure a more favorable showing.

On Saturday morning, April 14, confusion in the company was even greater than usual. At 07:30 the men were told that they would not participate in the funeral parade and that they were to put on their fatigue suits for regular training. Then, at 08:00 they were ordered to stand by for parade duty. At 08:02 they were told that the parade had been canceled, and at 08:10 they were ordered to put on Class A uniforms and to be sure that their rifles were clean. At 08:15, after most of them had changed, the company was ordered to fall out for training in Class C uniforms! Though they were accustomed to disorder, this was too much. Some just sat down on their footlockers and refused to budge until cadremen came in to chase them out. At 11:00 the company was marched back to the barracks, and at 11:30 it was announced that one platoon of Nisei would march in the parade and that it would be the 4th Platoon of A Company. At 11:40 an officer announced that a full company of Nisei would march, under the command of Nisei officers and NCOs. By this time, even the cadre did not know what to do, and several of them sat down beside the replacements to engage in speculation.

At 12:25 it was announced that all plans concerning Nisei participation had been canceled. Although most men agreed that dress parades were tedious, they were genuinely disappointed. This was a momentous event, a decisive occurrence of great historical significance; they had missed an opportunity to be part of something really "big." Some felt that this was their opportunity to show the *Keto* how well Nisei could march when they really wanted to excel. Several disclosed their willingness to give up their "free" Saturday afternoon in order to be in the parade.

"Maybe they left us out because we always fuck up when we march. Our staggered formation must really get them down."

"But, hell. The Buddhaheads can really march if they want to. If it's important like this, all the Boochies'll be on the ball. When the Boochies are on the ball, they can make any *Keto* outfit look sick."

"We'd get fucked out of Saturday afternoon though. Jesus Christ, we went on a night problem so we deserve a half day off. I heard we might get Monday off, but that's probably in the 10th Battalion. In this place they always fuck you in the ass when they get a chance like this."

"But at least we can tell our kids that we were in the funeral parade for Roosevelt. That would be something!"

The entire episode served to irritate everyone even further and to reinforce their views concerning the incredibly inefficient organization of the company. Although some argued that decisions concerning the composition of troops in a formal parade in honor of the president would be made at echelons higher than the A-9-3 orderly room, blame for the confusion was still focused there. Had the company cadre done its work effectively, all the Nisei would have been alert; then, the "brass" would not have hesitated to call on them to participate in the parade. Many reiterated their belief that the depot cadre was made up largely of misfits who were mentally or physically unfit for overseas duty, even though reports of the efficiency of the 10th Battalion cadre led some to question this. The Nisei concluded that most of their keepers were "dumb." They paid little heed to their orders, obeying only when it was unavoidable.

The general dissatisfaction and resentment in A-9-3 gradually developed into open hatred for four cadremen. The regular staff of the company consisted of three officers—the commanding officer, the executive officer, and Lieutenant Delaney—and thirty-four enlisted men. Most of the enlisted cadremen were assigned to specialized tasks in processing replacements, and the Nisei encountered them infrequently after their first week. Additional officers were assigned on a daily basis to lead the troops in various exercises. Most of the Caucasian NCOs experienced no difficulties with Nisei replacements, and some of them were very popular. Sergeant Thompson, platoon sergeant of the 2nd Platoon, and Sergeant Brown and Corporal Beaudette, who accompanied the troops on all hikes and field problems, were well liked. The latter two were wounded combat men who had reenlisted in the Regular Army. When in Class A uniform, both wore the coveted Combat Infantryman's Badge as well as the Purple Heart; they were respected combat veterans. The two not only sympathized openly with the Nisei but often warned them of traps being planned in the orderly room. Thus, the men usually knew in advance when a strict roll call was to be taken, and everyone would be present. Lieutenant Williams, the executive officer, tried to help the replacements whenever he could. Often there was little that he could do, and the men appreciated his unenviable position. There was general agreement that he was a "nice guy." The four on whom hostility focused were the company commander, the first sergeant, Lieutenant Delaney, and Corporal Salerno, the company barber.

The person who was held most accountable for the unpleasantness and to whom special malice against Nisei was attributed was the company commander. Captain Larsen was thoroughly disliked, not only by the replacements but also by the cadre serving under him. Many felt that he had probably been assigned to this post because he was too incompetent to be trusted with troops in combat. There seemed to be consensus, even among the depot cadre, that the captain had no sympathy for any enlisted man. One of the corporals in the company advised a gathering of Nisei: "If you're in dutch, the old man is the last guy you want to go to. I've served under a lot of tough COs, but inside all of them had a heart. They cared about their men, and they helped us. This son of a bitch is just looking out for his own ass. He don't give a damn for his men. He's bucking for a raise, and he's afraid that if he helps out his men he might make some small mistake, and they'll ship him overseas. He's chicken. He follows orders down to the last letter, even if it kills his men, because he's afraid they'll send him over." The few who did go to the captain with their personal problems discovered that the corporal was correct. The captain insisted that everything that went wrong was their own fault and refused to intervene in their behalf.

Of the company NCOs First Sergeant Mueller was without question the most unpopular. Although he often gave disagreeable orders, he was viewed by many more with contempt than hatred. A rotund, bespectacled man with a cherubic mien, his performance often elicited derision. He was initially dubbed "Humpty Dumpty." He made mistakes frequently; and since he was easily flustered, the men delighted in confusing him. When someone first commented that the sergeant reminded him of Oliver Hardy, others protested that this was an insult to the comedian. Within a few weeks, however, whenever the sergeant waddled through the company area, someone would whistle the Laurel and Hardy theme song—"The Cuckoo Song." Since they whistled through their teeth without pursing their lips, it was difficult for other cadremen to detect the violator of orders to stand at attention. Several rumors about him emerged: he had come to Fort Meade as a replacement and had been assigned as a company clerk after failing his physical examination; he had been promoted from clerk because no one else would take the position. The men thought nothing of talking back to him. The following exchange took place during one of the roll calls:

"I want Tanashi. Tanashi! TANASHI! It's spelled T-A-K-A-H-A-S-H-I. Tamaki? Tanaka?"

"Here!"

"Here!"

"Here!"

"Here!"

The sergeant was visibly irked. "What is this? Are there four different Tanashis? Now listen, fellows. I want only the man whose name I call to answer. None of you will answer for any of your buddies."

There was a brief interval of stunned silence; the men stared incredulously at one another. Someone giggled. Then, the entire company exploded in raucous laughter.

"What the hell is this? Can't you read?"

"Fuck up! Let's get on the ball!"

"How'd you ever get out of grammar school?"

"Who in the hell *were* you calling anyway?"

"Fuck up! Fuck up!"

The one man who was probably hated even more than the captain was Lieutenant Delaney. A tall, stern, sullen man, he always insisted on the strictest enforcement of all regulations and orders. Although he himself was quite clumsy, he was quick to castigate anyone who made a mistake. During an orientation lecture to one of the first groups of Nisei at A-9-3, he demanded prompt obedience and concluded with the threat: "Strict dis-CIP-line will be enforced in this company." Several experienced difficulty in muffling their laughter; they had never heard the word so badly mispronounced. For some time he was referred to as "Lieutenant Dis-CIP-line." The following conversation, initiated by a man who had almost knocked down an officer while sneaking away from a formation, is indicative of the manner in which he was viewed:

"Which looie was it?"

"That tall bastard. You know, the dumb one."

"Aw, you mean that guy who don't know how to talk English: Lieutenant Dis-CIP-line."

"Yeah, yeah. That's the one. He's a mean bastard. You know that?"

As increasing numbers became convinced of his churlishness, the jocular label was replaced by "von Rundsted." No one dared to defy him openly, but most men made little effort to conceal their feelings. They glared back defiantly at him.

Lieutenant Delaney did not confine his disagreeableness to Nisei, and there were several indications that he was despised throughout the battalion. He seemed to be completely indifferent to the welfare of the soldiers in his command, and the cadre sometimes cooperated in countermanding his orders. In one rifle inspection, for example, he insisted that he detected carbon on a piston head, only a few minutes after it had been polished with steel wool. Others in the vicinity who examined the piece agreed that it was shining like a mirror. Nonetheless, the lieutenant restricted the man. Corporal Beaudette, who saw the whole incident, happened to be CQ that night. When the time came for issuing passes, he "accidentally" gave one to the restricted man. Sergeant Brown also made no secret of his disgust, and he disclosed that Lieutenant Delaney was drunk almost every night. On one occasion, one of the officers assigned to a battle range commented, "That son of a bitch is so scared of going overseas that he's always browning his nose on the colonel's ass. He doesn't care what happens to his men; he just wants to stay here where it's safe."

Although Lieutenant Delaney never accompanied the troops on long hikes to battle ranges, he did go frequently on short and less strenuous exercises. Hence, the replacements knew him better than they did the other officers. Several incidents left an indelible impression on those who were present. One took place on May 4, when the company was at the creek for another foot-washing ritual. Lieutenant Delaney decided that the area was too dirty and ordered his charges to "police" it. He acknowledged, "I know that you men didn't get all this stuff on the ground; but whenever we leave an area, we want to make sure that it's perfectly clean."

The men were sluggish in obeying his order, and the lieutenant called for his "tech/sergeants" to supervise the work. When no one responded, he became very angry and screamed, "What's the matter with you tech/sergeants? Don't you know when you're being called?"

The undertaking came to a standstill. Everyone turned around to see what was happening. Then, to their dismay, he pointed his finger accusingly at T/4 Yamada, who was standing just a few yards away. The sergeant was stunned; he stood mute, staring about helplessly. The others had difficulty controlling themselves, and before long some burst out laughing. It seemed incredible, but the lieutenant was apparently unfamiliar with the various grades of sergeants! (See Plate 3.) For a raw recruit or a civilian to confuse a technical sergeant (second grade) with a technician/fourth grade was understandable, but for a lieutenant to do so was shocking. The men were utterly disgusted. For the rest of that week all NCOs who held the rank of T/5 were addressed in mock respect as "tech/corporal." As one private later remarked, "Jesus Christ! That's really sad! A first looie in the U.S. Army who don't know the ranks of noncoms! Lose fight!"

Another memorable incident occurred four days later. The schedule called for the foot hygiene routine. Although it was raining lightly, Lieu-

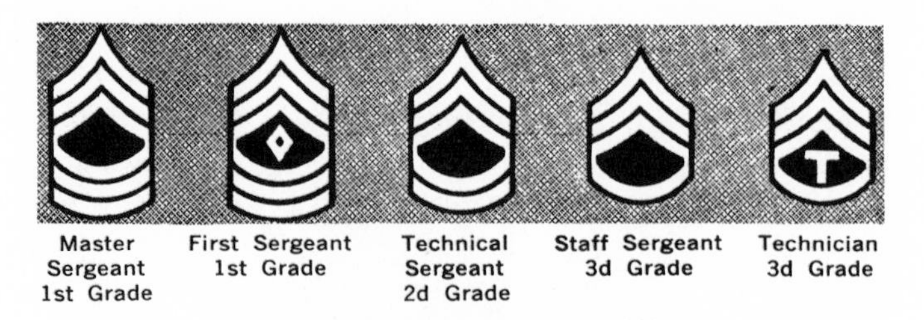

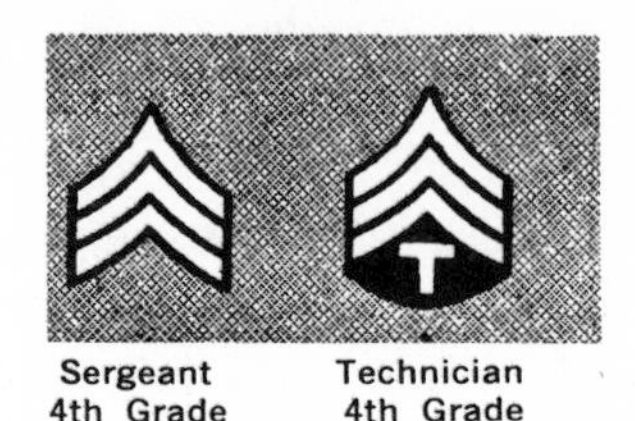

PLATE 3 *Insignia of Sergeants in the U.S. Army, 1945*

tenant Delaney ordered a road march to the creek. By the time the company arrived the sprinkle had turned into a downpour. When the lieutenant ordered a ten-minute break, someone remarked, "Well, at least he realizes it's raining," but Sergeant Brown retorted, "Don't be too sure. Maybe he just forgot his speech."

Though the area was familiar, somehow on that day it seemed remote and isolated. In the heavy overcast one could not see very far. Nothing else seemed to be going on in the fort; no other group had been encountered on the road, and no weapons of any kind were being fired anywhere. When the company was called back into formation, everyone expected to be marched back to the barracks. After all, a token effort had been made to follow the schedule. To their dismay the lieutenant prepared to lecture. The men stirred about restlessly but remained silent. Many hunched over to minimize the chances of water trickling down their necks as it cascaded off their helmet liners. As the lieutenant faced the group, the utter silence at this secluded point was disrupted only by the patter of raindrops on helmet liners.

"Men, today we are supposed to wash our feet, but I'm going to change the schedule somewhat because of the weather. It's raining." As he paused, a clearly audible stage whisper rang out in the silence from the rear ranks: "No-o-o-o shit!"

The lieutenant flushed and exploded. "Who said that? Sergeant, I want you to take that man's name!"

The men struggled to suppress their laughter. From time to time an uncontrollable giggle burst forth. The lieutenant pulled himself to his full height and ranged back and forth, glowering at the troops. He was obviously very angry. When Sergeant Brown's search proved unsuccessful, the lieutenant focused his ire on a Nisei corporal in the first row who could not stop snickering. He ordered that his name be taken.

When the commotion had settled down, the lieutenant continued, "Many times on the front you will not have the facilities that you have now. In spite of the importance I want to explain to you the reasons why we cannot wash our feet today. First of all, it would be impractical to take off our shoes because they might get wet inside. In the second place, our stockings might get wet, and that would not be good for the feet. In the third place, the foot powder would melt before it did any good, when you sprinkle it on your feet. Therefore, we won't wash our feet today. I want you men to remember that many of the common diseases come from wet feet. Athlete's foot comes from not wiping between the toes."

As the lieutenant droned on for almost a half hour, the men became more and more restless. He had to call "At ease!" on several occasions to regain their attention, but toward the end no one was taking any notice of anything he said. When he finished, the lieutenant ordered them to "police up" the area, even though they had had no opportunity to drop anything. As they scattered over the terrain to appear as if they were picking up waste, several voiced their feelings:

"Jesus Christ! Just because he's dumb, that's no excuse for giving us a crappy lecture like that. Shit, anybody knows that stuff, and even a dope knows that it's raining."

"That's the point. That guy is so damn dumb, he can't talk about anything that takes intelligence. He don't know anything else. I'll bet he memorized that speech last night and made us listen to it so that he won't feel he did that work for nothing."

The cadremen were no less annoyed. Corporal Beaudette smirked, "You guys call him von Rundsted, but I think you ought to apologize to the German General Staff."

The fourth unpopular cadreman, Corporal Salerno, was disliked primarily because he always delayed issuing passes on one pretext or other whenever it was his turn to be CQ. Unlike the officers, he was vulnerable and was frequently threatened with violence. On one occasion he demanded that those desiring passes present their rifles to him for inspection; when they protested that the rifle racks were locked, he sneered and pointed to his two stripes. One man became so enraged that he had to be restrained by his friends. Just then, the battalion duty officer (DO) —an officer from another company—appeared unexpectedly. After hearing the arguments he ruled in favor of the replacements. On the following morning, as some Nisei were sitting about waiting for a call to formation, the corporal ordered one of them to empty an ash can. The man glared back, "Kiss my ass!" Since the others in the vicinity appeared equally belligerent, the corporal began to walk off. As he approached the door, someone tripped him. He did not even stop to see who had done it. That afternoon, while the company was working with explosives, one man stole a dynamite blasting cap, and plans were laid for stealing some booby trap mechanisms. Although some argued that it might be dangerous, those planning to surprise the corporal argued that the blasting cap would only frighten him and was not likely to injure him too severely.

Thus, in the beginning disgruntled individuals had reacted to incompetence and indifference by responding sluggishly to calls to formation, by shouting from ranks, and by marching in their "staggered formation." Now, as hatred of the four cadremen crystallized, some began to defy orders more openly. As such insubordination became more widespread, the captain resorted to collective punishment. Instead of taking firm disciplinary action against individual offenders, he retaliated against all Nisei by holding them in formations in inclement weather or otherwise inconveniencing them, thereby arousing the resentment even of those who had been obeying orders. At the same time the more aggressive ones realized, much to their surprise, that they could disobey orders with relative impunity. In IRTC recruits were afraid to disobey; they rarely talked back to NCOs and never to officers, who were regarded almost in awe. At first a few had yelled in moments of intense exasperation or had sneaked off cautiously, but as it became apparent that individuals would not be held accountable for their deeds, increasing numbers joined them. The discontent in A-9-3 was also intensified by invidious comparisons. B-9-3 was reputed to be efficiently run, and the entire 10th Battalion was viewed longingly as a paradise in which necessary duties were performed promptly so that everyone could leave on pass as soon as possible.

Once they realized how difficult it would be for the cadre to locate any particular individual, men began going AWOL in droves—missing exercises, walking off of fatigue details, disappearing in the midst of field problems, and returning from pass several hours late. When tactical problems were being executed, several simply vanished, hiding in the thick vegetation until it was time to return to camp. Many Nisei saw themselves as being involved in a contest with the four cadremen and some of the officers occasionally assigned to them. They began fighting back by doing whatever they could to embarrass or inconvenience them. They stopped saluting officers and addressed them as "sir" only when such deference was explicitly demanded of them. Whenever an unpopular officer made a mistake, someone in the ranks would call attention to it in a loud voice; there was little the flustered officer could do, for he was in error. Such conduct further aggravated the already strained situation. At first such insubordination was spontaneous and individual, but before long the more audacious began vying with one another to see who could go the farthest without getting into serious trouble. To harass the cadre became the clever and daring thing to do. As the troops became increasingly unruly, Lieutenant Delaney was ordered to give another lecture on military discipline. When, as he approached the conclusion of his talk, he yelled, "You must obey orders regardless of how stupid or absurd they may seem," the entire company burst out in laughter. The cadre and the Nisei in A-9-3 became so preoccupied with fighting each other that they forgot about the war.

As the repetitious exercises came to be viewed as a nuisance, increasing numbers began to report for sick call. Since the company was gone by the time the indisposed returned from the dispensary, they were able to relax for the rest of the day. But these men were not just being lazy; they knew that an excessive number reporting sick from any single company would result sooner or later in an investigation. Such malingering was part of a deliberate effort to embarrass Captain Larsen. A comparison of morning reports for the month of April reveals that forty-two replacements from A-9-3 were hospitalized; none from B-9-3. In May, twenty-three were hospitalized from A-9-3 during the period when Captain Larsen was company commander; only four from B-9-3. The difference in the numbers reporting for sick call and not hospitalized was even greater. Suspecting that something of this sort was happening, Captain Larsen ordered everyone reporting for sick call to some kind of fatigue duty—generally cleaning the latrines or helping in the kitchen. Although the men realized that many were just loafing, considerable resentment arose during a cold epidemic, when several were actually ill. They also knew that the captain was aware of the epidemic, for orders had been issued from battalion headquarters to install shelter halves between the beds. The issue came to a head on April 9, when one of the sick men, who had been forced to work as a KP, was found to have contracted measles! He had to be hospitalized and did not rejoin the group for over a month. That evening, when the men learned of this incident from the malingerers, indignation ran high.

"What the hell! That fuckin' CO must think that all guys who go on sick call are goldbricks. Sure, some of them don't have to go, but most of those guys have something wrong with them. Even if you're sick only a little bit, you really feel like hell when you're on the field because you have to work so damn hard out there. If you get caught out there, it's really tough."

"That fuckin' CO is afraid of getting his ass chewed for not putting the shelter halves up on time. He's never on the ball. He figures that if we have too many guys on sick call, they'll investigate to see what the hell's up, and they might find out that he's not on the ball. He's just chicken shit, that's all."

"Goddam! What a chicken shit outfit! How can those dumb bastards tell if a guy's really sick or if he's goldbricking? It's none of their goddam business anyway. If the medic reports that there's nothing wrong with a guy, then they can clamp down on him, but not before. A guy might look damn healthy and still be sick—like that guy who had measles."

A clear understanding developed among the Nisei, however, that everyone could not disappear at the same time; a respectable number had to appear in each formation to go through the motions. Although no roster was kept, the men took turns in "goofing off." Those who disappeared too frequently were sometimes cautioned by others to wait their turn. Those who reported for sick call too regularly were accused of "goldbricking," and pressures were placed on them to appear for their share of the disagreeable duties. Private Onishi, an assimilated college graduate who was apparently unfamiliar with Nisei mores, was a frequent violator. He left whenever he thought he could get away with it—going on sick call or vanishing whenever a strenuous exercise had been scheduled. Since he took more than his share, he was accused of being selfish, labeled a "goldbrick," and ostracized. He was surprised that the others seemed to dislike him so much, when he excelled in what everyone else was doing.

The constant griping became increasingly acrimonious. Barracks conversations not only reflected but also reinforced the general state of demoralization:

"Goddam! This company is waste time, though. I never been in such a fucked up place in my life. That fuckin' von Rundsted and the first sergeant really get me."

"How'd that old pansy ever get to be a first sergeant? Either he's the captain's fairy or else he kisses his ass good every day. When I see guys like that get ahead, I lose fight. That guy don't know nothin'. I don't think he can count over fifty. Notice the way he counts us? He counts one at a time. It would be easier if he counted the number of guys in each row and multiplied by three."

"Yeah, but that's too intelligent. I think I'll lend him my fingers some day. Then, maybe they can count us once and get it over with."

"The guy who really gets me is that fuckin' von Rundsted. He's a first-class son of a bitch if there ever was one. He's dumb and fucks up all the time, and then we get it in the ass so he can cover up his mistakes. I don't like guys like that. I don't mind if a guy is tough, but when he takes it out on his men to cover up for himself, then I hate him."

"Boy, I'd like to go over with those guys. The first time we get into action I'd shoot 'em."

"Aw, you wouldn't have a chance. Somebody'd push 'em off the boat before they got there."

"All these guys around here are like that, though. They're all bucking for that other stripe or bar. They're all chicken shit."

"Yeah! Remember da day we go out in da rain? We have our raincoats on our belts, but dose guys no let us put 'em on because da schedule no call for raincoats. Den da major come and chew deir ass. Goddam! Orders is orders, but you got to use common sense, no? If you got to have a major to tell you a seemple t'eeng like putting on a raincoat, den what for we got officers lower dan a major? All you need is a major and a bunch of buck privates, no?"

"It's like them guys rushing our ass off to fall out in time and then making us wait for an hour while they decide what to do with us. Why don't those damn officers shag ass once in a while?"

"Everything around here is waste time. I don't see why they're so damn strict about us taking training. We had all this crap before. You don't learn anything here. If we learned something new, I wouldn't mind if they did run us a little, but this is waste time."

"They don't care if we learn anything or not. They're just afraid that they'll get their ass chewed if we don't show up."

"They think we're as dumb as they are. God, they treat us like we was a bunch of dumb kids. If we were as dumb as that von Rundsted, then they could talk to us like that, but Buddhaheads don't come that dumb."

"Goddam! I hope we get the hell out of here pretty soon. This is a pretty easy life, but I don't like this fuckin' company. If we were in the 10th Battalion I wouldn't mind, but this fuckin' place gets me down. It might be worse overseas, but I don't care. I want to get out of here."

Reprisals for Differential Treatment

For most Nisei in A-9-3 the last week of April turned out to be one of the most exhausting that they spent in the U.S. Army. Long afterward, when recalling their miseries, they referred to it as "the battle of B ranges" or "the week we marched 100 miles." For five consecutive days they not only hiked long distances to various training grounds but were also required to run through infantry assault problems when they arrived. They were aroused as early as 03:45 in the morning and had to work late every night. Since they were still required to maintain their barracks as well as to keep themselves and their equipment clean enough to pass inspections, there was little time for sleeping. This was also the week during which discipline disintegrated.

On both Monday and Tuesday mornings A Company was on the road at 07:30, marching to distant ranges to run through problems of infantry attack, village fighting, and close combat tactics. On Tuesday afternoon the men participated in a large-scale combined-operations maneuver—involving some tanks and artillery as well as infantry—and did not return to their barracks until 18:20. The total mileage covered in getting to and from these exercises was estimated by the cadre as thirty-four miles. On both evenings after supper, showers, and the cleaning of rifles and equipment it was too late to go anywhere on pass. Yet there were relatively few complaints, for it was assumed that after two such days in succession there would be a respite. They looked forward to Wednesday, when they expected to remain near the barracks for an easy day of orientation lectures, movies, and sports.

But on Wednesday morning the Nisei platoons were aroused at 03:45. Several remarked pointedly that Caucasian replacements had not been required to fall out. It turned out to be a wet, miserable day. At 06:00 the platoons set out for range B-2, nine miles from the company area. The drizzle continued for hours, and by the time they arrived most men were already wet from sweating under their airtight raincoats. Then, just as one of the units was starting through the assault course, the drizzle became a downpour. Since those performing had removed their raincoats, they were thoroughly drenched. The problem required their falling repeatedly and firing at various targets from a prone position, and by the time they staggered off the course they were covered with mud. Most of them deliberately exposed themselves to the rain in hopes of washing off some of the slime, but before long the wind rose in velocity to a point where everyone had to run for some kind of shelter. As the chilling gusts penetrated their rain-soaked uniforms, the men huddled under their raincoats, cursing their misfortune. A night problem had also been scheduled, but because of the weather the lieutenant in charge decided to bring in the troops. It was not until 19:00 that the grimy men returned to the company area. By this time some of the rifles were beginning to show specks of rust; all the pieces had been fired repeatedly and then soaked in water. Not many riflemen would have been so foolish as to leave their weapons in that condition; but the company was formally restricted, and everyone was required to work on his rifle until it passed inspection. Almost everyone was at work until midnight, and some did not finish for another hour after that.

Bitter resentment arose from the conviction that Captain Larsen was using these exercises to punish all the Nisei. According to the cadre NCOs he received three different schedules each day from battalion headquarters and was authorized to send any platoon in his command to any of the programs. One of the schedules was invariably light—made up of athletic contests, close order drill, and training films—and one consisted of field training. In previous weeks the schedules seem to have been alternated, but this week the captain sent the Nisei out to the field every day—presumably to get even with them for their misconduct. That Caucasian replacements in the company who had completed their processing—enough to make up half a platoon—had been excused from night problems and long hikes seemed to support this view. Furthermore, as they ate together in the mess hall, Nisei learned from the Caucasians that the latter had spent each day playing baseball and watching movies.

Since only the Nisei were being given disagreeable assignments, they became even more belligerent and uncooperative. When the Nisei platoons were again aroused at 03:45 on Thursday morning, some of the men refused to get out of bed. They announced that they were prepared to face a court martial to expose the captain, but at the last moment they were persuaded by their friends to get up. As increasing numbers reported for sick call in order to escape the long hikes, the size of the platoons dwindled noticeably. Corporal Beaudette disclosed that the captain had been asked by his superior officers for an explanation of the large-scale malingering in his company. On Thursday morning, when forty reported that they were ill, Captain Larsen required each of them to report to him personally to explain what was wrong. This order only led to further resentment, even though the men knew very well that most of the forty were in excellent health.

"Poot'a Christ! What the hell is this? We've been getting fucked all week! This is the last straw! I think I'll go on sick call too. Fuck this shit! I was up until 01:15 this morning cleaning my fuckin' rifle, and I don't feel like getting it dirty today."

"You better not. I hear the CO wants to see all the guys who want to goldbrick personally. He wants to know why they go on sick call."

"What the fuck does he have to do that for? He knows fuckin' well why everybody goes on sick call. It's *his* fault! He's a son of a bitch, and he's got it in for us, that's all."

"Hey, how come the fuckin' Haole don't have to go with us? They sit around and goldbrick all day. That's why no Haoles have to go on sick call."

"Ah, them fuckin' bastards! They think their shit don't stink. Fuck them! They can't take it! They'll be pushing up daisies as soon as they get over there anyway. No *shimpai* [worry]!"

"Son of a bitch! I'm not so sure I can take it either. My ass was really draggin' yesterday, hey. That mud is really the shits. You take a step forward and then slip back two. Boy, my ass is worn out."

After running through some infantry assault tactics, the troops returned to the company area at 16:30. Nothing was scheduled after supper, and on Thursday evening the barracks were almost deserted. On the following morning, however, the Nisei platoons were again aroused at 03:45. Two rumors developed that did not lighten their surly mood. One was that a night problem had been scheduled for that evening; since the one on Wednesday had been canceled, this seemed plausible. The second rumor was that the strenuous week was just the beginning of a special thirty-day training program. But why should the Nisei be singled out for additional infantry training? The fighting in Europe was almost over, and it seemed unlikely that they would be sent to combat zones in the Pacific.

Again the Nisei were on the road at 06:00 to hike over eight miles to run through yet another infantry assault problem. Because of the ingenious organization of the problem, its execution required vigorous effort. Since most of the men were already ragged from fatigue, they had a difficult time. They struggled through the course, puffing and cursing; on finishing, many just flopped to the damp ground—utterly exhausted. About 16:00 a lieutenant announced that the company would eat supper from a field kitchen and would then proceed to another range for a night problem. It was at this point that discipline collapsed. Howls of protest erupted everywhere. Some swore at the officers, and others made uncomplimentary remarks within hearing range expressly for their benefit. Since none of the Caucasian replacements had been required to come, they complained openly of discrimination and spoke bitterly about having to fight for an army that "treats us like a bunch of dogs." Several threatened to take up the matter with the regimental inspector. Corporal Beaudette, Sergeant Brown, and other cadre NCOs were just as angry as the Nisei and spoke up in their behalf. The officers, realizing that the men were in an ugly mood and unable to refute their charges of discrimination, made no effort to exercise their authority. A few tried to plead with the men, but most of them endured the insults without responding.

That night the officers assigned to the company had considerable difficulty maintaining order. As the troops started for another range where they were to participate in a night attack, the marching formation quickly deteriorated. The road was muddy, and it was difficult to remain in line even if one tried. A mass of men shuffled and staggered in the general direction of the next range; and the efforts of the officers to exhort, cajole, or threaten the troops were of no avail. The men just glared back at them and accused them of discrimination. Private Nakanishi, acknowledged to be the company comic, was just quipping that "this is our new staggered diamond formation," when an inspecting colonel drove up unexpectedly in a jeep. Those who saw him tried to jump back into formation, but it was too late. Some argued that the officers deserved to be embarrassed, but others pointed out that they themselves might lose their weekend passes for the next several weeks. The colonel called together all the officers and cadre NCOs and reprimanded them severely, but to everyone's surprise he said nothing to the replacements.

The night problem called for an assault on a hill by the entire company. The troops were deployed into four platoons, and after waiting for complete darkness, they advanced tactically behind their officers. Most men did their best. The presence of the colonel raised the specter of possible punishment. When they were told that a general was also present, they agreed that they should perform effectively to show the "brass" what Nisei soldiers were capable of doing. A sense of ethnic pride overrode all other considerations, and the men tried hard to remain concealed during the long tactical approach to the hill. Orders were passed back quietly, and each was alert and ready to do his part. It was so dark that no one could see more than a few yards. The officers led the way with compasses, and the others could only follow blindly, being as quiet as they could. When the flare signaling the attack was fired, however, the 3rd Platoon appeared unexpectedly abreast of the 1st Platoon, supposedly in the lead. In the scramble that followed, the "enemy," having located the attackers, opened fire. Two officers had led their platoons to the wrong place. Furthermore, since they had neglected to orient their men, some of the troops charged in the wrong direction, not knowing the exact location of their objective. No one doubted that the company had been annihilated. When the observing general later commented that "the problem was strictly snafu," the men had to agree. They were sorely disappointed. They had tried so hard in spite of their fatigue, but two lieutenants had made costly errors. Several expressed concern that the general left with the impression that Nisei troops were incompetent. They began to grumble about *Keto* officers.

Several Nisei became infuriated when they learned that the "enemy" detail consisted of Caucasian replacements from A-9-3. As all the troops mingled at the termination of the problem, the Caucasians indicated that they had been excused from all duties until evening, when they marched to the hill and simply waited. The Nisei had been "killed" by fresh and rested men from their own company. As open resentment against all Caucasians became more noticeable, many of the latter were astonished at the bitterness with which some Nisei addressed them. They could not understand why anyone should be so hostile toward those who were only obeying orders. Some Nisei were so bitter that they declared they would have tried even harder had they known the identity of the "enemy." "Hell, if I knew that Haoles were going to be up there, I would have used my bayonet without the scabbard. I could always say it slipped off."

On Saturday morning Corporal Beaudette and a cluster of men tabulated the number of miles they had been required to hike during the week and were astonished at the results. They had covered almost ninety miles in five successive days! They had been on duty for sixty hours in addition to their routine work. They had run through six assault problems in the rain and mud. They had worked four nights out of five. (See Table 2.) Most men were confronted with serious laundry problems, for every fatigue suit issued them was filthy. Although they complained bitterly about sore feet and not getting enough sleep, most were not so bothered by the physical strain. Because of daily exercise and well prepared meals they were in excellent condition. In spite of noisy lamentations about being exhausted, very few remained in the barracks to rest or to sleep. What upset them most was what they regarded as unwarranted

Table 2
A-9-3 Training Schedule for the Week of April 22

Day	*B Range & Exercise*	*Round-Trip Mileage*[a]	*Hours of Work*
Monday	B-9 Infantry assault	14 miles	07:30 - 12:30 13:30 - 18:30
Tuesday	Village fighting and close combat	4 miles	07:30 - 10:30
	Combined operations	16 miles	11:30 - 18:20
Wednesday	B-2 Infantry assault	18 miles	06:00 - 11:30 12:30 - 19:00
	Rifle cleaning		20:00 - 01:00
Thursday	B-4 Infantry assault	16 miles	06:00 - 11:00 12:00 - 16:30
Friday	B-3 Infantry assault	17 miles	06:00 - 11:00 12:00 - 17:30
	K-4 Night attack	4 miles	17:30 - 01:00

[a] Estimates of mileage were provided by the cadre accompanying the troops and were confirmed by the range officers.

incursion on their "free" time. Actually they had been free to leave on pass on Monday, Tuesday, and Thursday evenings; however, since they were kept busy until almost 21:00, they were not able to get away from the depot. Their resentment was aggravated by the feeling that the training was useless; they did not want to give up their "own" time if they were not learning anything new.

Another practice that was deeply resented as differential and punitive treatment was the issuing of monthly pay to Nisei soldiers or alerting them for overseas duty only during off-duty hours. When Caucasian replacements were on orders, the entire company was aroused—sometimes as early as 04:30—for the reading of a few names. Then the Nisei were sent out for field training. When some Nisei were to be alerted, however, their names were read in a formation called after supper, when most Caucasian replacements had already left on pass. Pay formation on May 1 was handled in a similar manner and led to serious difficulties. When the Nisei returned from the field, they learned that the Caucasian replacements had been paid at 16:00—on "government time." Indeed, they had been excused from all other duties that afternoon so that they would be present when the pay officer arrived with the money. Nothing was said during retreat formation about the Nisei being paid. This occasioned no surprise; the men recalled the chaotic manner in which they had been paid the month before, when from time to time small groups were called unexpectedly to the orderly room. In fact, there was no hint that any Nisei were scheduled to be paid that night until large numbers had left the company area—to the post theater or PX, to the boxing matches, or to visit friends in other companies. It was not until a man asked for a pass to leave the fort that Corporal Salerno, who was CQ that night, informed the applicant with unconcealed glee that all Nisei were restricted to the company area until everyone had been paid. Since the corporal gave each applicant a different reason, much confusion arose until Lieutenant Williams arrived with the money. He was surprised when told that the Nisei did not know about pay formation, but he confirmed the corporal's contention that no passes could be issued.

It was not until 19:00 that all Nisei were ordered to fall out in front of the company recreation room, where a temporary pay station had been set up. It was a chilly night, and the men were required to stand in the cold wind as Corporal Josey, one of the company cadre, slowly read the names of those to be paid—four or five at a time. Many whose names were called were absent; since this was a pay formation, requiring signatures on receipts, their friends could not answer for them. Even after he had been told repeatedly that half the company was missing, the corporal adamantly refused to call additional names until those not answering had been located. After several prolonged delays, Lieutenant Williams had to come outside to order the corporal to skip those who did not answer. From the thickness of the receipt pad it was apparent that only about a third of the Nisei would be paid that evening. To expedite matters the men tried to find out who was wanted, but the corporal refused to allow anyone to see the roster. The men then tried to guess from the names already read, but they could find no rationale for the listing. The names were not in alphabetical order; they were not arranged by the IRTC camps from which they had come; they were not in the order in which they had arrived at Fort Meade; they were not listed by platoons. Had the rationale been ascertained, those who knew that they would be near the bottom of the list could have run to the boxing matches, the PX, or to the other companies to fetch the others. But there was no way of telling whose name would be called next. As the wind began to blow harder, the added discomfort intensified the general sense of frustration.

There were numerous delays as the corporal, disregarding the lieutenant's order, continued to hold up the reading as absentees were being sought. The men began to mill about, huddling in small, angry groups.

"We fuck around, and we fuck around! What the hell is this, anyway? Goddammit, it's cold out here! If they're gonna pay us, why don't they get it over with? They can have my four bucks, if they want it so bad."

"We always get fucked up this way. I bet that fuckin' CO stays awake every night trying to think up some new way to fuck us up. How come those Haole boys got paid on government time? They're no better than we are. We work harder than them bastards anyway. That CO always thinks of some way to restrict us."

"I heard that one of the boys went to see the CO about something, and the CO really chewed him out. He told him that the reason we didn't get the breaks was because we didn't cooperate with him."

"*We* don't cooperate! How in the hell *can* we cooperate?"

"Shit, it's not us that fuck up all the time. They're the ones who're always fucking up. He's got a lot of guts trying to blame it all on us. If that's the way he feels about it, fuck him!"

In another circle the conversation took on a similar tone:

"You mean to tell me that the Haoles got paid on government time? That's dirty!"

"Aw shit! Wake up, will you! They don't like the slant of your eyes. You always get the ass end of the deal. They want you to die for your country, but when it comes to giving you something, you get fucked in the ass. That's the way these fuckin' *Keto* are. They show their Pepsodent smile when they want something, but after they got it, they want you to go to hell."

"I'll bet that fuckin' von Rundsted had something to do with this. Him and that angel first sergeant. They're always fuckin' everything up. I don't think they're smart enough to fuck things up like this on purpose, though."

"You can't tell. Those bastards got it in for us. They think we'll see things their way if they make us suffer a little, but they've got us wrong. The more you push a Buddhahead, the madder he gets. I want to think up one way to fuck them up in the same way."

In another gathering, made up largely of Hawaiians, the men muttered in the same vein:

"Son of a beetch! I t'eenk I go see da regimental inspector tomorrow. Dees is race discrimination, and da fuckin' army is against it."

"Aw, what's da use? Da inspector, him Haole too, no? Maybe to your face he say somet'eeng nice, but he take your name and fuck us up after, no? Waste time, dat kind! Best way—get dees bastards alone in Baltimore some night. Den, we bust up, yeah?"

Darkness fell, and the temperature continued to drop. As several ran into the barracks to get heavier clothing, the grumbling continued. Corporal Josey, apparently insensitive to the seething anger of those around him, continued to read the names at a leisurely pace in a heavy Southern drawl. He insisted on each person's answering smartly by yelling his first name and middle initial. Whenever someone answered, "Here," the corporal refused to allow him to enter the dayroom until he had responded properly. Finally, he stopped one man too many. There was a hot exchange and a scuffle. The infuriated man swung at the corporal and had to be restrained by others who pounced on him. Lieutenant Williams came running out, but when he saw the ugly mood of those milling about the door, he turned to the corporal and ordered, "Carry on." About fifteen minutes later the last man on the list reported, and the restriction was finally lifted a few minutes before 21:00.

Discrimination on an ethnic basis was also charged in the assignment of replacements to KP duty. Being sent to the kitchen was not as onerous in A-9-3 as it had been in IRTC, for the cooks treated their charges with consideration and did not work them too hard. In addition, there were a number of labor-saving devices—such as the automatic potato peeler—which eliminated much of the drudgery traditionally associated with KP. Assignment to Sunday KP was coveted, for it meant being free on pass on Monday and thereby missing a day of training; KP on Saturday was regarded as least desirable, for it meant losing a half day of "free" time, and there was no work on the following day anyway. KP on other days was viewed as desirable, if it came when the company was sent on a strenuous exercise at one of the B ranges. Although Sergeant Mueller insisted that men were assigned to various work details in regular rotation from the company roster, one could not help noticing that all Sunday KPs were Caucasian and all Saturday KPs were Nisei. Furthermore, whenever the company went to one of the B ranges all the KP assignments went to Caucasians. This became so regularized that the men were able to anticipate the degree of exertion they would face on the following day merely by examining the duty roster. If *all* fatigue details were assigned to Caucasians, this meant that something exceptional—probably including a night problem—was coming, and a larger number reported for sick call.

By this time the situation had become so disagreeable that even the shared joy of victory in Europe could not be celebrated. Toward the end of April the collapse of German resistance appeared a certainty, and each change in routine was interpreted in terms of reorientation of the war effort toward the Pacific theater. Official news of Germany's surrender came on May 8, as the men were sitting about in the barracks. The captain made a terse announcement over the public address system. But before anyone could cheer he added sternly, "This does not relinquish your responsibility of training." Everyone realized that this was so, but being reminded of it at that moment only dampened their initial reactions.

Further difficulties arose when sixty Nisei were assigned to a special construction crew. After retreat formation on the evening of May 9, Sergeant Mueller announced that a work detail of sixty would be sent to B-9 on the following day to build a new infantry-tank assault course. Since the rest of the company was scheduled for an easy day indoors, the sixty were counted out immediately from the 2nd Platoon. On the following morning all Nisei were aroused at 04:30, even though only sixty of them were needed. After an early breakfast all Nisei were ordered into formation. The work detail was short fifteen men. Since the first sergeant had neglected to take down the names of the sixty he had selected, he had no way of ascertaining who was AWOL. Fifteen protesting men were taken from another unit, and amidst a downpour of rain and hail the group was loaded on trucks. They arrived at B-9 at 07:30; since the trucks were needed elsewhere, the men had to get out. No one was there. Aside from the slim branches of a few trees there was no shelter of any kind in the area, and before long everyone was drenched. The men decided to build a huge bonfire. They tore apart all the empty ammunition boxes they could find, and when the wood proved too damp to burn easily, they poured on lubricating oil. Everything loose and combustible within a radius of fifty yards was collected and added to the fire, which was kept roaring until five officers assigned to the range arrived an hour and a half later.

The captain in charge of the detail was astounded when told that the crew had been waiting since 07:30; he sympathized openly with them and referred to A-9-3 as a "fucked up organization." After explaining what needed to be done, he permitted the men to set their own pace. Most of them worked diligently and without complaining in spite of the continuous driving rain. They built tank traps and machine gun emplacements, dug foxholes, filled sand bags, laid cables, and set up "enemy" dummies until about 15:00. All the officers worked with them and were obviously pleased, although some became quite upset whenever any of the men joked about being used as a "labor battalion." Finally, at 15:30 the captain complimented the men, announced that he had arranged to have them ride back in trucks rather than marching as scheduled, and then asked them if they would be willing to try a "dry run" on the new battle course with the tanks. He also promised that a different group of sixty would be brought out on the following day to correct any deficiencies uncovered in the trial.

Working closely with tanks was a new experience for most of the men, and they worked hard running through the assault course of approximately a mile and a half of rough terrain. Nisei who had trained in armored units at Fort Knox were permitted to work in the tanks along with regular Fort Meade personnel, and the others were deployed as riflemen. Many were thrilled by their first ride on the back of a tank—something they had seen in movies. Because of the uneven terrain and mud the tanks became stuck, and the exercise was delayed from time to time. Since all the riflemen were already thoroughly drenched, they fell into the mud with reckless abandon, much to the astonishment of the delighted officers. Several deficiencies were detected, but the captain made it clear that they were not the fault of the crew, whom he complimented again. By the time the trucks arrived most of the men were quite cheerful, even though they were covered with mud. Doing the "dry run" had made them feel like infantrymen again, and several indicated that they had learned many unexpected things about working with tanks.

On Friday morning further difficulties developed. At 06:40 the company was called into formation, and sixty men were counted off from the 2nd Platoon and ordered to board trucks for B-9. Approximately fifty of those chosen had been on the work detail on the previous day. When they howled in protest, Captain Larsen ordered them to be "at ease." When one of the Nisei NCOs tried to recount the promise made by the captain at the range, he refused to listen. Even those who were not ordered to the trucks joined in protesting the unfairness of the procedure, and several loudly called attention to the fact that not a single Caucasian replacement had been chosen. Again the trucks arrived at B-9 more than an hour ahead of the officers. As soon as they arrived, they recognized immediately those who had been there the day before. When they learned of what had happened, they were utterly disgusted. One lieutenant blurted out, "Listen, fellows. I don't like to say anything like this about another officer, but there isn't a single officer in the 9th Battalion that gives a shit for your CO. We know what it must be like to be in his company." After expressing a similar view in a more restrained manner, the captain decided that instead of making any of the planned changes the group would give the course its first trial using live ammunition. He indicated that the changes could be made at some other time.

Although they knew from their experiences the day before that the course was an arduous one to negotiate, the men appreciated the change ordered by the captain. After two more "dry runs" in the morning to be sure that everyone knew his assignment, the "wet run" was ordered soon after lunch. Seven tanks were brought in, and the men helped load them. Then, each rifleman was issued all the ammunition he wanted, and most helped themselves to two or three bandoleers. Before long most of them were wishing that they had not been so greedy; they had forgotten how heavy bullets were. Nonetheless, they trotted over the rough terrain. They were under constant pressure, for the tanks kept a steady pace, and the riflemen had to keep up. Except for a temporary confusion of signals that almost led to one platoon firing on a hill occupied by another, the trial was successful. When it was over and the group assembled for an evaluation, the sixty were in excellent spirits. They were elated in having been the first to do something different and exciting. They were proud that no one had faltered in spite of the grueling pace. They were especially pleased when the armored division personnel and the officers, who had run without arms or ammunition, expressed amazement at the endurance of the short statured Nisei. Among themselves they joked about their *Yamato damashii.* They agreed that they had learned many new things, especially about assaulting antitank guns. Since most men had fired several hundred rounds of ammunition, they commented that their rifles now deserved cleaning. Those who had worked with mortars joked about the number of war bonds they had sent flying that afternoon. Several laughingly noted that for once they had been "on the ball."

Their gaiety vanished as soon as they returned to the barracks. Everyone else in the company had gone to the warehouse to draw summer uniforms and had viewed a required War Department film on the discharge system. The sixty were ordered to fall out at 19:25 for the movie and to draw their khaki uniforms on Saturday afternoon—both on their "own" time. Their resentment was further intensified when they saw on the bulletin board that once again all Saturday KPs were Nisei, while those for Sunday were all Caucasians. Furthermore, the sixty had dirty rifles. Since they had fired so many rounds, their weapons had become burning hot, and all the oil and carbon were becoming caked.

They began cleaning their pieces at once, but at 18:30 they were interrupted when the platoon was ordered to "GI" the floor. If the sixty refused to participate, the few other Nisei left in the platoon would have had to do all the work; hence, everyone reluctantly took part. Additional tension arose at 19:25, when Captain Larsen ordered a man who had sprained his ankle at the assault course to march in formation with the others to the movie. By the time the contingent started for the recreation center the men were so angry that they no longer resembled a military unit. They smoked in ranks, refused to stay in step, and whistled loudly at every woman they saw. When the corporal in charge tried to restore order, several immediately challenged him to a fight. They made it clear that they did not care if an officer caught them; they were so disgusted that they did not care what happened. Most of them reiterated their determination to get even with the company commander.

The first sergeant aggravated the situation not only by ruling consistently against the Nisei but by being unable to conceal his feelings. An incident during a lunch hour added to the growing resentment. According to the bulletin board the 2nd Platoon, now made up entirely of Nisei, was scheduled to eat first. When some were slow in getting out, a few Caucasians in the 1st Platoon ran to get into line ahead of them. This started a good natured race for the mess hall line, and in a few minutes everyone was mixed in the melee. Most of the participants were still laughing and joking, when Sergeant Mueller wobbled out of the orderly room and ordered all Nisei out of the line. He insisted that each platoon had to eat together as a unit, that there was to be no mixing. When the Nisei protested that they were scheduled to eat first, he decided that they had to go to the end of the line "for running." The Caucasian replacements protested that everyone had run, but the sergeant was adamant; he continued to wave his finger toward the end of the line. A few Nisei obeyed, but most of them remained where they were and turned their backs on the sergeant. When it became apparent that most Nisei had no intention of obeying the order, the Caucasians also turned around, and the sergeant returned fuming to the orderly room. That evening the sergeant appeared before the 2nd Platoon with a broad grin. One of the men cautioned, "Oh! Oh! We're going to get fucked again!" Sergeant Mueller, who continued to grin, announced that all Nisei had to get up at 03:45 the following morning. That he was so delighted in making such an unpleasant announcement made the men angrier than ever.

As it became obvious to everyone that the Nisei were being subjected to collective punishment on an ethnic basis, a sense of group solidarity developed. The Nisei in A-9-3 were by no means united in the beginning. Many factions had formed, and some of the men disliked one another intensely. Outwardly, Hawaiian AJAs and mainland Nisei were on good terms; they addressed one another politely. But to their personal friends some on each side expressed their grievances. Those who were highly assimilated to American life found Little Tokyo standards a bit strange and often got together among themselves to complain about the pettiness of some of their buddies. Now the various factions united in their opposition to the company commander. A high premium was placed on protecting Nisei interests, and even those who were not ethnocentric participated. This sense of unity and determination to retaliate arose from their realization that the sole basis for differential treatment was ethnic identity. It did not matter whether one were a good soldier or a poor one, or whether he conceived of himself as Japanese or as American. All Nisei were treated alike, and they closed ranks to resist.

As the belief that they were being mistreated crystallized, the change in outlook became cumulative, for suspicions were reinforced through selective communication. Some Nisei had been so embittered by the evacuation that they distrusted all Caucasians. They were quick to interpret everything in the least favorable light, but at first their views had been discounted by the others. Sometimes such individuals encountered hostile reactions, especially from Hawaiians who asked them why they did not return to Japan; they were forced to keep their opinions to themselves and their intimate friends. As increasing numbers became convinced that they were being discriminated against, however, the few who were bitterly antiwhite found a congenial setting for their resentment. They argued that the *Keto* would never treat Nisei as equals. They insisted that the captain was a Haole who despised Nisei and wanted to make things "tough"; at the same time he did not wish to harm his "own people" by punishing the entire company. Old stories of discrimination were revived. Even the men who disliked Nisei and got along better with Caucasians joined in condemning discrimination. Highly assimilated Nisei had previously defended the Caucasian replacements, pointing out that the latter were only obeying orders and had no choice in the matter of taking training. But those who had argued that the Caucasians were going overseas in a few days and deserved special consideration now became silent. Even these men became so resentful that they joined in the general reorientation. Some conceded that their own conduct had contributed to the difficulties, but even they placed the blame primarily on the cadre.

"I hear the officers say they're going to be glad when the Boochies pull out of here."

"*They'll* be glad! How about *us*? Did we volunteer to come to this fuckin' place?"

"I guess they are kind of griped about us, but what the hell! They always fuck us up so we just fuck them back a little harder."

The climate of bitter resentment that prevailed was reflected in the following discussion of a news item concerning a heroic mission of the 442nd:

"Those guys sure fight like hell though, huh? Boochies are like that, though. They fight like the fuckin' Japs."

"Yeah, we fight like hell, and the fuckin' Haoles will put us in the fuckin' concentration camps again. They say that we fight good against the Germans because the Germans are white, and we like to kill whites. Jesus Christ! Lose fight, huh? We fight like hell, and they don't even appreciate it."

"Lose fight is right. We always get it in the ass. We always get the ass end of the deal. Boy, I bet a lot of guys got killed there too. They're just suckers."

"What the hell do you think we are?"

"When I get stuck with a fuckin' CO like this son of a bitch, I begin to wonder."

As their resentment rose in intensity, some of the Nisei turned against all Caucasians. After V-E Day increasing numbers of Caucasian replacements were stranded at Fort Meade. These soldiers were mocked for obeying orders, observing military courtesy, doing their best in various exercises, and making no attempt to disappear when they were needed. They were resented for the cooperative manner in which they addressed the depot cadre. Some Nisei were so angry that they forgot that they themselves had performed in a similar manner only a few months before. In field exercises they overlooked their own errors and called attention to those of the *Keto*, constantly reiterating their view that Nisei were superior infantrymen. Some of the Caucasians, puzzled by the hostility they faced, made polite inquiries, only to be rebuffed or ignored. Most of them remained friendly, but even the Nisei who had previously mingled with them were reluctant to appear too amicable. Nor were all the Caucasians blind to what was happening. One man from Tennessee remarked, "You boys is always gettin' fucked. Hell, I don't like to go on night problems at all, but I'd ruther go catch a little hell an' have ever'thing even-steven."

Although some Nisei insisted that all Caucasians were against them, most of the men focused blame on the A-9-3 cadre. Their continued faith in the U.S. Army was revealed in their repeatedly lodging complaints with the regimental inspector. Convinced that they were being victimized, the Nisei took measures to protect themselves and to fight back. Common understandings emerged that amounted to a campaign of obstruction and retaliation, and even the few isolated "loners" who were not embittered were forced to cooperate. The captain and the first sergeant remained the prime targets of hostility, but other depot cadremen also became targets, for a premium was placed on insubordination, especially the flouting of orders in contexts in which it was difficult to inflict punishment. Whenever an officer was inexperienced, slow, or clumsy, the men deliberately made mistakes in order to confuse him. No one even bothered to salute the company officers, and even NCOs as popular as Sergeant Thompson sometimes had to appeal to them to "play ball." If someone obeyed an order in a conscientious manner, this was almost invariably followed by loud sucking noises and taunting jeers about "brown nosing" or "bucking." To "fuck up" without being caught became the thing to do, and the men competed with one another in inventing ingenious ways to do so. They laughed openly at frustrated cadremen who could not single out anyone for punishment. Any error was immediately pointed out, and the embarrassed officer would only have made his own situation more ridiculous by inflicting punishment.

One retaliatory tactic, which soon became controversial, consisted of wearing down the opposition in speed marches. Especially when Caucasian replacements were participating in a long hike, Nisei platoon guides would set an unreasonably fast pace—perhaps 140 steps per minute—so that those at the end of the column would have to trot to keep up. One by one the Caucasian replacements would drop out; since they had been at the depot for only a short time, they had not yet developed their stamina. But this was thought to be even more embarrassing for march officers who were unable to keep pace, for they did not carry the packs and arms borne by enlisted men. Some of the lieutenants simply called for a jeep, but others attempted to keep up. The men were overjoyed on one occasion, when a lieutenant was puffing so hard at the completion of a race that he dismissed the company in a voice that was barely audible. This practice, though used with increasing frequency, did not enjoy the wholehearted support of all the Nisei. They also felt the strain, and many wanted to drop out, though they did not dare. Any Nisei who dropped out would have been ostracized; hence, they hung on in spite of blisters, colds, and other impediments. But the social pressure was resented, especially by those who had not previously lived among Japanese. They pointed out that the officers who finished on a jeep were not only not extended but perhaps were not even embarrassed. They argued that the Nisei were making unreasonable efforts in vain, since Caucasians did not regard dropping out a disgrace. In spite of the disagreement, however, it was understood that the speed marches were conducted for the purpose of retaliating, and all Nisei had to comply, regardless of their personal views.

Finally, when the discrimination became unbearably flagrant, several proposed filing a formal protest with the regimental commander; this effort disclosed deep divisions within the group in spite of its united opposition to the cadre. Soon after lunch on May 13 it became known that all Nisei would have to participate in a night problem but that all Caucasians would be excused. Inclement weather compounded the indignation, for it meant that part of the company would rest indoors and later go on pass while the rest had to work in the rain. It meant further that the Nisei would have to stay up for at least an extra hour to clean their rifles. The situation was further aggravated by a rumor that the Nisei would be aroused at 03:45 the following morning to hike to one of the B ranges.

During supper virtually all conversations centered on the formal protest. The men agreed that discrimination rather than inefficiency should be stressed, for this was the point on which Captain Larsen would be most vulnerable; it seemed unlikely that such a flagrant violation of War Department policy would be tolerated. A committee of college graduates was selected to draw up the petition, but disagreements soon arose concerning procedure. Some argued that a petition bearing the signature of every Nisei replacement in the company would be sufficient, but others felt that one man should request a court martial to be a test case. Several volunteered to be the test case, providing all the others would testify for them. Most men indicated their willingness either to sign a petition or to testify, but opposition developed. One corporal, obviously frightened, rushed from group to group, charging the men with plotting mutiny. He contended that mutiny was punishable by death and that organized protests were not tolerated under military regulations. When others countered that several of them had already gone through channels to the regimental inspector in vain, the corporal still insisted that the group was inviting serious repercussions. But the issue did not come to a head. The discussion became less heated when it was learned that the Caucasians had gone out on a night problem earlier in the week, even though they had completed their work by 20:00. As the rain let up, the resentment was further attenuated. When virtually all the Caucasian replacements were alerted for overseas duty on the following morning, resentment against them vanished.

But the abortive petition made many more conscious of some of the cleavages among themselves. A number were especially angry about the efforts of the corporal to block all attempts to organize. They did not know the regulations concerning mutiny, and they conceded that he might be correct. Nonetheless, they felt that they had a legitimate grievance to air. Among themselves they charged that the corporal was only thinking of himself. Since he outranked most of the replacements in the company, he would probably have been questioned before the others. Many concluded that he was a coward. Bitterness also developed among many Hawaiians, who pointed out that the mainlanders who complained most loudly about being mistreated were the first to back down when asked to sign a petition. Even mainlanders muttered among themselves that the very men who were the first to charge discrimination were the ones least willing to participate in an organized protest. Several commented on the unexpected manner in which the group had split during the controversy. Those who were most assimilated and felt no particular animosity toward Caucasians were among the leaders of the move to submit the petition; those who declared openly that they hated all Caucasians were the most reluctant to take action. The very individuals who derived the greatest enjoyment from showing up white men in speed marches held back from an attempt to have Captain Larsen reprimanded by his superior officers.

Confrontation with the Company Commander

By the beginning of May the disorder in A-9-3 was attracting the attention of outside parties. Some of the more sober Nisei in the company were becoming apprehensive. Although they continued to participate in harassing the cadre and in helping to cover for troublemakers, they felt that some were going too far. Among their intimate friends they expressed their disapproval, even though they felt obligated to conform. The men were also cautioned periodically by friendly cadremen that they were courting serious trouble. On May 4 considerable discussion arose after a warning from an athletic director, a former all-American football star and one of the few officers who commanded their respect. Just before a period of organized sports the lieutenant admonished them: "You fellows had better be on the ball from now on. I just learned that the battalion CO is plenty pissed off about the way you march around here and the way you've been conducting yourselves generally. Sure, you might have good reasons for fucking off, but remember this: you're still soldiers, and as such you'll have to take orders no matter how stupid they may be. I'd recommend that you ought to try to look good no matter how pissed off you are inside because they could very easily change one hour of sports into one hour of close order drill."

The long expected showdown came on May 16, an unusually sultry day on which the company was sent to B-2. The first call was at 03:45, and the schedule called for departure at 06:00 for the nine-mile hike to the range. When they realized that they faced a long hike on a muggy day, thirty-six Nisei went on sick call; it appeared likely that some kind of trouble would arise. It was not only hot but very humid. Many were already sweating after being on the road for only an hour, even though it was still dawn. By 10:00 the temperature was well over 85°, and the humidity had risen to the point where everyone was complaining. After spending a miserable morning running through the now familiar infantry assault course, the men were further frustrated at lunch time when the water supply proved inadequate; the lister bag was empty ten minutes after it was unloaded from the mess truck. Because of the heat and the endless complaints, the march officers allowed the troops to take a long rest after lunch. By the time the company started its return march the road had been blocked off; the route cut across the firing line of another range and had to be closed whenever it was in use. Although most men did not object to sitting about, some grumbling was heard from those who wished to go out on pass that night. The schedule called for "free" time from 15:00 to retreat, and they were anxious to get back to the barracks to finish various chores.

Once the return march got under way a number of men started clamoring good naturedly for more speed, and before long a grueling race among the various platoons got under way. Whenever the lead platoon was not moving rapidly enough, the others would catch up and begin yelling for it to move faster or to get out of the way. Then the lead platoon would set a very swift pace, and those following had to trot double-time in order to remain in formation. There was much yelling and joshing, as the units taunted one another about not being able to "take it." On several occasions, when the lead platoon had been goaded into an exceptionally fast pace, the march officer had to order the platoon guide to cut down his cadence. In spite of the oppressive discomfort almost everyone joined in the gaiety, although some who were older complained loudly of the stupidity of the whole affair. No matter how they felt, however, no one dared to fall out. The march officer appeared very tired, and this delighted the ones who wanted more speed. They were overjoyed when the lieutenant finally fell so far behind the rear guards that he had to ride in on the jeep following the column to pick up stragglers. Although the speed march began as an effort to return to the barracks on schedule, it continued despite the misery of the participants because of the competition. As one man later remarked in disgust, "That's one thing about Boochies. They don't want to be last. They'd try to keep up with the next guy even if it kills them." The nine miles were covered in an hour and forty minutes! As they entered the company area some were so exhausted that they appeared dazed; but on the whole the men were in a gay mood, proud that they had returned on time in spite of the two long delays after lunch.

As the troops stood in formation waiting to be dismissed, Lieutenant Williams came out of the orderly room to distribute cigarette ration cards. After a glance at their haggard condition he ordered: "Sergeant, dismiss these men. We can pass out the cards later inside because we won't have anything else to do until retreat anyway." Just as the sergeant was about to do so, however, Captain Larsen appeared and yelled, "Wait a minute! There's something I want to do."

He then ordered the men to open ranks and to remove their helmets—for a haircut inspection! It was so hot that some were reluctant to take off their headgear, and those who were slow in doing so had their names taken for extra duty. The men became sullen. There was no yelling from ranks; they stood silently at attention. Many felt that they deserved to be complimented for working so hard to return on time. They had learned to expect little from the captain, but this reception was worse than anything they had anticipated. After several had been "gigged" for inadequate haircuts, the captain left them standing at attention in the scorching sun for almost ten minutes before giving the order to put on their helmets. After that he paced back and forth, glowering at the men as they stood rigidly at attention.

After several minutes the captain began, "I want all the Japanese-American noncoms to fall out of ranks." When they were slow in responding, he bellowed, "Well, do you hear me? I want everyone—corporal and up—to fall out!"

When the Nisei NCOs had lined up before him, he proceeded to reprimand them. He then turned to one sergeant and demanded to know why the Nisei were so sluggish in obeying orders. When the sergeant replied that their morale was low, the captain screamed, "I don't give a damn about morale! Why is morale low anyway? Anyway, even if it is low, that's no excuse for your not being good soldiers!" Then, he turned toward the entire group and continued:

Men, I'm sick and tired of putting up with the way you've been acting around here. This is going to stop right now. We've tried it our way, and you won't play ball. Now I'm tossing the whole thing into the laps of your own noncoms, and I'm holding them personally responsible for your conduct. I'm through with this dillydallying. These noncoms will be personally responsible to me and will get all orders directly from me. There'll be no more excuses of any kind. And if you noncoms can't keep these men in line, you won't be noncoms very long. You don't deserve your stripes if you can't handle your men. I'm throwing the whole thing back on you. It's up to you.

You men are getting three-day passes like all other garrison soldiers. You should be happy to get them. The other boys around here [presumably the Caucasians] don't get any because they're not around long enough. You men have been getting privileges that the others don't get. The pass policy in this battalion is to let one-third of the eligible men go at one time, but I've had as many as one-half to three-fourths of the eligible men go out at once. You have no complaints to make.

Another thing. We had thirty-six men on sick call this morning. How do you explain that? Last week we had over sixty on sick call. How do you account for that? Why is it that every time we get up early we have such a large number of men on sick call? We've had so many men on sick call that the battalion sent an inspector down to check the sanitation of our barracks. This kind of thing will stop. I've tried to discourage you men from going on sick call, but if you persist, I'll have to take other measures.

I want to tell you once and for all. Low morale is no excuse for poor soldiering. If your morale is low, raise it! If you don't, there'll be no more three-day passes. There'll be no more weekend passes. There'll be no more overnight passes. Believe me, men, I can make it tough on you, and I will. As for you noncoms, if you can't get the men on the ball, there'll be no more stripes.

It takes you men twenty minutes to fall out in the morning. There's no excuse for that sort of thing. It will stop right away!

By this time Captain Larsen had lost his composure. Flushed and shaking in fury, he continued to scream at the top of his voice. As his mane of blond hair quivered, he appeared like a raving maniac. After more than forty-five minutes in the blistering sun the company was finally dismissed.

The initial reaction was one of anger, but more sober considerations were also brought up. Although there was complete accord on their estimate of the captain, the men disagreed about their own deportment. Heated discussions continued for the rest of the day.

"I had a feeling this was coming. Son of a bitch! Sure, we fuck up. We admit it, but he fucks up even more. In the army, though, rank counts, and I knew that sooner or later we'd get into Dutch."

"Yeah, but what can he do? He didn't have to tell us that we're slow getting out in the morning. We know that. We do that on purpose. Hell, if he'd leave the Boochies alone and just tell us when he wants us out there, we'll all be there."

"Shit, yes! No *shimpai* [worry] about Boochies. Nobody goes AWOL in a Boochie outfit. All he has to do is say what time we're supposed to be there, and we'll be there."

In another circle the captain's view on morale came in for discussion:

"He says all we have to do with our morale is raise it. Shit! What the hell does he think this is? He wants us to be happy when he spends his nights thinking up some way to fuck us up. Fuck him!"

One of the depot cadremen joined in, "Sure morale is low. Why? We come in from a hot hike, and he keeps us out there for a fuckin' shit inspection! That's why morale is low. If that bastard had any decency in him, he'd have let us all come in and then chewed us out later on."

"He did that on purpose. He wanted to make us suffer out there. He wanted us to pass out out there. Boy, I wish I fainted or something. I'd tell the doctor what he made us do. I'll go tell the fuckin' inspector about the time he made a guy with measles work in the kitchen as KP because he went on sick call. I'll tell those son of a bitches plenty! I hope he really hurts one of us. Then, we'll get him. We'll get *him* court martialed."

The Nisei NCOs were in a difficult position. It was obvious that they would be unable to enforce orders, and under the circumstances it was difficult for them even to request cooperation. The others commiserated with them.

"Well, I guess from now on we'd better be on the ball. It's going to be tough on the Boochie noncoms from now on. I feel sorry for those guys. There's no use fucking them up."

"It sure is tough on them, though. Hey, Yoshida, what do you think about it? You going to get us guys on the ball?"

"Fuck that shit! The CO can take my stripes and shove 'em up his ass!"

The sergeant who had borne the brunt of the captain's wrath was highly respected. He was quiet, unassuming, and clearly competent. Since he had been sympathetic and helpful, the men agreed that it would be unfair for him to be punished for their misdeeds. Some of the other NCOs were less belligerent.

"How about you Sarge? What do you think?"

"Well, whatever you fellows decide is good enough for me. If you guys want to get even with him and really fuck up, I'll go along with you. To tell you the truth, though, I don't care about the stripes, but I like the extra pay that comes in. It's almost fifty bucks a month, and I can use it. But you guys can do what you want."

"Fifty bucks a month, huh? That's a lot of dough. I don't blame you."

"Aw, shit! We may as well be on the ball. Shit, it'll be too tough on the noncoms. There's no use fuckin' them up. We can get even with that son of a bitch some other way."

Although the cursing and grumbling continued, by evening there was general accord that everyone should henceforth cooperate. None of the Nisei NCOs had ever exercised the prerogatives of their rank; they had stood in line along with the others and had taken the same abuses. They had worked diligently for several years to earn their position. Precisely because they had not asked for preferential treatment and had accepted the same fate as all other replacements the privates agreed to work for them.

On Thursday morning, May 17, reveille in A Company was a marked contrast to the previous routine. Almost everyone was out of bed with the first call. The NCOs were among the first to dress; although they aroused some of the privates sleeping near them, they made no effort to awaken late risers elsewhere. The tardy were pushed along by their neighbors who reminded them that a new regime had been instituted. Everyone was in formation well ahead of time. During the day two rumors emerged—that Captain Larsen was to receive a discharge and that all Nisei would soon be given furloughs. Most were skeptical, but soon after their return to the barracks an announcement was made that several replacements would begin their three-day passes that night. Contrary to the captain's assertion, this was the first time that a sizable number had been authorized to leave. At the same time the cadre NCOs reported jubilantly that the captain was leaving the company. Although they doubted that he was being discharged, he was being given another assignment. That evening, during the usual grouping and regrouping of personnel for a night problem, Corporal Beaudette disclosed that Captain Larsen was being replaced by Captain Jennings. Word spread quickly through the ranks and everyone was elated, for Captain Jennings was one of the few officers they did not regard as "chicken shit." He had led them on several problems during the past month, and he was respected as the kind of officer who went out with his men even in inclement weather and never asked anyone to do anything he would not demand of himself. The cadre NCOs were so delighted at the prospect that they went from group to group, gleefully discussing anticipated changes in company routine.

Before long unit after unit of strange troops marched into the A-9-3 assembly area, and by 19:00 it was apparent that all replacements in the 9th Battalion were to participate in a gigantic night problem. Nisei NCOs, most of them from B-9-3, were assigned to each platoon that was predominantly Nisei. The tension level rose, for the men felt that they were participating in something "big." One of the B Company sergeants told a large group of Nisei, "From now on you'd better be on the ball. Something good may be coming up, and you might fuck yourselves and all the rest of us by horsing around. There may be a good deal coming up." When questioned about what the "good deal" might be, he refused to say any more. Some speculated that Nisei infantrymen might be discharged; others conjectured that they might be placed in ERC; still others considered the possibility of forming an all-Nisei combat team for action against Japan. The sergeant's remarks were relayed from platoon to platoon, and before long the entire contingent of Nisei was buzzing in excitement.

The evening was pleasantly warm as the battalion got under way. Platoon after platoon marched out in a column that seemed to stretch endlessly. Once beyond the barracks area the lieutenant commanding several of the A Company platoons tried to raise their spirit by leading songs, and before long the singing of the jubilant men got out of hand. The Nisei in the 2nd Platoon, who probably had more reason for hating Captain Larsen than the others, were in especially good humor, and they sang lustily. They realized that their voices would probably carry for several miles in the stillness. At first they sang what the lieutenant suggested, but soon they turned to lyrics improvised by one of them:

The first platoon is always late, parlez-vous,
The first platoon is always late, parlez-vous;
The first platoon is always late, they stay in bed and masturbate,
Indi, dinki, parlez-vous.

Another song that they sang with special gusto was about their first sergeant:

Sergeant Mueller has plenty of class, parlez-vous,
Sergeant Mueller has plenty of class, parlez-vous;
Oh, Sergeant Mueller has plenty of class; he's always kissing the captain's ass,
Indi, dinki, parlez-vous.

As the battalion approached the problem area, the slight drizzle in which they had been marching turned into a torrential downpour. By 21:00, when the approach march for the attack got under way, everyone was dripping, and the area had been transformed into a quagmire. Most of the replacements had never before participated in such a large-scale operation, and they disregarded their discomforts. Once the problem got under way, talking was no longer permitted. It was an unusually dark night, and there were times when each had to cling to the cartridge belt of the man ahead to keep from getting lost. When A Company finally lined up for the attack, orders were whispered down the line: "Fix bayonets!" There was no joking about making a "*banzai* charge"; each had to be careful not to stab a buddy in the darkness. As they were crawling up a hill toward the "enemy," a flare was fired, indicating that the problem was over. Tension was suddenly dissipated. The officers and NCOs made frantic calls on their walkie-talkies and held hurried consultations. It was not until most of the units had reformed on a road that it became apparent that A Company had lined up on the wrong road! As the colonel in command kept the troops standing in the rain to criticize the lack of coordination among companies, some grumbling began for the first time. The privates pointed out that they had had no access to maps, compasses, or portable radios; indeed, they had not even been briefed on the plan of attack. They had done all that could be expected of them, and it was not their fault if their officers had gotten lost. As one exasperated Nisei muttered, "Jesus Christ! I lose fight! For once the Buddhaheads were on the ball, and the looies fucked everything up!" It was not until 23:45 that A Company returned to its barracks, and at least an hour had to be devoted to cleaning and oiling weapons. But most men remained cheerful. The few who were downcast were reminded that unpleasant days were now numbered.

On the following day it was announced that all Nisei would move on Monday evening to B-9-3. The explanation given was that all Nisei in the battalion were being placed together to meet repeated complaints to the regimental inspector that some companies had more lenient pass policies than others. It soon became apparent that Captain Larsen had made things as difficult as possible. All Sunday KPs and all other weekend details were made up exclusively of Nisei, and those on Sunday KP were notified that they would not be issued passes on Monday. All three-day passes had been canceled. But the greatest indignation arose when it was learned that men who had earned three-day passes by participating in battalion boxing matches had been placed on KP for Monday. Some of them had taken considerable punishment, and they were now being excused from duty only on Saturday morning! When the infuriated boxers stormed into the orderly room to demand an explanation, Sergeant Mueller, who appeared very frightened, insisted that he had merely followed Captain Larsen's orders. He then promised that he would take up the matter with the B Company orderly room in hopes that the passes could be issued there. Nor were the Nisei the only targets of Captain Larsen's parting shots. He had previously recommended three cadre NCOs for discharges; then on his last day he had notified battalion headquarters that the special skills of these men were in critical demand and that they could not be spared. On Monday morning, however, company procedures were suddenly transformed. The transfer of 127 Nisei to B Company was made on "government time"; the move got under way at 13:00 and was completed in a half hour. The men recalled that they had been told to be ready to move in the evening; they assumed, therefore, that Captain Larsen had planned to have them move on their "own time" but that Captain Jennings had changed the order.

Nisei replacements had been assigned to B-9-3 from the middle of March and had accumulated there until they made up an oversized platoon. They included a number of NCOs from Camp Robinson, several in the top three grades. There had been no difficulties; the conduct of these troops had been exemplary. On May 21, all Caucasian replacements were transferred to E-9-3—making B Company a segregated unit. The 127 new arrivals were mixed with the Nisei already there. All privates were lined up by height and grouped into squads of twelve, and an NCO was placed in charge of each squad. Four squads were placed in a platoon, each with a higher ranking NCO as platoon sergeant. There were five platoons in all. In the barracks each squad was assigned a sector, and within each squad the men were assigned to bunks alphabetically. This resulted not only in the disbanding of aggressive cliques but the separation of close friends. Some were delighted to get away from individuals they disliked, but many complained that they missed their buddies. What facilitated the reinstitution of military discipline, however, was the termination of the anonymity enjoyed in A Company. Since each individual was assigned to a definite squad and known to his squad leader, responsibility was easy to fix. It was no longer possible to miss formations without being caught, and those who were caught could not give false names.

After supper all newcomers were called to an orientation meeting. The company commander, Captain Parmentier, told them that he had long desired to have an all-Nisei unit and that he had requested one some time ago. He also indicated that he was familiar with their record in A Company but wanted to forget about it. "We can start all over here," he urged. "I'll expect you men to be on the ball. I don't want you to wreck the fine record that the Nisei who were already here have made in this company." He then went on to list some company regulations—standard operating procedures regarding the appearance of the barracks. He emphasized that he wanted to see nothing on the floor—not even shoes or wet *geta*. The men drew several conclusions; the captain knew about the record of the 442nd; he had had previous contact with Nisei; he was favorably disposed toward them. On the other hand, he also had high expectations, and this aroused mixed feelings.

Most Nisei NCOs did not try to enforce regulations; they worked diplomatically and barked out orders only when they were absolutely necessary. They tried to reason with subordinates who were reluctant to carry out some assignment. When confronted by a difficult or dirty task, most of them did their share of the work. When they caught privates violating regulations, they either pretended not to see or only cautioned them and did nothing more. They did not report infractions to officers, nor did they bawl out anyone in front of the others. Several NCOs were embarrassed on Wednesday, May 23, when they were required to give close order drill to their respective squads. It became apparent at once that some of them had won their stripes as cooks and supply men. Most of the privates were riflemen, accustomed to being drilled by competent IRTC cadre, and some of the squad leaders did not even know how to give commands. How much authority each NCO enjoyed depended on the respect he commanded as a person. Those who had been IRTC cadremen had relatively little trouble. For example, on June 5, a foot inspection was ordered at 05:30. Since it was quite cold, complaints arose at once. One discontented man yelled, "Hell, we could do this just as well inside!" Similar cries were cut short when Technical Sergeant Matsui replied calmly, "Yes, it could be done inside. But we're doing it out here, aren't we?" The men had seen the sergeant perform; he was one of the most proficient infantrymen they had ever encountered, and they admired him. Some grumbling continued. Since most of the NCOs were regarded as "nice guys" who did not "pull rank," however, nothing serious developed.

An incident on Tuesday evening, May 22, did much to win over many of the newcomers who had been ambivalent. As he had promised, Sergeant Mueller sent a note to B Company, naming the boxers as well as those who had been on KP on their final weekend in A-9-3. Since the fatigue details had been performed elsewhere, B Company was under no obligation to give the men a compensatory day off. Nonetheless, they were called into the orderly room and asked on what night they would prefer to take their pass. The first sergeant told them that they could leave immediately if they wished but that they would stand to gain more by delaying their pass for a day or two. A night problem was scheduled for Thursday, and everyone would be restricted on Friday night to "GI" the floor. Therefore, he recommended that the men select one of these days—to miss one of these unpleasant duties. This was a sharp contrast to the policy in A Company, where it appeared that every effort was made to get all Nisei into night problems and special duties. Since those eligible for passes were now assigned to different platoons, the news spread quickly throughout the company. The others were quick to appreciate the significance of this gesture. One man commented, "Here the CO, the cadre, and the noncoms are all working *for* you. They try to help you out. In A Company they were always trying to fuck you up." Those within hearing range concurred. The spirit in which this orderly room was run was clearly different, and ill feeling began to die down. The men no longer felt that they were at war with the depot cadre.

Although some griping continued on a low key, the transfers from A Company were suddenly transformed into components of a unit with the sharp military bearing that characterized most Nisei troops in World War II. There was a conscious effort on the part of many to do exceptionally well in order to make Captain Larsen appear unfit. Since Captain Parmentier had asked for an all-Nisei unit, these men wanted to make the contrast between the two officers as conspicuous as possible. Not all the men cared, but those who did urged the others to be "on the ball." Whenever grumbling began, the more sober newcomers quelled the incipient protest by urging agitators not to "fuck up" their fellow Nisei. When this appeal did not work, the plea to help get even with Captain Larsen usually stopped the complaints.

Many remained ambivalent about the transfer. Though exasperating, inefficiency had had its advantages; life in general had been easy. The NCOs in B Company were too efficient, and after months of loafing the discipline was hard to take. Some even admitted that part of the difficulty had been their own fault, although everyone agreed that the captain was primarily to blame. A-9-3 had been labeled as a "fuck-up outfit" by other Nisei; their friends elsewhere sometimes taunted them, and some had become quite sensitive about it. Although life was now more orderly, some were not sure that they were better off.

"Jesus Christ! This fuckin' place is too goddam strict! I wish to hell I was back in A Company. We got away with murder there. This place is too well organized. I don't like this shit, hey. They make us put our wet *geta* in the barracks bags. God, what a lot of chicken shit! We didn't know when we were well off, hey. We shouldn't have fucked around so much in A Company."

"This place is going to be tough, but at least they're on the ball."

"I think A Company was better. We got to fuck up there. Life was more *kiraku* [carefree] over there. Here, you've got to be on the ball all the time."

Most men were convinced that they had succeeded in their vendetta against Captain Larsen. According to official records the captain was assigned to Headquarters Detachment of the 3rd Regiment on May 22, and there is no indication that he had been demoted or even reprimanded. But the men assumed that he had been removed from his command for incompetence. Their belief was reinforced by reports from the cadre that he was crestfallen and that the first sergeant was frightened. They felt that their deliberate efforts to "goof off," especially in the presence of higher officers, had worked. They felt that Captain Larsen was not fit to be an officer in the U.S. Army, and they were proud to have played a part in exposing him for what he was.

Jungle Fighting on A.P. Hill

In the reconstituted B-9-3, attention during the first week was focused primarily on a forthcoming bivouac. Rumors of going on maneuvers had been entertained for weeks, and on Monday evening it was announced that the company would leave on the following week for the A.P. Hill Reservation in Virginia. Additional rumors developed. The distance estimated ranged from twenty miles to fifty. Some heard that the troops would ride part of the way but would have to march the final thirty miles; others understood that trucks would take them all the way. It soon became apparent that units of the 10th Battalion as well as A-9-3

—which had been replenished with fresh Nisei IRTC graduates—would also participate. Since so many Nisei were involved, many were ambivalent. High standards of performance would be established, and everyone would have to be alert to avoid being shamed.

By Saturday morning, May 26, it was common knowledge that the maneuvers would take five days, including travel time, and that the eighty-seven-mile trip would be made by truck convoy. Enthusiasm gave way to apprehension, however, when the list of GI equipment each man would be required to carry was posted on the bulletin board. When a properly made-up combat pack, cargo carrier, and blanket roll were placed on the supply room scale, the total weight was sixty-eight pounds. When strapped together, the three pieces comprised a pack that was bulky and awkward to carry. Those who were shorter, some barely over five feet tall, discovered that it reached from their head all the way down to their lower thighs. The men were told that they could take any additional items and personal belongings they wished—on their backs.

Since so many items had to be crammed into limited space, NCOs supervised the work closely. When the packs were finally made up, most men tried them on and were astonished. They agreed it felt much heavier than the portable flamethrower, which weighed about seventy-five pounds.

"Jesus Christ! How in the hell do they expect us to fight with these goddam things on? You can't even walk around with 'em."

"I saw a picture the other day where the guys were fighting on Peliliu. They had on packs like these when they hit the beach there. They were running like hell too!"

"Goddam! Lose fight, huh? I guess I'd run too if somebody was shootin' at me, but I don't think I could walk a mile with this fucker on right now. It's a good thing we're ridin', hey."

On Monday morning, May 28, the troops were aroused at 03:10. Breakfast was served immediately, and lunches to be carried on the trip were issued. About 04:00 orders came over the public address system to wear gloves and to place either a fatigue hat or a wool knit cap into the cargo carrier. This evoked loud protests, for it meant that each man had to tear apart his pack. Since the captain had ordered that the packs be left undisturbed after inspection on Saturday, many had endured the cold weekend with only one blanket and had gone without other items that had been packed. Now they had to reopen the packs just to enclose a small item that could easily be stuffed into one's pocket. For the next two hours a succession of contradictory orders were issued, and each time the packs had to be torn apart and reassembled. The confusion resembled that on the morning of President Roosevelt's funeral. Most men sat down in disgust; some started laughing; others yelled back at the loudspeaker.

The company was finally called into formation at 06:15. While the officers and NCOs checked and rechecked to be sure that everyone was present, several began to complain of leg pains. Although they were not held at attention, most were reluctant to sit down; the packs were so bulky that they were not sure they could get up again. The grumbling was attenuated temporarily when Private Nakanishi quipped, "Jesus Christ! It's a good thing this ground is hard, hey. Otherwise we'd go right through it!" After a twenty-minute wait the troops were marched to an entrucking point about 600 yards away. The packs felt heavier and heavier, and the straps began to cut. The men had become accustomed to carrying regulation forty-pound packs, and this one was so much heavier that their bodies had not yet adjusted to them. At the assembly area they were required once again to stand about and wait.

Finally units of the 10th Battalion began to arrive. As column after column of Nisei soldiers marched into the area, the somber mood began to lift. The men felt elated; they were participating in something "big." They began joking.

"Jesus Christ! I never seen so many Boochies in all my life! Where'n the hell'd they come from?"

"It's a good thing we're inside the fort, hey. If we were outside, somebody'd think the invasion started."

"Holy cow! Look at all the Boochies! No wonder the fuckin' war never ends!"

"Yeah! No wonder they want to deport us!"

The trucks began arriving after about twenty minutes. Since the convoy included four oversized companies—A-9-3, B-9-3, A-10-3, and C-10-3—each with its own field equipment and kitchen, it was quite large. As truck after truck, led by motorcycles and jeeps with screaming sirens, roared tactically through the small towns of Maryland and Virginia, the inhabitants turned out to see. All cross traffic was stopped by MPs, and the men became acutely conscious of being on display. They were at the center of the stage. They cheered, waved, and whistled. One man kept yelling to every group of civilians he passed, "Is this trip necessary?" After the first two or three towns the excitement began to wane. Many fell asleep, and others talked quietly. Several speculated about the reactions of civilians on seeing such a large truck convoy, filled with

Asian soldiers and bristling with weapons, roaring over an American highway. What did they think was happening? Many surmised that the civilians thought they were cheering Chiang Kai-shek's legions. On one truck Private Onishi, who had gone on sick call so frequently, became the butt of acid comments. Nakanishi reassured him that he would be cared for by a "mobile dispensary" and that he need not be overly concerned. Since they had not eaten since 03:30, most of the men became hungry. They had been ordered not to touch their lunches until noon. Several had had the foresight to steal some extra lunch boxes, however, and the spare sandwiches were shared.

After a five-hour ride the convoy arrived at the A.P. Hill Reservation. Everyone was relieved to learn that the campsite would be nontactical; this meant that it would not be necessary to camouflage the area or to dig foxholes. Still, a number of irritating tasks had to be done. The 3rd Platoon, made up largely of men who were in the troublesome 2nd Platoon in A-9-3, was ordered to dam up a creek to get water for bathing and to dig the company latrine. The creek damming was completed in a few hours, but installing the latrine turned out to be a formidable undertaking. Two huge boxes approximately twenty feet long, on which a row of toilet seats had been mounted, had been unloaded, and holes large enough for each had to be dug. The 3rd Squad, which carried the brunt of this task, had to toil for much of the afternoon, and others went about sniggering that "the turd squad of the turd platoon has to dig the turd house." As in a number of other details it was the members of the work crew rather than the NCO in charge who decided when to stop working. After several hours of digging and hauling away dirt, one of them yelled, "Hey, Sarge, this is good enough, huh? Hell, if the CO don't like it, he don't have to shit here."

For the replacements in B-9-3 Tuesday, May 29, turned out to be one of the most arduous days they had in their years of service. The jungle area on A.P. Hill was approximately 3,000 yards square, and the problem consisted of one company finding its way through with a compass to attack another company dug in at the other end. The original plan called for the attacking force to negotiate the dense jungle three times—a reconnaissance patrol in the morning, a combat patrol in the afternoon, and a final assault at night. Each of the four companies was to take turns in attacking and defending, and B-9-3 was the first to attack. When the officers saw the condition of the troops after they had gone through twice, they realized that they had underestimated the difficulty of penetrating the jungle and decided to require the others to do it only twice, eliminating the afternoon patrol. B Company was the only unit that had to go through three times.

When the company was aroused at 05:30, the men were told that a night problem had been scheduled and that they would probably be out until 06:00 the following morning. Nothing was said of the character of the night problem. After breakfast they were marched to a starting point about a mile and a half away from the bivouac area. The exercise was labeled a "reconnaissance patrol," and each of the five platoons was assigned a different lane with an azimuth to follow on a compass. None of the privates had ever seen a jungle before, and they had no conception of how difficult it would be to negotiate such terrain. Although the platoons moved very slowly, most of the men felt worn out by the time they were half way through. The clinging vines, the thorns, the swamps, the unexpected holes, the mosquitoes, the chiggers, and the everpresent danger of copperheads—all combined with the stifling heat to make the operation extremely enervating. Some of the squads encountered snakes and killed them without bothering to ascertain what kind they were. One person in each platoon was designated a first-aid man and carried emergency medical equipment instead of weapons; they were kept busy caring for a variety of minor wounds. Officers did not accompany the troops on this patrol, and as the privates became more tired and disgusted, the NCOs had increasing difficulty in keeping their charges under control. After about 1,000 yards the "breaks" became more frequent; one man would sit down and refuse to move, and the entire unit had to stop so that no one would get lost. During one of the breaks a rumor developed that the company would have to attack over the same territory that night. It was not implausible. If they had to reconnoiter the area, sometime during the week they would have to attack. But the thought of going through again on the same day was disturbing. Many declared that if this happened, they would "get lost." Since the growth was so thick, it would be impossible for any officer to find an individual who was hiding. "Fuck this shit!" was a common cry, and several wondered aloud how they could last until 06:00 the following morning when they felt so ragged already. Covering the 3,000 yards required more than four hours, and when they reached the road on the other side, most men felt so drained that they just flopped to the ground. When the NCOs tried to get them to march back to the bivouac area, they refused to move. After considerable delay, by collective agreement they started their trek "home"; their canteens were empty, and they were hungry. It was so hot that salt tablets had to be issued, but the water supply was low, and more had to be ordered. By the time the company was called into formation again most of the men were exhausted.

Since they had had such a rough morning and were scheduled for a long night problem, the men expected to have an easy afternoon. Thus, when they were marched back after lunch to the same starting point, groans filled the air. When told that they had to go through the jungle again, this time on a "combat patrol," many indicated that they intended to get lost. But the B Company officers were ready; they had been warned by the A-9-3 cadre. It was made clear that no platoon would be allowed to leave the jungle area for supper until every member had been accounted for. Since no one wanted to inconvenience his entire platoon, each realized that he had no alternative to sticking with his unit. To the participants it did not matter that the afternoon exercise had a different label from that in the morning; it still meant struggling painfully through the jungle for three or four hours. It consisted largely of picking out places to step, hanging on to avoid slipping, wading through swamps, swatting insects, trying to keep one's balance on precarious terrain, crashing through thorns steel helmet first, holding back branches for each other to crawl under, disengaging a buddy's equipment from vines, and above all striving to keep his rifle clean and dry in spite of everything. By common agreement they took frequent breaks. It was impossible to find the same trail that had been used in the morning. Realizing that they might have to go through again, most of the men remained alert, looking for anything unusual that might serve as a landmark and trying to remember passages that turned out to be relatively easy to negotiate. NCOs could not order their subordinates to do anything. When they wanted to go on, they had to plead and reason with the privates. The 3rd Platoon encountered some 10th Battalion men bathing in a stream. Some started taking off their clothes to join them; reluctantly they rejoined their platoon, however, when Sergeant Matsui protested that he did not want to stay in the jungle all day. Two platoons got through the jungle at about 17:00 and deployed for a raid on the "enemy" position. A friendly rivalry was developing among the platoons. Since riflemen took pride in their ability to "take it," after a brief rest, they went on a spirited bayonet charge up the steep hill. Much to their disappointment, the position had been abandoned. The "enemy" had left for supper.

The temperature had become oppressively hot, but only a few were unable to march back to camp. Friends offered to carry the rifles of those who appeared completely exhausted, but most of them were too proud to give up their arms. Since it had been an exceptionally debilitating day, the friends insisted that no one would lose face on this particular occasion, but they hung on doggedly. On returning to camp, the men learned that the night problem consisted of going through the same jungle again! Since most of them were already staggering, they were unable to see how it could be done. Furthermore, they felt that the whole exercise was "waste time" after the first trip, since it was simply a matter of somehow getting through the jungle. Supper consisting of fish, rice, and hot coffee did not help morale, and several were too ill to eat anything. Considerable resentment arose over the water shortage, and supply men were cursed for their inefficiency. A mobile PX unit came into the area, but the waiting line had over 200 in it from the beginning, and some had to wait as long as an hour and a half for a bottle of Coca Cola. Since everyone had been ordered to be back by 19:20, most B Company men were unable to get anything.

B Company was called into formation at 19:45. Several had reported to the dispensary during supper, and some of them—about ten according to rumors—had been excused from the night problem. When the company assembled, however, there was no appreciable reduction in numerical strength. As long as the others were going, most Nisei felt that they could not drop out, even when given permission to do so. An officer was assigned to accompany each platoon on the night problem; since the troops appeared so dejected, several of them tried to boost morale by leading songs. The men were aching, tired, disgusted; at first no one would sing. Several confessed to their friends that they doubted that they could finish on their own strength. About 20:00 a lieutenant, who identified himself as a Greek, told some unusually lewd jokes, and everyone within hearing range started laughing. Then, someone in the 1st Platoon yelled that the 3rd Platoon couldn't "take it." With that the 3rd Platoon suddenly snapped up and began singing:

The 1st Platoon is always late, parlez-vous,
The 1st Platoon is always late, parlez-vous;
The 1st Platoon is always late; they stay in bed and masturbate,
Inki, dinki, parlez-vous.

As the 3rd Platoon sang with gusto to show the rest of the company that it was not yet prostrate, the other platoons also came to life. The lead platoon set a faster cadence and taunted those unable to keep up, and a contest started to see which platoon was the "most rugged." Stragglers were heckled, unless they were really tired or elderly (over 25).

By the time the company arrived at the assembly point, most of the men appeared in good spirits. Once the platoons separated, each going to its respective starting point, however, the prevailing mood once again became somber.

The attack was launched at 21:00. It was an unusually dark night, and soon all five platoons were lost. Anticipating such difficulties, members of the 4th Platoon had tied strips of toilet paper on branches along the trail they had cut that afternoon, but it was so dark that their scouts could not find the paper! Since the attack was tactical, flashlights could not be used; this rule could not be overlooked, for this time an officer accompanied each platoon. The main barrier was a large swamp about one-fourth of the way through, and each platoon tried to cross it at a narrow point. Two of the squads in the 3rd Platoon had found relatively easy crossing points during the afternoon, but their squad leaders were unable to find them in the darkness. Trying to find an easy crossing, the scout for the 3rd Platoon—known affectionately as "Cyclone" Nakasone —had to go off the azimuth assigned to the unit, and by 22:00 the platoon was lost.

It appeared that the 3rd Platoon began to cross the swamp at one of its widest points, perhaps 100 yards, and the men had to wade through hanging on to one another's cartridge belts. It was so dark that one could not see a piece of white paper held more than six inches from his eyes. Once physical contact was broken, the men had to whisper to locate each other. They followed the scout blindly, holding on to the belt, hand, or rifle butt of the man ahead. Private Nakasone managed to find a path through the swamp that was generally not more than knee deep, but from time to time someone in the chain would lose his balance and fall into deeper water. Whenever anyone lost contact with the person ahead of him, a plaintive whisper was passed from man to man: "Hold it up!" The whole platoon had to wait until the break was closed. As they stood in the mud, each could feel himself sinking deeper and deeper into the slime. Mosquitoes were ubiquitous, but the men could not be bothered with them; they had to concentrate on the task of getting through the swamp. It took more than an hour to reach the other side. Since "enemy" patrols might be roaming about, even the lieutenant could not use his flashlight, although he confessed being tempted. Occasionally, the "enemy" fired flares, and these gave the attackers a fleeting opportunity to orient themselves.

Just when the 3rd Platoon had crossed the swamp and stopped for a rest it began to rain. Those who were not already wet became thoroughly drenched. After a brief respite the platoon started up the first of three hills that had to be scaled. It was 01:30. Boots and socks were soaked, and they squished with every step. Most rifles were wet, and many wondered if their weapons would fire. Some shooting broke out ahead. This was disheartening to those still mired in the jungle; a firefight ahead meant that at least one of the other platoons had already gotten through. The 3rd Platoon had not even reached the midpoint.

Yard after yard they advanced. The men could not think of how tired they were, for they had to concentrate on maintaining contact with one another. Whenever a low branch, sharp thorns, a hole, or a slippery spot was encountered, warnings had to be passed down the line from person to person. Similarly, orders could not be shouted, and these likewise were whispered down the chain. If the platoon was to operate as an efficient unit, every single man had to be alert. Though many felt miserable and wanted to sit down where they were to go to sleep, they plodded onward, helping each other over barriers and pitfalls. Although the individuals assigned heavier weapons such as the BAR did not request it, others voluntarily took turns carrying them so that no single person would have to suffer more than the others. Orders that this be done were not issued by an officer or NCO; the privates did it on their own. At first one man volunteered to carry the heavy weapon; after he had had it for a while, someone else would volunteer to take his turn. Confronted with a challenging situation, the men became close and cooperation was effective. They pooled all their resources to accomplish the arduous task.

The 3rd Platoon moved rapidly over the last two hills and at 02:30 reached the road that was to be the point of departure for the final assault. There the men lay down to take their first extended rest, and many fell asleep at once. The lieutenant and the NCOs held a conference, but the others paid little heed to what they were saying. Then, the privates were awakened and marched down the road a few hundred yards and were told to remain quiet. A small "enemy" patrol came through at this point and was seized. When one of the startled captives yelled "Corporal of the guards!" those nearby had difficulty stifling their laughter. But most of the men were nearly asleep; they heard the commotion, but they were too weary to care. As the platoon was again aroused, a rumor developed that the problem was over; it had been called off because everyone was too tired. Just when most of them were beginning to relax, they were ordered to deploy for a bayonet assault. They were puzzled. No one had bothered to tell them the plan of attack; in fact, they did not even know that an attack was still scheduled. They had simply followed their NCOs in the dark.

The final bayonet assault was a dismal failure. Even in A-9-3 they had never made an assault that was so ineffectual. So many fell behind the pace set by the leaders that a skirmish line could not be maintained. Once one exhausted individual fell behind, others faltered in droves, and the formation quickly became staggered. When it became apparent that the skirmish line had collapsed, some just stopped running. The "enemy" was supposed to retreat and then counterattack, but there was so much confusion in the darkness that no one knew who was on which side. The men were unwilling to fire blanks when others might be at such close range. No one in the attacking force was in the mood to continue masquerading as soldiers. Apparently no one in the defending unit cared either. When the attackers reached the top of the hill, they sat down and chatted with the defenders, who were curious about the jungle. They could see how ragged the B Company men were. They knew that they would be going through on the following day, and they wanted all the advice they could get.

When the exercise was over, the 3rd Platoon was ordered to form on the road. Everyone expected to be marched back to the bivouac area, but the lieutenant insisted that it was still too early to return. The problem was not supposed to end until 06:00. The platoon was taken to a secluded spot, and the men were told to get some sleep. Several objected that this was unwise; conditions were ideal for catching cold. Everyone was soaked from the rain, the swamp, and sweat; their shoes were wet; they were haggard and weak; and the ground was cold and wet. As increasing numbers began to cough and sneeze, some intentionally, Sergeant Matsui kept after the lieutenant until he finally agreed to return to camp. Once they were back most men changed immediately into dry clothing and fell asleep. A few waited for breakfast, which was scheduled for 05:00. As the other platoons returned, those who were still awake discussed the relative difficulties of negotiating their respective lanes. When breakfast was delayed, however, they too gave up in disgust.

On the following day the company was not awakened until 11:00. Many complained of still being sleepy, but most were ready for another day's work; months of conditioning had made them sturdy and resilient. Since everyone was expecting a "soft" day, however, protests erupted when they were ordered to put on their combat packs. When told they were to defend the hill, however, the objections died down. At 14:00

the company marched into the road behind the jungle, and each platoon was assigned a sector to defend. Being familiar with the terrain, the men agreed readily on the kind of places that could be defended most easily and dug in at these points. Little labor was needed. Foxholes dug on the previous day were still there, and only a few new holes had to be dug and camouflaged. The only problem was the placing of automatic weapons. All those who were interested, regardless of rank, participated in drawing up the defensive plans. Once this was done, everyone took turns bathing at a nearby creek. The rest of the afternoon was spent relaxing. One or two remained awake to patrol the hill ahead and to give warnings; all the others, including the NCOs, slept.

The raid by the "enemy" combat patrol came at about 17:00, just when the company was getting ready for supper. The defensive positions had been set up in such a manner that it would have been extremely difficult for any frontal assault to succeed. When those in the 3rd Platoon recognized their old friend Corporal Beaudette leading the charge, however, they let up their fire to let his unit through. There was much joking about "shooting your ass off" and dire warnings about what would happen that night. Learning from the corporal that A-9-3 had not done any reconnaissance that morning, many were resentful. When he explained that plans had been changed because the officers thought three trips were too "tough," however, they agreed. Still, some complained of being "robbed."

That evening most members of the 3rd Platoon did not sleep. Since their rifles were already dirty from the afternoon firing, there was no point in avoiding a firefight; the weapons would have to be cleaned later anyway. Most men got all the blanks they could, loaded, and waited anxiously for Beaudette and his platoon to come again. What they were hoping for was that some of the A Company officers would accompany their men. Captain Larsen was no longer with the company, and Lieutenant Delaney rarely joined the troops in strenuous tasks. But with a new company commander things might be different, and there was a possibility that the lieutenant might come. There were a few other A Company cadremen who were not especially liked, and plans were laid to "get" them. The unit was alert. To be sure that they would not be caught off guard, large sections on the next hill were wired with booby traps so that sentries would not be caught napping. Sentries were doubled later in the evening, and patrols were sent out periodically to check. These moves the men made on their own. They were not ordered by the officers or NCOs, although the latter were delighted that for once the privates were taking their work seriously, using all the skills they had acquired in many months of infantry training.

But all the preparation turned out to be in vain. The attacking force lost its way in the darkness and assaulted a position being defended by another platoon. The firefight in the next lane did not concern members of the 3rd Platoon; they were still focusing on their own sector. Corporal Beaudette then ordered his unit to attack the 3rd Platoon from the rear. The defenders did not know what was happening; they had assumed that these men belonged in the next lane and were returning to camp after having completed their work. There was much grumbling about unfair tactics and sarcastic comments about infantrymen who did not know how to read a compass. The new men in A-9-3 then disclosed that most of them had trained in an armored division at Fort Knox; they were not riflemen. No A Company officer accompanied this platoon, and many were disappointed that they did not even have a chance to frighten Lieutenant Delaney. It was not until 03:30 the following morning that B Company returned to the bivouac area.

The remainder of the week was uneventful. Amidst complaints about the inadequate food and water supply, the troops went through a variety of exercises in a lackadaisical manner. On Thursday, B Company was ordered to fall out at 13:00 carrying no equipment other than soap and towel. They were to be taken on trucks to an engineer's camp four miles away to take their first shower that week. The grueling schedule was now beginning to take its toll, and as the 3rd Platoon straggled toward the trucks, the following exchange occurred:

"Jesus Christ! My ass is draggin', hey."

"*Your* ass is draggin'! Shit, *mine's* all red and blistered!"

"Mine's bleeding. It's been bleeding ever since Monday."

"What are you guys moaning about? I ain't got no ass left!"

"Honest to God! Look at this platoon! They must be teaching us how to do a disorganized retreat."

Even though little hot water was left, most men were delighted at the much needed shower. Then, the captain led his company to the camp PX, where they were able to purchase soft drinks, beer, and candy. When the NCOs told them that the captain had not been authorized to do such a thing and might get into trouble if caught by an inspecting colonel, the privates were even more appreciative. Complaints stopped, as those who were unhappy about anything were told by the others, "No *monku* [complain]."

Although they had to march back to the bivouac area, most men were in high spirits. They began to sing, and as they passed the A Company campsite, their voices boomed out:

A-9-3's a bunch of jerks, parlez-vous,
A-9-3's a bunch of jerks, parlez-vous;
Oh, A-9-3's a bunch of jerks; they're always fuckin' up the works,
Inki, dinki, parlez-vous.

When given the evening off, most men spent the first few hours cleaning their weapons. As they sat about, a rumor developed that everyone would receive furloughs on returning to the fort. In the 3rd Platoon area Private Onishi continued to be the butt of caustic remarks. Private Nakanishi often caricatured the first sergeant whenever someone yelled to him, "First sergeant, report!" He snapped smartly to attention, saluted, and announced, "Sir, all present or accounted for! As you were! T/7 Onishi temporarily incapacitated in the mobile dispensary!"

The troops left A.P. Hill at 11:30 Friday in a forty-truck convoy. Again motorcycles and jeeps with screaming sirens cleared the way, and again civilians flocked to the streets to see what was happening. Most of them waved, and again the men speculated about what these people were thinking. On one truck, on which a man at the rear was wearing his steel helmet and holding his rifle while the others were relaxing in fatigue caps, someone commented, "Christ, I bet we look like a bunch of POWs and fuckin' Fujimoto looks like our guard." There was much speculation on what was to happen to the group. Several rumors developed. One was that the company would go to a beach on the following week to learn about landing crafts and LST boats. When someone remarked that he did not care where he went as long as he could "fuck the dog," Nakanishi answered with a straight face, "Yes, I hear that some of us are going to be transferred to the Canine Corps." Several wondered why so much money and effort was going into training Nisei, who were no longer needed in Europe for combat duty. Much of the talk also centered on the first day in the jungle. In the misery, in sharing water, and in helping each other through, many new friendships had been formed. Those who had previously been in B Company had gotten to know the transfers from A-9-3. They expressed pride in their "toughness," their ability to "take it" like infantrymen, and disparaged the armored men as "soft." Several noted that they were now better trained than other IRTC graduates, most of whom had never seen a jungle. The trucks arrived at Fort Meade at 16:30.

Analytic Summary and Discussion

The administration of A-9-3 stood in sharp contrast to the efficiency of IRTC units. Although everyone was familiar with the absurd situations and bureaucratic bungling that arose periodically in the armed forces, none had encountered so many blunders, some of them approaching the grotesque, in a single company in such a short period of time. At first the men viewed the cadre with astonishment, then with humor, then with irritation, and finally with bitter resentment.

It is possible that the replacements were somewhat hasty in imputing foul motives to selected members of the cadre, but they had no way of understanding some of the difficulties involved. They did not know how much the personnel roster changed from day to day. For example, the morning reports of A-9-3 show that during the month of April, 32 newcomers arrived in dribbles of 1 to 6 a day; 48 came on April 20. Various ailments hospitalized 42 during the month; 4 were AWOL; 2 were transferred to another battalion; 7 went to a transportation company. In addition, 69 were sent to POE on April 9, and 98 more shipped out on the 20th. Few could appreciate the difficulties involved in keeping track of transient personnel. Furthermore, the similarity in the confusion that preceded both the funeral parade and the maneuvers suggests that someone in battalion or regimental headquarters had trouble making up his mind.

The pattern of insubordination and resistance developed gradually through trial and error. At first just a few yelled in moments of utter exasperation, and some inadvertently missed a formation. When they went unpunished, others began to take chances. When they too escaped undetected, increasing numbers began to avoid onerous duties. Those who could not escape began to feel foolish. Thus, several behavior patterns not commonly found among fresh recruits took shape. Shouting insults from ranks and missing exercises became a spontaneous reaction to any inconvenience. The authority of cadremen was initially challenged because they were regarded as totally incompetent. Why should anyone take orders from people who were so stupid? A distinctive pattern of reacting to frustrations crystallized in efforts to cope with a succession of exasperating situations.

Captain Larsen's inability or unwillingness to recognize individual differences among Nisei then consolidated an otherwise heterogeneous group. Rather than punish specific offenders, the captain elected to treat all Nisei alike—whether or not they had engaged in derelictions. This reinforced their sense of ethnic identity. As the men came to regard themselves as a disadvantaged minority involved in a contest with the orderly room, they united to fight back. Insubordination that was initially spontaneous took on some measure of organization; the men cooperated in devising stratagems for embarrassing the cadre. The ethnocentrism fostered by the captain's policy also reinforced various Nisei norms, and even those who were relatively unconcerned felt compelled to participate. The more the captain retaliated, the more he was viewed as unjust. When he was defined as a bigot who was violating army policy, his authority was questioned. When the men were subsequently segregated, their sense of ethnic identity was even further enhanced.

Not much reinforcement was needed, for most of the men were still acutely conscious of their tenuous position in American life. Regardless of the context, whenever they became aware of being on display as Nisei soldiers, their sense of ethnic pride overrode all other considerations, and the quality of their performance was suddenly transformed. They were especially concerned that officers outside of A-9-3 recognize their competence as soldiers, once they were beyond the control of the A-9-3 cadre. Thus, on the final day of the strenuous week in April they were utterly exhausted; yet, when they learned that a general was among the observers, they did their very best and were infuriated when the mistakes of a few officers made them appear inept. Similarly, the sixty who were in the construction crew performed diligently when their audience consisted of outside officers who appeared to understand their plight. In B-9-3 those who were still discontented became model soldiers, for they saw this as an opportunity to strike back at Captain Larsen. The more effectively they performed, the more likely it was that the regimental commander would realize that the captain was primarily responsible for their miserable record. Indeed, their performance on the first Monday at A.P. Hill reveals both the aptitude of the men when confronted with a task that they recognized as essential as well as their stubborn determination to do their full share and not to fall behind other Nisei.

5. «» the disposition of surplus personnel

On 26 April 1944, a few months after the reopening of the Nisei draft, the War Department announced that soldiers of Japanese ancestry would not be used as combat troops against Japan because of possible enemy retaliation and the "increasing hazards of enemy infiltration." Although this was published in the *Pacific Citizen*, most of the replacements did not know of it. They had heard countless reports of this kind, and they could no longer remember which were official and which were not. Furthermore, none of the officers at Fort Meade knew of the policy, for they answered questions put to them in surmises. Most Nisei assumed that infantry service in the Pacific was unlikely. They had seen enough of ground combat to appreciate how considerable confusion could arise—on both sides.

V-E Day brought the unusual position of Nisei infantrymen into sharp relief and tended to isolate them psychologically from other replacements. They had been drafted as replacements for a combat team fighting against Germany, and now the Germans had surrendered. What now? In speculating about their future several possibilities arose. One was that all Nisei, most of whom were in the Army Ground Forces, would be transferred to the Army Service Forces for rear-echelon work. Some joked about replacing WACs for active duty, but others were not particularly amused at this prospect. Another possibility was that they would be sent to Europe for occupation duty, replacing combat veterans who were entitled to discharges. This seemed a reasonable alternative, and most spoke in favor of it. A third possibility was that they would be retrained for service in the Pacific. This would mean transferring to Fort Snelling to prepare for intelligence work. This also seemed plausible, but most spoke against it; MISLS was labeled as being even more "chicken shit" than B-9-3. Since there seemed to be nothing left for them to do, a few suggested hopefully that Nisei might once again be placed in ERC. This was most attractive, but most of them had been in the army long enough to realize that this was extremely unlikely.

The fate of the 442nd Regimental Combat Team was watched with special interest, for as long as Nisei infantrymen remained segregated their future was tied to that unit. It was known that the regiment was back in Italy, for its part in the 5th Army's final drive had been publicized widely. But there was no assurance that it would remain there; it had once been detached from the 34th Division and sent to France. Several contradictory rumors emerged. One was that the 442nd was going to China, that some of the advanced elements had already arrived in Chungking. It was even claimed that someone in A Company had seen a letter from a friend en route to Asia. Another rumor was that the 442nd was being ordered back to the United States; another, that it had been attached to the 15th Army for occupation duty in Germany. Still another rumor was that volunteers would be called to form a new all-Nisei combat team for action in the CBI (China-Burma-India) theater. All the rumors were attributed to a variety of "reliable" sources—in most cases men with friends or relatives in the regiment—but most remained skeptical.

The few who had entertained seriously the possibility of being released early were jolted rudely on May 11, when they viewed a motion picture explaining the discharge system. Most agreed that the plan to give points for length of service, dependents, wounds, and decorations and to give those with the most points priority was eminently fair (Stouffer et al., 1949: II, 520–48). While heartily approving the procedure, they realized that they themselves would remain on duty for some time.

"What the fuck you counting your points for? No *shimpai* [worry]. You won't get no fuckin' discharge. Look at old Beaudette. He's been in the fuckin' army since Pearl Harbor. He was with the 1st Division, and he's got over 100 points, but he's still here. He was telling us that when we left this place he'd probably still be here to lock the gates."

"Yeah, it don't make no diff how many points you got. If they want to fuck you up, they can do it. They talk about that 'military necessity' shit. Remember? They evacuated all the Boochies from the Pacific

Coast on that kind of crap. If they can do that, they sure as hell can keep us in the fuckin' army if they want to."

"How many points you got?"

"Seven. Naturally. Shit, you know I haven't got any more than you."

"You're married, ain't you?"

"Sure, but that don't cut no ice. Ya gotta have kids."

"Why don't you get off your ass and make some?"

"The fuckin' points don't count after May."

"No lie! That's the shits. I thought if I stayed in this fuckin' army for about seven years I'd get out, but if the points don't count any more, no hope, huh?"

"In your case the best hope is Section 8 [psychiatric discharge]."

"Stay in this fuckin' camp very long, and we'll all go nuts."

"No lie, boy! I lose fight in this fuckin' place."

Everyone would have to serve out his time. Wherever it was, it meant continued regimentation. Furthermore, it was not likely to have any promotional value for the Nisei cause. The men felt useless.

Recruitment for the Intelligence Service

Not long after their return from A.P. Hill, Captain Parmentier announced that all replacements would be issued three-day passes, one platoon at a time. Throughout the following week, during the desultory performance of routine duties, the central preoccupation was with getting maximum benefit from the passes. Then on Thursday, June 7, the company was upset by a rumor that everyone was on orders. The immediate concern was with passes, for only two platoons had been let out, and it was believed that Army Ground Forces regulations permitted only 20 percent of any command to be absent. That evening the captain cleared up the matter: "Men, I have some news for you. I don't know for sure what's coming up, but I can't let you leave camp after next Monday. This means that we'll have to revise our schedule on three-day passes. I hadn't counted on your going on orders this quickly, but it's one of those things that can't be helped. In order to keep my promise to you, all men who haven't had passes will leave tonight. You must be back here by Monday noon. I'm sorry that I have to cut it down, but I can't extend your time any longer. This doesn't mean that all of you are on orders, but I want to play safe and make sure that all of you have your passes before the clamp is down. I want to play safe and give all of you what is due to you." When some grumbled that their passes were good for only two and a half days, others pointed out that the captain's hands were tied; furthermore, he was risking censure for allowing 60 percent of his command to be absent at one time.

On returning from their passes on Monday, the men learned that a team of recruiters from Fort Snelling was interviewing prospects for the military intelligence service. There were a number of rumors—that all qualified Nisei would be drafted for MISLS, that those who did not qualify would be sent to Fort Riley for reassignment, that 150 Nisei would be sent to Camp Ritchie for a special assignment.

At a mass meeting preceding individual interviews Master Sergeant Nishida, an older Hawaiian who was introduced as the battalion sergeant major at Fort Snelling, announced that Nisei replacements faced three alternatives: (1) About 250 would go to Fort Snelling for training in intelligence and civil affairs work. Three different courses of study—of three, six, and nine months—would begin about October. (2) About 150 would be sent to Camp Ritchie to demonstrate Japanese arms and tactics. Although the sergeant did not elaborate on this, a United Press report of June 2 that was reprinted in the *Pacific Citizen* on June 9 had disclosed that Nisei soldiers were to participate in the training of combat divisions now being redeployed from Europe to the Pacific. They were to demonstrate Japanese uniforms, weapons, and tactics. This alternative was quickly labeled as "permanent Saipan Joes." (3) The rest would be sent to Europe for various assignments.

Members of military organizations are seldom consulted about their preferences, but this group was being given the opportunity to select a branch of service from a limited set of alternatives. Everyone immediately became completely preoccupied with the problem. The men talked of little else. All relevant information was pooled, evaluated, and discussed. Many points were considered, and many questions were raised. Each man had to make a decision, but it had to be made in an atmosphere of utter confusion.

There appeared to be several advantages to going to Fort Snelling. It meant that one would remain in the United States longer. For those who thought their future was in some line of work involving the Orient, it meant an opportunity to learn the Japanese language at government expense. Learning Japanese might turn out to be helpful in many other ways; if nothing more, they would be able to communicate more effectively with their parents. It meant assignment in rear-echelon work, where casualties thus far had been very low. It also meant getting faster promotions, for all graduates thus far had been given ranks of at least T/5. This would bring more pay.

On the other hand, several disadvantages were cited. The European theater was now inactive, and occupation duty was not hazardous. MISLS graduates, however, might be sent overseas before fighting ended in the Pacific. As increasing numbers of Nisei became available for combat intelligence duty, they would become more expendable and might be sent closer to the front lines than they had been in the past. Furthermore, translators and interpreters might be needed for a long time for the military occupation of Japan; most had heard that Germany had been occupied for five years after World War I. Although they constituted but a small minority, some just did not want to fight against Japan. Those who were proficient riflemen were apprehensive about transferring to a branch of service in which they would be judged by different criteria. Gifted athletes were generally agile riflemen, and these men confessed to their friends that their standing might be jeopardized were they to be evaluated by intellectual rather than physical standards. Men who were more assimilated also had reservations; they wondered how they could possibly keep up with those who had grown up in segregated communities, had attended as much as twelve years of Japanese language school, and had used the language almost exclusively at home and at work. Some who had no objections to going to Fort Snelling hesitated lest they miss out on an even better "deal," such as the remote possibility of being placed in ERC; others noted, however, that in the past some useless Nisei had been placed in labor battalions. But the most frequently cited objection was Fort Snelling's reputation for "chicken shit" and discrimination. The general distrust of the MISLS administration by the troops already there had spread throughout the Nisei community.

The general effect that arose from the discussions was that no one wanted to go to MISLS. Those who wanted to volunteer were afraid to say so. Some confessed quietly to a few intimate friends that they would not mind going but otherwise continued to speak out against this alternative. Its reputation for pettiness was stressed over and over. Some indicated that they had previously volunteered and had been turned down; now they considered it beneath their dignity to go. The pattern of acceptable conversation was clear; everyone wanted to stay in the infantry—in a *man's* world. They were "rugged"; they sneered contemptuously at rear-echelon assignments; they did not want to become "chair-borne." Only the effeminate who could not "take it" would accept "male-

WAC" duties. Since those who had something favorable to say about MISLS refrained from speaking while those opposed were free to give vent to their views, a predominant impression was created that did not actually match individual views. Many hesitated to volunteer because everyone else insisted that he was not going. They wanted to stay with their buddies, all of whom kept insisting that they wanted to have nothing to do with MISLS. Furthermore, everyone endorsed the enlisted man's dictum: *Never volunteer for anything.* If one were ordered to an assignment that turned out to be disastrous, this could be written off as one of the misfortunes of war. But if he had volunteered and then suffered, the responsibility was his own; he had asked for it.

Although the troops marched to the interview station in platoons, each man was interviewed alone. Each was closeted with a recruiting sergeant in a room in which no one else could hear what was said. Thus, each person had a chance to decide for himself. On the first day some who had gone in the morning claimed that the recruiters had tried to trick them into volunteering for intelligence work. Resistance developed immediately. Unsavory reports about Fort Snelling were revived. Several declared flatly that they did not wish to have anything to do with an organization that was "slimy," "tricky," and "underhanded." Afterward several insisted that they had told the interviewer that they knew nothing. One commented, "Aw, that was all fake! The bastard talked to me in Booch, but I just said I didn't know what he said." Although several claimed that they had even intimidated the recruiter, there was no way of finding out just what each candidate had done.

The final weeks of training were conducted lackadaisically, and the men justified sloppy work on the grounds that they were no longer riflemen. Psychologically, many had already made the transition to rear-echelon duty. For example, on Tuesday afternoon, June 12, three platoons were selected to defend a hill while two platoons assaulted. Claiming that they would soon be "pencil pushers," no one bothered to pick up blank cartridges that were issued. The defenders picked cool, comfortable spots and went to sleep. The officers, who apparently felt the same way, napped with them. When the attackers came thrashing noisily and carelessly through the woods, they were greeted by shouts: "Bang! Bang! Goddammit, simulate a dead position!" Everyone laughed, and the defenders commiserated with those who had had to walk through the woods while they slept.

Even serious incidents that would otherwise have evoked considerable comment passed almost unnoticed. On one cold night B-9-3 was assigned an area to defend. After a detachment of about twenty Caucasian replacements from E-9-3 had joined them, the men were told that another company had been deployed to attack, and each was issued blank cartridges. It soon become so dark that they could identify one another only by their voices, and those who did not want to dirty their rifles threw their shells away. Instead of waiting for the attack most Nisei found comfortable places, chopped down the vegetation with their bayonets, and spread out their blankets to sleep. Nor did they bother to sleep tactically—one man sleeping while his partner stood guard. Suddenly, at 23:15 everyone was startled by an explosive burst of rifle fire. Unmistakable in the outburst were three sharp cracks overhead, followed by a fading swoosh. Live ammunition! Even without combat experience they were sufficiently familiar with infantry work to be able to recognize rounds fired by various weapons just by sound alone. Most were up instantly; many were frightened. They got together quietly in small huddles to confirm their impression that live rounds had been fired. Why was real ammunition being used in a night problem in which trainees were shooting at each other? The few who thought the "enemy" might consist of Caucasians who were "gunning" for them were especially concerned. Within minutes it became apparent that somehow the "enemy" had infiltrated the final protective line and was wandering about among the defenders. As word was passed along, everyone became alert. Action suddenly erupted. There was shooting, running, and yelling. When the first few prisoners had been taken, the Nisei realized for the first time that the "enemy" force *was* Caucasian, and the cry went up: "It's the Haole!" After that, anyone who was not Nisei was seized. Excitement ran high. A captured rifle came to symbolize a prisoner, although no announcement to this effect had been made, and most men did their utmost to capture as many rifles as possible. Those who fired blanks had to be very careful; their friends were at close range, and no one could see. For those who hated *Keto* the problem provided an opportunity to strike back; only their inability to distinguish between friend and foe kept down the number of cuts and bruises. In about fifteen minutes the fighting was over, and everyone, though shaken, went back to sleep.

On the following morning it became apparent that the "enemy" consisted of the small detachment from E Company. The "enemy" had marched with them into the area to be defended, knew the defensive alignments, the location of outposts, the password and countersign—

everything! They could not be distinguished in the darkness, and were it not for the quick realization that the line had been drawn on an ethnic basis the defenders would have had no way of telling who was on which side. Many wondered aloud about the point of the exercise. There had been definite risk of serious injury. What could anyone have possibly gained from it? What were they supposed to have learned? And the live rounds remained a mystery. Caucasian replacements who were questioned confirmed hearing the live rounds, but they were equally puzzled about them. Could the officers have fired the shells purposely to arouse them into more realistic action? Did some especially embittered Nisei fire them on learning that Caucasians were on the other side? Did some Caucasian replacement who did not know any better fire them for the fun of it? No explanation was given, and the officers disclaimed any knowledge of the bullets. But attention was soon directed to other matters, and the men stopped talking about it once they returned to the barracks.

On the last day of training a Nisei was shot accidentally during a village fighting exercise. Since his injury was serious, under other circumstances it would have become the center of considerable concern. But the men were so preoccupied with shipping orders that the incident elicited scant attention outside the victim's circle of friends. Everyone learned of the accident as soon as his platoon marched into the company area. Each checked to see whether the injured man was a friend, but those who did not know the man expressed their regrets and went on with their tasks. A rumor developed that he had been shot in the back. Since it was difficult to maintain a skirmish line in village fighting and since bullets frequently ricocheted, this was easy to understand. Some commented that it was astonishing that more trainees were not injured in that exercise.

The training routine was virtually ignored after Wednesday, June 13, when about forty replacements were alerted immediately after lunch. Considerable speculation arose over which of the three destinations they faced. The initial guess was that this group was going to Fort Snelling; then it was discovered that two of the men did not know any Japanese at all. When it was suggested that only volunteers would be transferred to MISLS, everyone denied vehemently that he had volunteered. On successive days additional men were alerted. Those on orders remained in the barracks for an inventory of clothing and equipment, while the others continued their field training. The depot inspection was regarded as a waste of time, for only a fool would leave Fort Meade without all the items he had been issued. By June 17 everyone was on orders. There was no longer any question of which men were going overseas; they had to listen to the reading of the 26th Article of War about desertion in the face of hazardous duty and had to sign a statement that they had heard it. The other two contingents were being prepared for transfers, and the cadre indicated which of the two was going to Fort Snelling. The latter were especially irked on learning that they would have to carry gas masks! Combat serviceable masks that had been fitted and tested in the gas chamber had been turned in; now they were replaced by old, leaky gadgets that would only have gotten in the way in the unlikely event of a Japanese gas attack on the troop train. Even the officers shook their heads in disgust.

Those who were going overseas appeared delighted, and the others congratulated them on their good fortune. Those who had been ordered to Fort Snelling frequently expressed their envy of the others. They continued to insist adamantly that they had not volunteered for MISLS, that they had been "Shanghaied." Many contended that going to Camp Ritchie as "permanent Saipan Joes" was the best alternative, for it appeared likely that these men would be discharged before the others. In spite of the deprecatory statements about Fort Snelling, however, there was a noticeable increase in the use of simple Japanese phrases and expressions. For example, in referring to the medical examination of genitals—commonly called the "short-arm" inspection—one man commented, "They are going to look at your *mijikai ude*." This expression, a literal translation of "short arm," was quickly adopted. That so many were on orders for Fort Snelling occasioned no surprise. Since no one had volunteered when personnel was badly needed, selections had been made on some other basis. Those who were needed had been drafted, whether they liked it or not. After all, this was the army. Since most men known to have high AGCT (Army General Classification Test) scores were on this list, some wondered whether this was the key criterion (Cf. U.S. Department of the Army, 1948:6). A rumor that some veterans of the 442nd were also being sent to MISLS evoked bitter comment. It seemed so unfair. One man remarked, "It's sure going to be tough on them! They're combat men and not used to garrison life. They won't stand for all this chicken shit." Many of those transferring to G-2 insisted that they would continue to wear the blue braids and crossed rifles of the infantry.

On June 18, 19, and 20, 1,235 Nisei replacements departed from six companies of the 9th and 10th Battalions for the three destinations.

Camp Ritchie received 149; Fort Snelling, 337; and the rest went to Europe. B Company contributed 17 men to the first order, 75 to the second, and 109 to the third. As the various contingents were ordered to the trucks, there were many fond farewells, exchanges of addresses, and promises to keep in touch.

Ex-Infantrymen on the Turkey Farm

By the summer of 1945 the Military Intelligence Service Language School was well established, both in the army and in the Nisei social world. Many of the practices of the early volunteers had become institutionalized. Although their sense of mission was largely gone, the practices they had initiated were the accepted pattern, and they were maintained—even in the face of constant lamentation. Nisei soldiers had acquired a favorable reputation in the Twin Cities area. Their study habits were becoming legendary; rumors of men being admonished for studying under their blankets after "lights out" were recurrent. Since AWOL was virtually unheard of, the pass policy was lenient and flexible. Among servicemen the school had a reputation for strict discipline. In spite of the general atmosphere of distrust and resentment, flagrant insubordination and violence were unthinkable. All this changed in the late summer of 1945, and the men who were credited with smashing these patterns arrived at Fort Snelling on the morning of 21 June 1945.

The newcomers started complaining as soon as they entered the fort. As they were ordered to march in formation, carrying handbags, they began to grumble.

"What the fuck is this? Why don't they bring trucks if they expect us to go so far? Ain't they got no sense in this place?"

"They just want to get you hungry for breakfast. You're lucky they didn't make you carry your duffle bag. No *monku* [complain]!"

"This damn bag's gettin' heavy. That chow'd better be good, goddammit."

"Yeah! We're *chair*-borne now. We can't do no more heavy work."

After breakfast the new arrivals gathered in small circles around their baggage, which had been unloaded at a firebreak. About 07:00 a Nisei first sergeant called the newcomers together and asked them to be "at ease." No one paid any attention to him. Finally the sergeant, obviously enraged, stalked up to one of the groups—made up largely of men who had been in A-9-3—and ordered them to be quiet.

"What the hell you pickin' on me for? All them other guys're talking. Go yell at them!"

"Look, soldier, don't talk to me like that! When I tell you to be quiet, you be quiet."

"Yeah? That's what *you* think!"

The dumfounded sergeant looked around at the sea of hostile faces. Then he walked away.

It was not until 10:00 that a general formation was called for orientation. Again difficulties arose. As the 337 men gathered in the firebreak around a wooden platform, the cadre could not get them to stop talking. Finally, in desperation, the sergeant began to yell as loudly as he could. Even then he could not be heard. A captain climbed on the platform, and the men nudged one another to pay attention.

Captain Schmidt announced that he was the company commander and then proceeded to harangue them: "All right, men, this kind of stuff's going to stop right now. When I or any of the men appointed under me tell you to be at ease, you'll be at ease. You men have been in the army long enough to know that noncoms are to be obeyed. If any more of this continues, I'll see to it that it stops. You're making so much noise that you can't hear the sergeant. He's got an important message for you. Now quiet down, or the sergeant'll lose his voice." Someone in the rear sniggered, "That's tough shit."

The captain became livid with anger. "Who said that? So it's tough shit, is it? If that's the way you men feel about it, there are some things I can do, too. Now let's have it quiet."

When some semblance of order had been restored, the newcomers learned that they had been assigned temporarily to Company E, the casual company in which they were expected to perform fatigue details for the entire School Battalion, pending more permanent assignment to one of the school companies. Since members of the other companies were busy studying, they did not have time enough for KP or other maintenance duties. When objections arose, the sergeant argued that when members of this contingent entered school, someone else would do their fatigue details for them. These tasks included among other things KP for the nine mess halls in the battalion, cleaning the latrines, working on victory gardens, moving chairs in and out of the field house, repairing equipment and buildings, and mosquito control—soon designated the "kamikaze detail." Some jobs, such as clerical work, were assigned on a more extended basis, since it would be inconvenient to have different typists reporting to an office each day. A few tasks, such as boiler-room duty, required working at night, and arrangements were made for these men to be free on the following day. They also learned that the area in which they were billeted was the farthest removed from the main gate of the fort, had by far the poorest facilities, and was called the "Turkey Farm"—an object of contempt.

During the orientation talk a mimeographed pamphlet entitled *Information Bulletin for Enlisted Men Attending the MISLS* was distributed. It began, "You have been especially selected from all American citizens as a man possessing a knowledge of the Japanese language, and because of this knowledge you have been ordered to this school to secure approximately six to nine months specialized training in order that you may perform valuable services for your country."

This paragraph was followed by a long list of the regulations of the School Battalion; the rules filled up ten pages, typed single space on legal size paper. This confirmed the reputation of the organization. Among the regulations were the following:

> Barracks will at all times be kept in spotless condition.
>
> The complete authorized uniform will be worn at all meals from breakfast Monday through dinner Saturday except by men on fatigue details. Only Class A uniforms will be worn about the grounds after Retreat on any day, except by men on clean up details.
>
> Breaking into the mess line is prohibited.
>
> Loud talking and vulgar language will not be tolerated in the mess hall.
>
> No card playing, games, or dice shooting of any kind will be tolerated during duty hours or after taps.
>
> In order to attain familiarity with new regulations, notices, and other miscellaneous matter, *you will be required to examine the bulletin board located outside your company orderly room at least twice daily.*

These were not suggestions; they were orders. Most of the men read them carefully, and the initial reaction was one of curiosity. Do they really mean it? Are they serious, or is this just propaganda? Many were skeptical of the possibility of enforcing such orders.

"Jesus Christ! What're they tryin' to do, scare us? Who's gonna check up to see if we're gonna do all this shit?"

"I hear they really do it though. My brother was here, and he said they were chicken shit as hell. I hear they court martial guys around here for being out of uniform."

"No shit!"

"Yeah. Ask any of these guys around here. They're really chicken shit in this place. When they yell, they expect you to snap shit."

"Yeah. I hear in this place you gotta figure on gettin' screwed."

Then a Nisei cadreman began to assign men to their bunks—six to a hut. The assignments were in alphabetical order. Although some complained that they wanted to be with their buddies, most were anxious to get settled. The huts were dirty. Although no order was issued to clean them, before long everyone was hard at work washing and scrubbing. Most of the afternoon was spent drawing mattresses, pillows, footlockers, and other equipment from the supply room. Then each had to search through piles of barracks bags for his property. Once settled, friends sought each other out to exchange hut numbers.

In the next few days additional troops arrived, and by July there were 567 enlisted men in E Company. Not enough work was available to keep everyone occupied. Night details and clerical positions were given to qualified volunteers; these men were excused from KP and other menial tasks, but they had to report for work every day. The others lined up daily in platoon formation, and duties were assigned at the rate of approximately two platoons a day. The remaining platoons were excused, though restricted to the post. Each day some were restricted to the company area on the pretext that they might be needed later. Those not assigned to some detail were free to do their laundry, play cards, shoot dice, write letters, visit friends, or sleep. Before long, however, they learned that it was unwise to remain in the barracks. They ran the risk of being assigned to some other task. Furthermore, after the first week they ran the additional risk of being required to do calisthenics or close order drill. Later on, road marches were instituted—with full equipment. But these boomeranged on the officers, who were accustomed to leading student companies composed of men in flabby condition. The newcomers from Fort Meade or fresh out of IRTC centers did not tire easily. The short road marches were nothing more than minor inconveniences that only served to irritate them. Only the officers became tired.

The work was done lackadaisically. Although some tasks, such as KP, were recognized as essential, many others were obviously unnecessary. In some work details the men discovered that they had been ordered to spend a whole day doing something that could be finished in a few hours. Even those who did not mind working, since they had nothing else to do, found such duties irksome. Since those who were not on assignment were free to do their laundry or to while away their time, they saw such unnecessary details as encroachments on their "own time." After performing such tasks resentfully for a few days, increasing numbers adopted the evasion patterns that had been developed in A-9-3. At first the others were astonished to see the former A-9-3 replacements "disappear." When these men went undetected and unpunished, however, they began to compete with one another in the extent to which they could "goof off." As this attitude was adopted by an increasing proportion of those sojourning on the Turkey Farm, incredulous mess sergeants and company commanders lodged complaints with the E Company orderly room. Mess sergeants reported that the KPs refused to obey orders; attempts to discipline them only elicited challenges to fight. Those on other details frequently hid behind bushes and went to sleep until dinner time. Walking about the post, one frequently saw groups of five or six soldiers in fatigue uniforms stretched out comfortably under a tree. When the physical conditioning exercises became more irksome, even larger numbers began to "disappear."

The Nisei cadremen were placed in an awkward position; they were virtually helpless. Sensitive to charges of being "chicken shit," they were unable to enforce regulations, except when a Caucasian officer was present. All they could do was to appeal to the men for their cooperation. Privately they agreed that many of the duties were unnecessary, but they had their orders and had to execute them. Those who insisted on enforcing an order on their own were threatened with a beating. This made their position even more difficult. Although many who made such threats were only bluffing, others were genuinely angry. The ex-infantrymen looked well conditioned and had been trained to fight; many of them appeared capable of inflicting severe injuries. Should a fight occur, their difficulties would be compounded even further. Win or lose, they could not "squeal" on another Nisei. They could not even report the names of minor offenders—those who had disappeared or had refused to obey an order—to the company commander. They had to cover up for many flagrant violations. Thus, after one week, the entire work program broke down.

The general impression created was that the tough infantrymen were annoyed by the petty regulations of garrison life. Virility was stressed constantly. Some were arrogant; they disdained clean uniforms and the fastidious observation of military courtesy as effeminate. After all, they were riflemen. They were not "softees" like most of the students, who had had only six weeks of basic training. They had been in the infantry from five months to over two years. Many had received training far beyond what most soldiers learned in IRTC. All their past efforts had been oriented toward the day when they were to go into combat; now they found themselves ordered to become desk soldiers—the kind of troops on whom they had heaped so much scorn in the past. The appearance of toughness was constantly reinforced in loud complaints about "chicken shit." Anything unnecessary was "chicken shit." All inconvenient orders were "chicken shit." Any interference with a man's enjoyment of private affairs was "chicken shit." Could such men be fitted into the academic routine of MISLS? Those familiar with their background were frankly pessimistic; they expected the worst. Outwardly, the men showed a stubborn determination to resist all efforts of G-2 officers to transform them into effeminate garrison soldiers.

Although the cadremen could not control their charges, restraint was exercised through informal arrangements that arose spontaneously among the men themselves. While "goofing off" was a widespread practice, it was kept within reasonable bounds. The men realized that they had to maintain at least a minimal appearance of compliance to enable the cadre to cover for them. Those who "goofed off" skillfully were admired for their ingenuity; others felt themselves to be "suckers" for being caught and forced to do work they considered stupid and unnecessary. On the other hand, those who were absent too often and especially those who "rubbed it in" by taunting comrades who had been trapped came in for censure. Each was expected to do his fair share of what was regarded as necessary, and the presence of enough bodies to make each formation appear respectable was recognized as essential. It was felt that at least half the members of each platoon had to be present in order not to arouse the suspicions of the officers. As in A-9-3, men who "disappeared" too frequently, such as Private Onishi, became the targets of curt remarks and occasionally were told bluntly to do their share. No one kept score, but each day a substantial number managed to get their siestas.

Although the men resented having to waste time in useless tasks and exercises, they were pleasantly surprised by the liberal pass policy of the School Battalion. In the casual company anyone not on duty was free to leave every night and on weekends, and Fort Snelling was ideally situated between the Twin Cities—eight miles from Minneapolis and seven miles from St. Paul. Street cars to both areas passed through the fort at intervals of about twenty minutes. Since enlisted men of the first three grades and married men with wives in the area were issued passes good to 07:30 on Monday and Thursday mornings, wives of MISLS soldiers moved there in large numbers. Many young, unattached Nisei women were also drawn to the region. The War Relocation Authority had opened an office in Minneapolis in February 1943, and volunteer workers had helped to find housing and jobs for resettlers. No one knew exactly how many Japanese civilians lived in the Twin Cities; WRA estimates ranged from 2,000 to 4,000. Most Nisei, civilian and military alike, tended to congregate in certain centers. Two Japanese restaurants that served Chinese and Japanese dishes as well as several Chinese restaurants were popular meeting places. The weekly dances held at the Minneapolis YWCA attracted large numbers, and dances held from time to time at the field house also provided opportunities for meeting girls. Old friends who had not seen one another for years met on the streets or at one of these gathering points, and information was exchanged avidly. Thus, soldiers stationed at the fort were inevitably drawn into the Nisei social world. Personal appearance assumed greater importance, and most of the men became more concerned with Nisei norms.

Since MISLS was part of the intelligence service, no one was surprised at first at the emphasis placed on security. The *Information Bulletin* read:

> As this school has been classed by the War Department as a "Restricted" activity, the following regulations must be observed carefully. Any reports of violations will result in immediate disciplinary action.
>
> There will be no mention of the name of the school or any activities thereat (including courses studied) in or on the cover of any personal correspondence.
>
> There will be no mention of the number or composition of the personnel.
>
> There will be no discussion or talk of the name of the school, activities thereat (including courses studied) or personnel thereat, outside of the limits of the Post. This will apply not only to casual acquaintances but also to close friends and relatives as well as other persons stationed at the Post.

Once the men went out on pass, however, they quickly realized that few military installations were more accessible to the public—or to enemy agents. A main street ran through the fort. There were no gates guarded by sentries. Although military police were stationed at various places, there was no checkpoint at which persons without passes could be stopped. This made the security regulations appear ludicrous and contributed further to their conviction that MISLS was a "chicken shit" organization. The mail regulations, which prohibited using "MISLS" in the return address, elicited raucous laughter.

"Crazy, huh? People outside know more about this place than the guys who are here. Yeah. They know all about this place. They can tell you how many guys are going to school here, what they study, how long it takes, and what they do when they graduate. They know all about it.

If any Boochie spy wants to find out about this place, all he has to do is go to camp or Chicago. Those people know all about it."

The men's negative view of MISLS and their dissatisfaction were increased even further when they visited other companies in the fort and discovered that the regulations they had scoffed were indeed enforced. As free time enabled them to see friends and relatives in the Twin Cities, they also learned of the reputation they were acquiring in the Nisei community. Most Nisei were proud of the performance of MISLS soldiers; they pointed with pride to their sacrifices while training, their excellent record of discipline, and their gallant service overseas. In their eyes the "men from Fort Meade" were disgraceful—the one rotten apple that could spoil things for everyone. The newcomers had been labeled a "rowdy bunch"—rough, good-for-nothing "bums." Some reported being snubbed on admitting that they had come from Fort Meade. Some were proud of this reputation; but others, who got their orientation primarily from friends already at the fort, felt the pressure of their unanimous condemnation. They also learned that the Turkey Farm was an object of contempt because it was the bottom of the status ladder at Fort Snelling. It was for "rookies," for those who could do only menial work. The facilities were much like an afterthought—an appendage for poultry. Their friends told them that they were fortunate that it was still summer, for in the winter the walls of the wooden huts did not provide sufficient insulation to keep out the bitter cold. A frequently repeated account was that a man had to sleep alternately on his back and on his stomach because the side of his body facing the stove would burn while the other side would freeze. One glance at the huts convinced most men that this could well be true; the mainlanders had seen similar structures in the relocation centers.

Much of the general resentment over their lot came to focus on the company commander. While the newcomers managed to get along with most of the Nisei cadre, Captain Schmidt could not be approached. He soon became the object of special hostility and was generally referred to as "that chicken shit bastard." The men regarded him as incompetent and petty, and they resented his frequent threats. No matter what they did that displeased him, he threatened them with restriction. Since the pass privilege was the one thing that they prized, they felt that he was continually hitting "below the belt." The belief grew that the captain had managed to rise in rank through "brown-nosing" the colonel and that he was "bucking" for another promotion by being overly strict; he was seeking personal gain at the expense of his subordinates. The more the captain threatened them, the more those who had been in A-9-3 spoke of schemes for getting even with him. The company's executive officer, Lieutenant Catalano, was no more popular than the captain. The few who had come into contact with him dismissed him too as "chicken shit."

The first crisis involving the entire group occurred on the Fourth of July. At the 05:30 reveille formation Captain Schmidt announced that the entire company was restricted indefinitely: "Last night some men from this company were on detail to deliver two cases of whiskey to the Officers' Club. It was found that two bottles were missing. The entire company is restricted until those bottles are returned or until the two men give themselves up."

This led to howls of protest from the ranks.

"Ah, shit! Why the whole company? Why don't you check the driver?"

"Hell, yes! You keep a duty roster, don't you?"

"Who was the driver? What the hell've we got to do with it?"

But the captain was adamant. The entire company was restricted until the two bottles were returned to the orderly room. Since everyone had anticipated a holiday pass, this was a severe shock. Only the KPs had expected to remain in camp.

At first the men joked among themselves. They pointed accusing fingers at each other and demanded, "O.K., get them bottles back there so we can get the fuck out of here!" But several had dates. The humor was soon replaced by resentment, and within an hour the prevailing mood became ugly. Why should the entire company suffer for the deeds of a few men whose names should be known? Nisei cadremen were questioned bitterly and threatened, but they insisted that they knew nothing about the matter. Since passes were not necessary to get out of the fort, many simply went AWOL. Late in the morning some men began running about the huts yelling excitedly that the restriction had been lifted. There was a rush to the orderly room for passes, and the line that formed was well over fifty yards long. As they stood impatiently in the hot sun, several rumors formed among them: the two men had confessed; they had to confess because they had consumed the whiskey; they had broken the bottles and could not even take back the empties.

"Who was it anyway?"

"You got me."

"So the guys confessed, huh? I guess it's pretty rough bein' responsible for the whole gang gettin' restricted."

"Yeah. That's a dirty way of gettin' guys. If them bastards in the orderly room were on the ball, they'd a known who it was to begin with, and we wouldn't be fuckin' around here like this."

No official announcement was made concerning the lifting of the restriction; the captain was not even in the company area. The rumors about the men having consumed the whiskey and about the broken bottles were never confirmed; no one was court martialed. It was simply taken for granted that this was what happened, and the incident was soon forgotten.

During the first few weeks the newcomers were constantly hounded for not saluting officers and for other violations. One frequently heard comment was that "Fort Meade spoiled us." Although the general atmosphere was one of suspicion and resentment, several admitted privately that their fate could have been worse. The physical facilities were the most comfortable they had seen. To be sure, the huts on the Turkey Farm were substandard, and some two-story wooden barracks were used in the induction and separation center area. But the School Battalion, except for two companies, was housed in large brick buildings that were steam heated. These buildings had adequate laundry facilities, soft drink and candy-vending machines, recreation rooms, and many other conveniences. On cold days it would not be necessary to leave these buildings at all. The mess halls were clean and efficiently run. Nisei cooks worked conscientiously, serving rice and soy sauce in addition to army rations. The food was well prepared and tasty in comparison to what they had had. Amidst bitter complaints about "chicken shit," one observant individual commented, "All these guys around here say they're pissed off, but I know damn well that a lot of 'em are glad to be here."

Taking the Classification Examinations

On the day of their arrival the group was addressed by Sergeant Nishida. As soon as he mounted the platform, the men recognized him and began to yell: "Yeah, Shanghai!" "Shanghai Joe!" "There's the guy!" In his orientation talk the sergeant emphasized several points: (1) there would be a six-month course for men who were already proficient in the Japanese language, a nine-month course for those who did not know as much Japanese, and a shorter "conversational" course of unspecified length for those who were so "hopeless" that they could not qualify for the "regular" school. (2) There would be no "washouts" from the school. There was nowhere else for Nisei to be sent. They could not be used as infantrymen in the Pacific, and the last contingent of

Nisei occupation troops had already been shipped to Europe. "Whether you like it or not, you're going to stay here, and you're going to graduate." (3) With *two exceptions* all the men who had said definitely at Fort Meade that they did *not* want to come to Fort Snelling were *not* included in the transfer order. (4) In the near future a classification examination would be held, and men would be assigned to different companies on the basis of their linguistic ability. It was clearly understood, then, that the fate of each soldier would rest on his performance in the classification examination.

Although almost everyone—including some who had lived in Japan for almost ten years—insisted that he did not know any Japanese, only a handful were actually ignorant of the language. A few had come from areas in which only a few Japanese lived—Michigan, New Jersey, Texas. They could not even pronounce simple Japanese words without a heavy accent. In addition, there were some from the Pacific Coast who had grown up apart from other Japanese; they barely knew enough to converse with their parents. These men faced no problem; they could not pass the examination no matter how hard they tried. But all the others had to make a decision. They could fail purposely, or they could study hard and pass with high marks. Which alternative would be more advantageous? Each had to make up his mind in an atmosphere of confusion. Some were placed under considerable social pressure.

Much of the pressure came from outside the company—from Nisei living in the Twin Cities area and from those already in school. The newcomers soon realized that the status ladder in MISLS was quite different from that in the infantry; in fact, the two scales contrasted rather sharply with each other. In the infantry strength, agility, and stamina had been the basis for prestige and self-respect. After a few weeks at Fort Snelling, however, it became clear that one's status here rested more on mastery of the Japanese language. In his talk the sergeant major had labeled those unable to qualify for the "regular" school as "hopeless" and "sad cases"; friends in other companies as well as cadremen shared these standards. This elicited some resentment, for it gave Kibei a definite advantage over Nisei. Men whose national loyalty some Nisei had reason to question enjoyed high rank. In the infantry openness was prized; in garrison duty men appeared petty. Furthermore, there was a different orientation toward rank. Infantrymen were ambivalent about promotions; NCOs were envied for their stripes, but each promotion brought additional responsibilities and greater exposure to danger. Many riflemen with natural ability deliberately held back, content to remain privates. But among Nisei civilians, rank was apparently of critical importance; many declared openly that each soldier should work hard for stripes—the more the better. Sissies and bookworms also found favorable opportunities. They had suffered in the infantry, but they now had a better chance in the scramble for ascendancy. Some previously respected men felt intuitively that they were moving downward; their ability as riflemen counted for little. Almost everyone connected with MISLS seemed to feel that *all* Nisei should master the Japanese language and that there would be a promising future for them in Asia after the war if they did so. Friends outside the company all urged the men to do well to "make the grade."

Expectations were built largely on rumors. Sergeant Nishida had mentioned a new course, to be offered on an experimental basis, for those who failed the examination. But he had been vague and indefinite. After a few days it was designated the "*kaiwa* [conversational] course" or the "three-month course," although he had not specified any particular period of study. What was in store for graduates of this new course? Estimates were uniformly negative. According to one rumor the course was for front-line duty; for that reason seventeen weeks of infantry training were also required. Returned veterans had testified that linguists rarely got closer to the battlefield than regimental headquarters. But the need for interpreters on the front lines was easy to understand; prisoners interviewed on the spot could provide valuable information. Other rumors were that the special course was for MPs or for those who were to guard Japanese POWs. Would graduates of the new course be promoted on the same basis as others? According to rumor, graduates would receive *no* ratings. Another rumor was about the fate of the failures while they were still at Fort Snelling; students in the "*kaiwa* course" would attend school from the Turkey Farm. Still another rumor was that those assigned to the nine-month course would study civil affairs; this meant that they would remain at the fort (and out of the battle zone) for a longer time to prepare for occupation duty. Individually and in small groups the men pondered these questions. They asked the advice of friends in other companies, and instructors and school officials were flooded with questions. The general effect of their replies and the rumors was that failing the examination could lead to unpleasantness and even danger.

But there were cross pressures. In the countless discussions that went on most men talked as if they were disappointed infantrymen, and the overall impression created was that only a few would cooperate with the MISLS administration. A few stated candidly that they intended to do their best in the examination. Some declared that everyone should try to do well but that they themselves did not intend to study. The vast majority, however, consistently maintained that they knew no Japanese or that they would fail deliberately so that the authorities would never find out what Japanese they knew. Most of the men had attended Japanese language schools; when reminded of this, they insisted that they had forgotten almost everything. "The only reason why I went to Japanese school was because my folks said so. I didn't want to go. I didn't study. If I learned anything, it was just an accident." The sergeant's contention that only two nonvolunteers had been included in the order evoked considerable comment. Several stated emphatically that it was a lie; they insisted that they knew more than one other person who had not volunteered. A few confessed that they had volunteered prior to induction or during basic training. "Sure I volunteered, but that was a long time ago. That's different. I volunteered from camp. But I didn't want to come here from Meade. I got pissed off when they rejected me before. Why should I come here when they didn't want me before? When I wanted to get in, they wouldn't take me. So the hell with them!" The prevailing stance was one of defiance. Most men insisted that they had been brought to Fort Snelling against their will; therefore, they would refuse to study. As one Hawaiian youth announced, "Maybe dey can bring me here, but dey no can make me study, yeah? Me, I get eight months training already. Infantryman! Yeah? I t'eenk we should all fuck around. Den, dey no can do naut'eeng." All disclaimed any responsibility for having been sent to Fort Snelling, and they reiterated their resentment. The publicly avowed values were still those of the infantry. Many meant exactly what they said; others apparently did not.

Many actually had no objection to learning Japanese, but they confessed this only to their closest friends. Even then, they were quick to give excuses: "I wanna take this conversational course. I don't give a shit if I never learn how to read or write, but I want to be able to talk to my folks anyway. Christ! I'll bet they'll fall over if I go home and talk Booch to 'em. I'd like to learn a little. It's embarrassing, you know, when you can't say nothin'." The only person who stated openly that he was eager to study the language was a youngster from Michigan who did not know a single word. The others tested him by calling him a *baka-tare*, a colloquial expression denoting stupidity; when he continued to smile, they concluded that he was telling the truth. Since he knew absolutely nothing, he was not condemned for wanting to learn.

Those who had lived almost exclusively in Nisei communities felt the outside pressure more than the others. Although they were reluctant to say so, they were very concerned over whether they would be able to "make the grade" and qualify for the "regular" school. They feared being left behind, for not being able to keep up was regarded as a personal disgrace. Some of these men wanted to pass the examination in order to escape the group. All their friends elsewhere viewed its conduct with reproach, and they were inclined to agree. Furthermore, discipline at Fort Snelling was not as lax as it had been at the replacement depot, and they felt that it might turn out to be costly to remain with it. There were too many ruffians who might get out of hand, and the entire group might be punished. But even some of these men felt ambivalent. They did not care to study as hard as their predecessors. They had heard so much of studying with flashlights under blankets, of sneaking into latrines to memorize *kanji*, and of widespread cheating in desperate efforts to get passing grades. This was not appealing. Was disgrace preferable to this?

When it was announced that the examination would be held on Monday, July 16, those who wanted to study became desperate. Some borrowed textbooks from friends in other companies and tried to study surreptitiously in the huts. Whenever they were caught, they explained that they had not wanted to come but now that they were here, they might as well make the best of it. Some of them took extraordinary measures to ensure sufficient privacy. On the weekend of July 14 and 15 a number stayed in hotels or in the homes of friends so that they could study without being taunted. Some remained with friends in other companies, although this was more risky. A few, who had nowhere to go, began to study openly in their huts. They came in for considerable harassment and had to stop periodically to justify what they were doing.

On the eve of the examination there was still much confusion and indecision. No consensus had developed on the wisest course of action. Those who failed would presumably go overseas sooner, where they could accumulate points for discharge at a faster rate. Furthermore, there was a possibility that civil affairs specialists might be "frozen" after the war. Suppose selective service were suspended at the end of the war? But would it be wise to go into an active combat zone? Many wanted most of all to stay with their friends and decided to do whatever their friends did. They had been together for a long time—sharing many hardships, covering for each other, giving one another a hand in tight situations. They did not want to break up now. Although everyone continued to insist that he would fail, a few observant individuals became suspicious that at least some of their comrades intended to do their best and feared that they themselves would be left behind. On Sunday evening many had not yet reached a definite decision. Those who had made up their minds kept their own counsel or confided only to their closest friends.

There were no fatigue details on Monday morning, July 16. "Academic necessity" had priority over almost everything else at Fort Snelling. During the reveille formation everyone was ordered to fall out at 09:00 in Class A uniform.

The examination took place in an atmosphere of bedlam. Soon after 09:00 the men sauntered into the various classrooms and talked noisily. There was much yelling and horseplay, and in most rooms the monitors had considerable difficulty restoring sufficient order to give instructions. They were shocked. Many of the teachers had been educated in Japan, and they expected the same deference traditionally accorded teachers there; they were not accustomed to such disobedience. Clashes took place before the papers were even handed out. The following exchange took place in one of the rooms:

"Will you men quiet down so I can tell you what you're supposed to do in this examination? How do you expect to pass if you don't know?"

"Ah, waste time! We're gonna flunk anyway. What's the difference? We're all gonna hand in blank sheets of paper."

"That's not the way to look at it. You have the wrong spirit."

"Look, we're not volunteers like you guys. We didn't ask to be sent up here. We were Shanghaied! I don't give a shit! I don't know any Booch from nothin'. If you told me to write my name in Japanese, I wouldn't know what to do. Just hand me the paper, and I'll write my name, rank, and serial number on it."

When threatened with discipline, the men laughed at the befuddled sergeants; those who persisted in trying to enforce order were threatened with violence. Finally, the instructors handed out the mimeographed forms and asked the men to do what they could. Immediately someone complained that he did not have a pencil, and pencils had to be provided for everyone.

Minutes after the examination started someone yelled, "Hey, Sarge, can we take off as soon as we finish the exam? Waste time sittin' in here." Soon thereafter men began to turn in their papers and walk out of the rooms. Instructors who pointed out that it was a two-hour examination were accused of being "chicken shit" and were reminded again that this was not a volunteer group. Within thirty minutes well over half the company was either drinking Coca Cola at the PX or standing around on the driveway outside the school building.

Almost everyone talked as though the whole affair were a joke. Most of them, however, wrote something; they had to do something to "kill the time." Inasmuch as the first section was a test on English vocabulary, almost everyone completed it. A few worked on the second section on Japanese vocabulary, answering the questions in Romanized characters. Only a handful worked on the third section, which required the translation of a Japanese passage of considerable difficulty. Some actually did what they said they would do. On their papers they wrote jokes and drew lewd cartoons. Some simply wrote their name, rank, and serial number; after a respectable amount of time elapsed, they departed. Throughout this period pandemonium persisted in all the classrooms; there was loud yelling and joking.

Those who were desperate to pass were confronted with a painful dilemma. Since they were allowed to sit in any room that they chose, most men found themselves surrounded by friends and acquaintances to whom they had been boasting for weeks of their ignorance of Japanese and of handing in a blank sheet of paper. The monitors were unable to prevent them from talking or from looking at one another's papers. Whenever an instructor cautioned them against talking, he received the stock reply, "Don't worry. We ain't gonna cheat. We don't wanna pass anyway." Only a few had the courage to try their best in the midst of open taunts from their comrades. Many waited until their friends had left the room before filling in their papers.

Those who were outside competed with one another in boasting of how they had "goofed off." Exaggerated claims of how poorly each had done were exchanged.[1]

"Hey, Johnny. How come you stay dere so long? You no like us guys, huh?"

"Hell, I had to stay there long enough to make it look respectable. I only did the first part. Hell, that was in English. I couldn't say I didn't know English."

"No shit! Was that first part English? I didn't even look at it. I just put down my name, rank, and serial number and walked out."

1. One of the instructors who graded the papers reported that they were unquestionably the worst that he had ever seen. However, he insisted that there were not as many lewd pictures and charges against MISLS as the men claimed.

"You should've seen old Slick. Jesus Christ! He sat there drawing a cartoon of a pussy. He might get into trouble."

"Aw, hell, what could they do to him? A lot of guys were drawin' pictures. What else could they do?"

"Well, as long as they ain't dirty, it's O.K."

"Hey, Joe. What were you doing in there so long?"

"I was waitin' for Wacky. I didn't know what the fuck was takin' him so long. I thought he was tryin' to fuck us up and really doin' the stuff. What the hell! He was writin' something about all the chicken shit in this place in English."

Some accused others of doing too well, although most of the loud, open indictments were addressed to those who had actually walked out early. The resentment against the minority who had really tried hard was silent.

By the end of the first hour all but a handful were outside, and cries arose with increasing stridency for the rest to hurry: "Hubba hubba!" "Let's go! Let's go!" The men were not anxious to get back to the Turkey Farm, where they would probably be assigned to some work detail, but resentment was mounting against those who were still trying to answer as many questions as they could. Only a few exceptions were made. Yoshihara, for example, was a Kibei with an extensive knowledge of Japanese. He had never denied this. Furthermore, he was 35 years old, married, and had two children. Infantry service had been very hard for him; those who had been with him at Camp Wheeler knew that he had lost consciousness in more than one strenuous exercise. He was quiet and unassuming and had always done his share of the work. No one blamed him for trying his best. But the others were seen as traitors. Several friendships were terminated abruptly, when those who had claimed they knew nothing were seen writing in Japanese.

A week of waiting and speculation followed, and a number of rumors emerged: that students in the three-month course would have to take more physical training than the others, that three Nisei linguists had been killed in Okinawa, that the three-month group would live in barracks in the separation center rather than the Turkey Farm, that all married men would automatically be assigned to the nine-month course, that the classrooms as well as the barracks of the three-month group would be on the Turkey Farm. Apprehension developed that those who failed

---

He did confirm that a large number filled out only the English section, leaving the rest blank.

the examination would not only get the worst facilities but would also go into combat. Some who had failed deliberately began to wonder if they had made a foolish mistake.

One rumor that elicited special interest was that graduates of the conversational course would be assigned to the 1st Army, currently being redeployed from the European to the Pacific theater. Its personnel was on furlough, and the unit was expected to leave about November. The rumor was that fifty interpreters would be assigned to each of its ten divisions—500 Nisei would be needed. The 1st Army was a veteran unit, and the prevailing belief was that it would not be assigned to occupation duty.

"I wouldn't mind. That 1st Army really did a good job in Europe. I wouldn't mind being in the 1st Army."

"Yeah. You'll have something to tell your grandchildren about."

"Shit, we ought'a get something good out of this deal. We got fucked. We got fucked all around. It's a good thing we're only gonna be around here for three more months."

"No sheet, boy! Dees place gonna be cold like hell, no?"

"You ain't shittin' there, boy! Look at them fuckin' walls. I can feel the damn wind blowin' through right now. God damn! I pity the guys who're gonna be here when it's twenty below."

"Twenty below! Whee! Me, I no like dat kind."

Another rumor was that the Nisei with the 1st Army would become an integral part of it, that they would train with the veterans and would come home with them.

"Boy, that's gonna be rough. It's bad enough going through all that infantry training again, but imagine going through it with all the veterans."

"You'll learn something though. They're all combat veterans. They'll teach you the real stuff. It'll be rough, but it'll be worth it. If I'm going into combat, I want'a go in with a veteran outfit."

"I figure it'll be worth it if we can come home with them. All them guys've got high points, and they'll come home as soon as it's over."

"Naw, dey no let us come weeth dem. Who say?"

"We go fight with them, and then we stay. Yeah?"

"Can't tell, though. They might let the whole bunch come home. They been fightin' a long time. They got a rest comin' to 'em."

"But what we get? Battle of da B-ranges! *Banzai* charge on A.P. Hill! Dat's all!"

"Yeah. But dey should geeve us ribbons for dat, no? Goddam! I sweat my balls off out dere. No sheet! Me, I sweat. I work hard up dere, and what I get? Naut'eeng!"

Two men from the Turkey Farm had been assigned as typists at the personnel office. When cutting stencils ordering personnel to various student companies, they looked for the names of their friends and quickly notified them of their fate. From this source further rumors emerged, placing various individuals into different units. On Monday morning, July 23, long lists appeared on the bulletin board announcing assignments to various student companies, along with orders to move on Wednesday. The largest number, 284, were assigned to a new company that was being activated—Company K. The second largest number was ordered to Company H, and the rest were assigned in smaller numbers to other companies in the School Battalion. Since the longest list contained the names of everyone who knew little or no Japanese, it was assumed that Company K would be for the "*kaiwa* course." Soon after the posting a rumor developed that the three-month men would not receive furloughs after they graduated; this was related with glee by those who had been assigned to the "regular" school. Thus, all failures were to be accumulated in a special company, where the worst was to be expected. Furthermore, it appeared that those in the conversational course would be sent subsequently to more dangerous assignments. Although there were exceptions, on the whole this group contained the better infantrymen and the poorer linguists, including most of the NCOs.

Word of what had happened at the examination spread quickly throughout the Nisei community. With the posting of assignments opinion crystallized: Company K was an undesirable place for leftovers—soldiers who were otherwise useless. Those who had passed the examination had difficulty in concealing their delight. There was a marked tendency on their part to look down on the others. Although some still spoke politely about the "conversational course," the "*kaiwa* course," or the "three-month course," others referred to it bluntly as the "dummy course." The negative evaluation was also clearly implied even in the remarks of those who were not outwardly insulting. Those who were uncertain of one another's assignments sometimes asked, "Did you *make* the nine-month course?" Some expressed condolences and assured the failures that things would probably not be so bad. Others consoled their friends with remarks such as, "Well, it'll be over with fast anyway," or "You can get out of this damn place in three months, anyway." When asked by others where he had been assigned, a Company K man replied, "Oh, I'm with the rest of the fuck-ups." By the evening of July 24 some of the men going to the "regular" school were referring contemptuously to those going to Company K as "turkey farmers."

Many who were moving to Company K were not especially upset about the assignment itself. But they resented the overnight change of attitude on the part of some of their erstwhile "friends." They felt betrayed, for it now became obvious that many of their comrades had not been honest with them. The haughtiness of some who had passed provoked indignation, and those who expressed sympathy with condescension were especially resented. Since most men were aware of how much Japanese various individuals actually knew, they realized that many who were competent had failed deliberately. On the other hand, there were others who must have studied very hard, for they knew barely enough to get by. Some who had failed on purpose to stay with a friend now discovered that the latter had written an essay in Japanese. Although linguistic competence was not irrelevant, most men had actually been classified by the extent to which they had complied with informal norms to be nonchalant about the test. Although few serious accusations of being "double-crossed" were expressed openly, many were deeply hurt and disillusioned. The more assimilated men did not care as much; they did not know enough to pass anyway and did not feel themselves disgraced. But those who were more involved in the Nisei social world were embittered. Men who had passed the examination and had cheerfully deserted them were labeled "stab-in-the-back Japs."

"Them fuckers piss me off. All the time they were making out that they didn't want to come, and they volunteered. They made out they didn't know any Japanese, and all the time they were studying. It was just a front. Now that they're in the regular school, they think their shit don't stink. I remember when we were down at Camp Wheeler taking basic. They were crying because they got turned down when they applied to come here. They were too yellow to go fight in Europe so they wanted to come here. Then they said they didn't know anything, and now they razz us because we got stuck in this *kaiwa* course. I didn't know Buddhaheads could be so low."

"Aw, I found out when I was back in camp. Back home I used to think all the Boochies were pretty nice, but in camp I found out they were just like everybody else."

"Some of these guys are really dirty though. When we were gettin' interviewed at Meade, Yasui and them guys were all saying that they didn't want to come here. That's why Nakanishi and Hoshino and them guys told the interviewers that they didn't want to come here. Hell,

Hoshino knows plenty of Booch, and he was accepted here once already. But he thought the other guys were going to Europe so he told them he wanted to go to Europe too. Look what happened! Those two guys went to Europe, and the rest of their gang came here."

"Yeah? No shit! Is that what happened? I'll be fucked. That's a Jap for you! Stab-in-the-back kind! They're just like Tojo."

"You can't trust them guys."

The Activation of Company K

Company K, School Battalion, was activated on 25 July 1945 with 258 student enlisted men, three cadre NCOs, and one officer. When the men learned that the company commander was to be First Lieutenant Joseph Catalano, who had been the executive officer of E Company, groans of despair arose. The NCOs, who had had more contacts with him, insisted that he was "not a bad Joe," but the privates contended that he was "chicken shit." Had he not enforced all the petty regulations on the Turkey Farm? When they learned that Staff Sergeant Endo, a Texan who growled orders with a heavy drawl, was to be acting first sergeant, their unhappiness was compounded.

"Them two bastards ought to get along pretty good. That Endo is the worst Boochie noncom I ever saw. He's a bastard! He thinks his shit don't stink! And that fuckin' Catalano! Jesus, what a chicken shit bastard! I seen him stop a guy for not salutin' him. He chewed his ass for five minutes and then made him clean the can. He acts tough as hell too. Two of a kind, and we gotta have both of 'em for three months!"

It was bad enough to be in the leftover company without having a petty commanding officer and a first sergeant who could not get along with Nisei. The other two cadremen were accepted; the company clerk was known to the men only as Ito, and the supply sergeant was called George.

Additional men were subsequently assigned to the company, eventually bringing its student personnel to 293—58 Hawaiians, 233 mainlanders, and 2 Caucasians. This included 31 who had been in A-9-3. (See Table 3.) Unlike most student companies a number of NCOs were included among the students, and they were given special responsibilities. The student personnel was divided into six platoons of four squads each. Each squad of 12 men was assigned a squad leader, and each platoon had both a platoon sergeant and a platoon leader. Since there were not enough NCOs, 22 privates were selected to serve as squad leaders. All privates other than the acting NCOs were assigned to platoons in alpha-

Table 3

Student Personnel of Company K
(October, 1945)

	Place of Induction				*Total in*
Rank	*Hawaii*	*Chicago Area*	*Elsewhere*	*From A-9-3*	*Grade*
T/Sgt	1		3		4
S/Sgt	1		3	1	4
Sgt			2		2
T/4			1	1	1
Cpl		1		1	1
T/5		1	1	1	2
Pfc	56	73	150	27	279
Totals	58	75	160	31	293

Source: Company Roster on October 1, 1945.

betical order. Thus, each man had a definite place in a tight organization, where he was known and could be identified easily. Duties would be more difficult to shirk.

On learning that Company K was to be billeted in the separation center area, the men were initially pleased. While the white two-story barracks were not quite as comfortable as the brick buildings of the "regular" school, they did have indoor latrines. The area was fairly close to the "dummy" street-car line; and a branch PX, a barber shop, and a telephone exchange were nearby. It was certainly better than the Turkey Farm. On arrival, however, they discovered that they had not been assigned to the clean white buildings; those were occupied by Company A—the Caucasian OCS men. Company K had been allotted a row of seven dilapidated, tar-papered barracks—the kind in which the mainlanders had lived in the relocation centers. There was little insulation against the cold; only one piece of wood and a layer of tar paper separated the occupants from the frigid Minnesota winter. In addition to the seven barracks there were three smaller buildings of the same type. One was used as the dayroom; the second was the latrine; and the other housed the orderly room and the supplies. The latrine door did not quite close, and several wondered aloud what it would be like taking showers in freezing weather. The first of the seven barracks was used to billet the cadre and the six platoon leaders; each of the other six housed a platoon. A public address system connected each building to the orderly room. A road separated this complex from the white barracks. A small latrine was located in the A Company area, set apart from the other buildings. Since it was seldom used by the OCS men, the newcomers decided that they might as well avail themselves of it.

Everyone was present on the morning of the move, and it was accomplished with a minimum of confusion. The efficiency was attributed to the planning of Nisei NCOs. The men moved to their new quarters by platoons. Barracks bags were taken by trucks and left near the building in which their owners were to sleep. Since the barracks had not been occupied for some time, they contained no footlockers. The men were therefore ordered to carry the footlockers they had been using in Company E. But the barracks were so dirty that they refused to take their belongings inside. They were empty but for dust laden double-decker beds and filthy mattresses.

"What! We gonna live in them things? Jesus Christ! I'm glad we're only gonna be around for three months. I hate to live in them things in the winter. We'll freeze our ass off."

"No shit! I thought we was gonna get them white barracks."

"Naw. I hear the Haoles got 'em."

"I thought they closed the separation center."

"They're almost closed. I hear they're almost done cleaning up now. Them barracks ain't for guys gettin' discharges though. I hear them Haole OCS guys are gonna live there."

"I'll be damned! How come they rate, and we have to live in this shit hole?"

"Well, them guys're OCS."

"Aw, no *monku*, no *monku*!"

Soon after their arrival the recently appointed platoon sergeants and squad leaders took charge. All mattresses were taken outside for airing, and the platoons took turns using the one hose available in the company. On their own initiative the men scrubbed the floors, washed the windows, and hosed down the ceilings, storm doors, porches, and steps. Since only one broom and one mop had been issued to each platoon, the units had to borrow each other's equipment and take turns working. It was not until late afternoon that the barracks were clean enough for occupancy. Most of the men were bitterly disappointed; they had not expected too much, but they did not think it would be this bad. (See Plate 4).

On 30 July 1945 Division F, the Oral Language School, was activated, and academic work began. Heretofore, emphasis at MISLS had been placed on writing and translation; but with increasing demand after

PLATE 4 *Company K and the "Regular" School*

V-E Day for interpreters in the Pacific, training had been intensified and shortened. Work in the new division was to be focused largely on conversation.

Over the weekend of July 28 and 29 a roster assigning each man to one of fifteen sections, presumably on the basis of linguistic ability, was posted on the company bulletin board. At first, questions arose concerning whether section F-1 or section F-15 was the most advanced, but most men concluded that F-1 must be the highest. It did not contain all the men reputed to know some Japanese, but sections F-14 and F-15 included most of those who knew very little or nothing. The men also learned that their classrooms were to be in the small, latrine-like brick structures between the separation center and the Turkey Farm. Each building had been partitioned into four parts, one room for a section.

This arrangement facilitated communication throughout the company. The men were assigned alphabetically to platoons; they were grouped on another basis into school sections. Thus, word of anything that happened in any platoon or in any section soon became common knowledge, for someone from each platoon was in each of the sections.

The first day of school was one of utter chaos. The company marched to the school area in the rain, and after some confusion most students were able to locate their classrooms. The doors were locked. When they.finally got inside, they discovered that there were no desks; each room contained only one or two chairs. Nor were there any instructors. The only things ready for use were the fluorescent lights on the ceiling and the blackboards.

"What the fuck is this? They make us come to school, and they tell us to hurry up so we can get overseas. And they got nothing ready."

"Jesus Christ! Why don't they clean up the fuckin' place? What do they expect us to do, sit on the floor?"

"Jesus Christ! It's cold in here. No stove! I don't see no radiator. Holy fuck! We're gonna freeze our ass off when it gets cold."

"Goddam! We always get the ass end of the deal. Boy, we been salty ever since we hit Fort Meade. I figured something like this would happen. Nothin' ready! How in the hell do they expect us to study in a dump like this? Look at them other guys. They got a nice warm building with Coke machines and everything else. They got nice desks and everything."

The teachers were almost a half hour late; they had been receiving instructions from the division chairman. They ordered the men to go to another building to get some tables and benches. Not enough seats were available, and during the morning several had to sit on the tables.

In all fifteen sections it became apparent from the beginning that discipline could not be enforced. When one of the instructors warned his students that if they did not do well, measures might be taken to make things more unpleasant, the reaction was hostile.

"Goddam! There you go already! Always threatenin'! What's the matter with this goddam place? All they do is threaten ya. No matter what you do they threaten ya. No wonder everybody's pissed off! That's why everybody wants to goof off. Lose fight!"

"What do you want to goof off for?" the sergeant retorted. "It won't do you any good. You can't buck the army. You been in long enough to know that."

"Aw, hell. We didn't ask to be sent here. We got Shanghaied."

"Not all of you though. Some of you guys volunteered."

"That was a long time ago. Hell, I volunteered from camp, and they turned me down. That's what pissed me off."

"Anyway, this place is chicken shit. All they do is threaten ya."

An instructor in another section found his students even less responsive. He wrote a simple character on the blackboard and asked, "Kumamoto, what's that?"

"You got me, teach. I don't know Booch from nothin'."

Looking at his class roll, the sergeant continued, "Honda, you answer the question."

"Aw, what for you peeck on me? How 'bout all dem other guys? Why me?"

"Listen! When I call on you, you stand at attention and recite. You're in the army."

"You like fight? Come, we go outside."

Teaching was impossible on the first morning. The instructor of section F-1 kept reiterating that it was the best in the division. Each time he said this the reaction consisted of guffaws and hoots of laughter. The sergeants were unable to control their classes. No matter what they said, no one paid attention. Anyone who tried to enforce his orders was immediately challenged to a fight. The men talked among themselves or slept.

"Hell, we might as well go over now. We ain't gonna learn anything in this fuckin' place. The main thing is you gotta know how to fight or you'll stop a bullet. If Okinawa was so tough, you can imagine what it's gonna be like when they hit the mainland. Goddam! I'll bet them fuckin' Boochies'll be out there with bamboo spears and everything else. We won't be able to trust nobody in that place."

Once school got under way, the daily routine became well established. The company was aroused between 05:50 and 06:00, and first call to reveille was at 06:10. Reveille was always held, except in inclement weather. Breakfast was served between 06:20 and 07:00. Each day everyone was expected to make his bed, sweep and mop the floor, clean the stoves, line up his clothing, and get everything else in order before leaving for classes. Inspection was held at 09:00 each morning, and violations usually led to restrictions or some other form of company punishment. The latrine and dayroom were "off limits" to all personnel from 07:20 to 07:45—to give the duty platoon of the day a chance to clean them. At 07:45 all men were in formation on the company street in the uniform of the day to march to school. The platoons were marched in formation by the acting officer of the day, usually one of the platoon leaders. Classes were held from 08:00 to 12:00 and from 13:30 to 15:30, with ten-minute breaks each hour. At 15:30 the duty platoon of the day returned to the company area for various fatigue details while the other five platoons were taken out for physical training—in most cases a road march, organized sports, close order drill, or an obstacle course. Retreat formation was held at 17:00, and supper was served between 17:00 and 18:30. On four nights a week (Monday, Tuesday, Thursday, and Friday) two hours of compulsory study began at 19:00. Instructors were present to answer questions and to take roll. At 21:00 the students were expected to clean up the classrooms—sweep and mop the floors, dust the desks, wipe the windows, wash the blackboards, and anything else deemed necessary. On Wednesday night study was optional, and passes were issued. Lights were turned off at 23:00. Sometime between 23:00 and midnight the company CQ—platoon sergeants and squad leaders in rotation—held a bedcheck, and everyone was expected to be in bed at that time.

Grumbling was especially noticeable during mealtimes, for mess hall conditions quickly became a major source of irritation. Unlike the rest of the School Battalion, Companies A and K ate at the Consolidated Mess Hall, intended primarily for the induction and separation centers. The mess sergeant and his permanent staff were all Caucasian, and the food there was much like that in other army kitchens; rice and soy sauce were not served. From the very first day complaints arose about unsanitary conditions and not getting enough to eat. Nor were the complaints limited to the Nisei; OCS men were just as critical. The Nisei, however, tended to explain the unsatisfactory conditions on ethnic grounds: white people were not as neat and efficient. Failure to maintain minimum standards of sanitation was a serious matter. The washing of silverware was often so inadequate that food particles were still left on them, and the dishes were frequently greasy. Furthermore, the food was not palatable. Many Nisei had been guests of friends in other companies and had served as KPs in the other mess halls. Hence, they made invidious comparisons.

Nor were the complaints unfounded. From the first week a number of students were inconvenienced by mild cases of diarrhea, and a few suffered severe stomach cramps. The "GI shits" soon became a serious problem at school. Diarrhea could not be dismissed lightly where men were involved in a tightly organized schedule in which they were not always free to run to the toilet. Of course, it did provide a convenient excuse for "goofing off" from classes. No instructor could refuse to allow a man with diarrhea to relieve himself, and from the second day of school the latrines were full of F Division students who stood about smoking and whiling away their time. But many of them were actually ill.

Complaints about hunger were so widespread and persistent that Lieutenant Catalano finally intervened personally. One evening he remained in the company area after retreat and went over to the mess hall with his troops. After speaking to the mess sergeant, he stood behind the KPs serving food and objected whenever he felt that a man did not get his full share. He remained there until the last Company K man had gone through the line. He then walked up and down the aisle, asking his men whether they had gotten enough to eat. Those who were still hungry were told to get up for seconds, and the lieutenant went with them to be sure that they got as much as they wanted.

A few days later the complaints resumed. At lunch on August 24 the men were pleasantly surprised. As they entered the mess hall, they saw the tall figure of Lieutenant Colonel Murphy, commander of the School Battalion, bawling out the mess sergeant, who stood rigidly at attention. As the colonel gave vent to his wrath, they chuckled in delight. By the time they reached their tables they learned that Lieutenant Catalano had objected so frequently and vociferously about his men getting diarrhea that the colonel had come down for a personal inspection. As the colonel walked up and down the aisle asking the men what they thought of their "chow," most of them replied, "Lousy, sir."

That evening they found out what had happened from the KPs,

some fresh recruits from the induction center. The colonel was so infuriated by the unsanitary conditions he had found that he ordered every piece of silverware, every plate, and every cup in the mess hall—including those that were never used—to be scrubbed in soap and boiling water. Furthermore, he remained in the mess hall all afternoon to see that his orders were carried out. The KPs reported that the colonel periodically examined the piles that had already been scrubbed; whenever he found a single speck on one dish, he ordered the entire pile to be done over again. Although most men expressed sympathy for the KPs, since it was not their fault, they were delighted. "That's tough shit. They pulled KP on the wrong day."

A second major source of irritation was the weekly parade. Every Thursday a dress parade was substituted for physical training; every company in the School Battalion had to pass in review in Class A uniforms. Most of the students detested the parades as more Fort Snelling "chicken shit." Since the performance of the battalion often left much to be desired, the companies were periodically ordered to practice parading on Tuesdays. This aroused even more resentment. The parades were disliked for several reasons. One was that everyone had to change into Class A uniforms; since August and September afternoons were often hot and humid, this created laundry problems. In addition, although the other companies were closer to the parade grounds, Company K had to march almost a mile to get there. Thus, on most Thursdays the men were not able to get back to their bunks until just before supper; on all other days they were free from 16:30 until retreat formation. Thus, they felt that part of the parade and the march back to the company area were incursions on their "own time." When parades were held in honor of combat veterans who were being decorated, they did not mind as much. For example, when a Silver Star was awarded posthumously to the widow of a man who had been killed in action, no one complained. Since so many parades were held, however, there were not enough people to honor. Hence, various officers at the fort took turns occupying the reviewing stand. This irked the men most of all. They resented having to go to so much trouble to pay homage to those whom they held in such low esteem.

Although the aversion to parades was almost unanimous, it was the one exercise that brought Company K men some measure of pride and drew them closer to their commanding officer. Since they had spent so much time in the infantry, marching in formation was relatively easy. Many of the students in other companies, however, had completed only six weeks of basic training and frequently appeared awkward. On August 23 a parade was held to honor seventeen heroes from the Minnesota area who were being decorated posthumously for gallantry in action. Before it got under way the other companies were drilled by their respective commanding officers, who were vying with one another for the honor of having the best marching company in the battalion. Much to their surprise Lieutenant Catalano marched his troops to a shaded area and allowed them to rest. He commented loudly, "Those guys need the practice. When you men march, you look like the Regular Army." Although the men laughed, they were obviously pleased. As the other companies marched by and made errors, they called each other's attention to the clumsiness of the linguists. They were supposed to be soldiers, but they did not even know how to march—and they were not even carrying weapons. September 6 was an especially hot day. All the other companies had to drill for an hour before the parade started, but Lieutenant Catalano again marched his company to a shady spot and allowed his men to rest.

Thus, the two sources of irritation led to a redefinition of two Caucasian officers—Lieutenant Catalano and Colonel Murphy—as decent men who were doing their best in an otherwise dismal situation. Contrary to initial expectations, the lieutenant was definitely not "chicken shit." He was sympathetic; he fought hard to get things for his men; and he was a "square shooter." He always told the group in advance what punishment he would inflict for various infractions, and he did what he said he would do. Even in the meting out of punishment he was fair and impartial. The men were also surprised that he had made a special effort to learn the name of each individual; after a few weeks, when he said that he knew every man in the company by appearance if not by name, they did not doubt him. Those who had never been in the orderly room reported encountering the lieutenant in Minneapolis and being recognized. The lieutenant also understood the expectations the Nisei had of their own NCOs, especially on matters of informing on a fellow Nisei, and he respected them. He did not ask anyone to violate any Nisei code. In a short while the men developed a strong sense of identification and loyalty toward their commanding officer.

One incident that endeared the CO to his men began during the final week in August. On Tuesday night several members of the 6th Platoon, who were singing after 23:00, neglected to turn out the lights. Colonel Murphy, who was touring the post, caught the offenders. Some Hawaiians had left the lights on, and the mainlanders had hesitated to ask them to turn them off; but the entire platoon was guilty and anticipated some kind of punishment. In a meeting on the following day the lieutenant announced, "As far as the 6th Platoon is concerned, you're restricted to the post over the weekend. I'm going to give you men the freedom of the post because I know that it was only a couple of men who are guilty. The colonel recommended that each man be required to sign in every hour; so you fellows just keep your mouths shut about this. You'll get *my* tail in a sling. Now you fellows behave because the pressure is on the whole company." After the lieutenant's departure the first sergeant explained that there was nothing more the CO could do about it, for the offense had been detected personally by the colonel. Everyone understood this, and resentment in the 6th Platoon turned against the Hawaiians.

At pay formation on the following day the lieutenant announced that Monday was Labor Day and that the entire company would be issued Class A passes from Saturday noon until midnight Monday night. When the rejoicing had calmed down, he added that he had lifted temporarily the restriction on the 6th Platoon. Since it would not be fair to hold the platoon on a three-day weekend, it would be restricted on the following weekend instead. This announcement drew appreciative comments even from those who were not affected. They knew that the punishment had been ordered by Colonel Murphy and that the lieutenant was jeopardizing his own position by countermanding it.

"Boy, the old man's really sticking his neck out. If the colonel finds out about it, the old man's gonna have his ass in a sling."

"That's no shit. He's really a regular guy—straight from the shoulder. He never fucks you from behind. If he's gotta do it, he does it from the front."

"You gotta hand it to that guy, boy. He's really soft hearted inside. He can get plenty tough, but he's all right."

"CO good guy, huh? Goddam!"

Later at the same meeting the lieutenant cautioned them about their conduct: "Now look, men. I can understand why you're pissed off. I would be too if I were in your shoes. But remember this. You haven't won your fight yet. The Nisei have a fairly good reputation, but we've got to fight for it. Don't let down the reputation that was made in blood by the men in the 100th and the 442nd. Those men gave their lives for the right of Nisei to live like Americans. You men are taking up where they left off. Don't let them down, men."

He also warned that "higher ups" were beginning to "crack down" on the company: "I don't have to tell you why things are getting hot around here. You know about it better than I do. You fellows haven't been doing too well in class." This evoked raucous laughter. "The other night Colonel Murphy caught one platoon with its lights on and found several men in the latrine. He took names, but some of the men gave names that weren't on our company roster. Now don't tell me that those guys came from some other company just to use our latrine! That's all right with me, men. If you can get away with it, it's O.K. with me. But don't let me catch you."

Some took the warning seriously. Even those who believed that their obligations did not extend beyond those of any other soldier did not doubt that serious reprisals could come, if they persisted in doing what they had been doing.

Before long the men of Company K began to take pride in their marching ability. Week after week they did their best and consistently won top honors. When the company guidon was brought out for the first time, several surrounded the guidon bearer to examine the colors. Once a parade got under way, most men worked conscientiously; they were irritated at the few who were careless. All horseplay stopped. They checked each other's uniforms. Men who were slightly out of step were cautioned by those behind them. Everyone marched at attention, and no one raised a hand to swat a fly or gnat. As the company passed in review, everyone snapped sharply on command to execute a smart "eyes right." Company K was the sharpest marching unit on the field, and the men knew it. They felt that this was natural; after all, they were infantrymen.

They not only took pride in their appearance but were delighted at the way in which the company commander responded. The lieutenant made no effort to hide his pleasure. He declared on one occasion that he heard Colonel Andersen say to Colonel Murphy as the company passed in review, "This is the best marching outfit." Those who had been on the right file confirmed the lieutenant's claim; they too had overheard it. They were jubilant.

"Boy! The old man was pretty happy today, huh?"

"No lie! Did you see that grin on his face?"

"Shit, he ought to be happy. This is the first time the company did anything right. Now he's got something to say the next time the colonel chews his ass."

On another occasion the lieutenant told them, "Now we have to go to a parade again today. I know how you fellows feel about it, but this is the army. I want to tell you what I think of you though. This is the best marching company in the whole damn battalion. I know, men. I was on the reviewing stand at that last practice parade, and I saw them all. When you did eyes right, you probably saw me smiling from ear to ear. Believe me, men, I was proud of you. After you went by, the colonel said to me, 'I think you have the best marching company.' So I turned to him and said, 'We're the best in everything sir.' Huh! He just laughed and said, 'Yes, you're the best fuck-ups.' "

Initial Adjustments to Academic Life

The instructors assigned to the new conversational course had been drawn from other divisions of the School Battalion, where they had become accustomed to teaching under conditions unknown in American educational institutions. The regulations concerning classroom conduct were among the items in the *Information Bulletin* that had been distributed on the first day, and by this time all the students knew that these rules were actually enforced.

The authority of the instructors was clearly defined: "Any orders issued by them will be accepted as orders of the Commandant." Regulations concerning classroom conduct were explicit: "At the last note of assembly commencing each school period, *each student will be in his seat* and ready for instruction. . . . When called upon to recite, unless given specific instructions to the contrary, the student will stand at attention until he has finished reciting and then promptly sit down. When one student is reciting all others will remain 'at ease,' refraining from talking or making other noise, and will be attentive." No smoking or eating was allowed at any time, and no unassigned reading matter could be brought into classrooms without the *written* consent of the instructor. The rule concerning cheating was similarly explicit: "Any student caught cheating in examinations or any other schoolwork will be summarily dismissed from this school." Tardiness or absences were treated as AWOL, and excuses could be granted only by the commandant, the director of academic training, or the post surgeon.

The regulations concerning studying reveal much of the expectations of the administrative officers. "Although there is no required study period Wednesday night, the primary purpose of this period is to permit voluntary study. Wednesday evenings are not intended to be general holidays. Any student reported as being unprepared on any Thursday will be ordered to compulsory study on the following Wednesday evening." "Students are not permitted to study after 23:00 any night of the week. The display of a spirit of earnestness by studying after 23:00

is gratifying, but studying beyond 23:00 results in more harm than good." "At any time that a recitation period, including examinations, ends before 'recall' is sounded, students will devote the remaining time of the period to study or review. No students are permitted to roam about the school area during class or study hours."

Although the instructors in Division F arrived expecting these regulations to be observed, they were forced to make an abrupt adjustment. Whenever tempers flared, they never won. In one section, for example, a Kibei instructor tried to persuade his students that they had little reason to complain, for they had things "easy." He then went on to describe the harsh conditions in the Japanese army and contended that infantry training in the U.S. Army was "sissy stuff" compared to what the Japanese had. This evoked loud protests. Several accused him of being pro-Japan, and one man demanded to know why he did not return to Japan if he liked it so much. The men continued to abuse him until the sergeant finally started to cry as he apologized to the class. When they saw tears in his eyes, they stopped. The period ended before anything more could be said.

Like the cadre in Company E the instructors were essentially trapped in a hopeless situation. If they accepted a challenge to fight, they might be beaten. But win or lose, they could not report their assailant. Assaulting an NCO was a serious court martial offense, and being responsible for the incarceration of a fellow Nisei could lead to unpleasant repercussions in the Nisei community.

The major problem confronting Division F personnel, teachers and students alike, therefore, was how to pass the time for four months without getting into serious trouble. Except for infrequent occasions in some of the higher sections, sustained instruction was out of the question. Somehow the men had to be kept occupied to prevent mischief that could lead to disastrous consequences. Enforcing the rules of the School Battalion was out of the question, but there was always the danger of being caught by an inspecting officer.

During the first few weeks the instructors and students worked out a pattern of accommodation, a day-to-day modus vivendi that enabled them to get by. The men smoked in classrooms, engaged in horseplay, drew lewd pictures on the blackboard, and whistled or sang when the spirit moved them. Since there was nothing else to do, the instructors carried on informal chats with their charges on a variety of topics. Many of them disclosed their own unhappiness and commiserated with the men. They agreed that MISLS was more "chicken shit" than the rest of the army, and several aired their personal grievances. They indicated that most of the teaching staff at the fort was demoralized and that they considered their assignment to Division F as punishment they did not deserve. They frequently left the classrooms unattended for long periods, thus giving the men an opportunity to give vent to their frustrations without becoming targets of their hostility. Whenever some high-ranking official was scheduled to visit, the instructors warned the students of it. Lookouts were posted, and as soon as the dignitary arrived, word was passed from room to room. Before long most men conceded that their teachers were for the most part "nice guys" and sympathized with their plight.

Although the teachers were constantly challenged and baited, they were under orders to maintain a heavy work schedule. As classroom discussions turned from time to time to the Japanese language, it became obvious that most students already knew considerably more than even their friends had suspected. Many could already read and write, and the few who actually knew very little found it impossible to keep up. When an instructor started teaching *hiragana*, one of the alphabets, someone yelled, "Waste time!" There was only one man in section F-1 who did not know *hiragana*. When the sergeant wrote some *kanji* on the blackboard, some objected that they were too easy; this aroused the ire of others who did not wish to learn any more than the minimum. Some held their pens in the Japanese manner and proceeded to write complex *kanji* on their tablets; others, while still protesting that they knew nothing, did likewise. A period devoted to singing popular Japanese songs elicited an unexpectedly enthusiastic response in most sections. Several unwittingly revealed extensive knowledge of the lyrics and were immediately asked where they had learned so much and what they were doing in Division F. The stock response among mainlanders was that they had learned the songs in relocation centers, but the Hawaiians could not talk themselves out of it so easily. The instructors decided to mimeograph some sheet music. When asked which songs they wanted, an almost unanimous request was for *Shina no Yoru*—which almost everyone seemed to know. Within a week most men had evaluated the linguistic capacity of those in their section. Some section F-15 students were obviously more competent than those in F-1.

The only exercise that won the undivided attention of the students was the showing of Japanese movies. Twice weekly they were taken by bus to the post theater to see a succession of commercial films. The purpose was to give them familiarity with correctly pronounced Japanese, customary usages, idiomatic expressions and gestures, and a general "feel" for Japanese culture. The first picture shown was *Die Tochter der Samurai*, a film released in 1937 to introduce Japanese culture to the German public. It stressed the traditional values of Japan and provided a striking depiction of the subordinate position of women, the poverty of some of the people, and the beauty of the terrain. The second movie, *Dan Ryu*, was about the love affairs of an enterprising young executive in a large Tokyo hospital. This showed Japan in its modern, westernized setting and led many to reconsider their beliefs about the country. Attention was focused on the beautiful actresses. Few had seen Nisei women who were so attractive. Although the movie discussion period was not scheduled until Saturday morning, the instructors found it impossible to get the men to stop talking about the pictures they had just seen.

This provided the instructors with one way of maintaining some semblance of order and taking up time. Considerable interest developed in the subordinate position of Japanese women. Asked if it were true that Japanese men simply gave commands and that women followed unquestioningly, one Kibei sergeant replied, "Men never walk behind women in Japan. When you guys get over there, don't take any shit from the women. You just tell them what you want them to do. If you treat them like American women, you'll just be sitting there, and they won't know what to do. Japanese women are trained to obey. You guys remember that, O.K.?" Instructors who had been in Japan were asked to recount their experiences in the brothels there, and some had to stretch their imaginations to take up as much time as possible. Whatever the subject matter on the schedule, instructors were continually interrupted by questions concerning boy-girl relationships in modern Japan.

Those who had seen Japanese movies regularly before the war whetted the appetites of the others: "Hell, you ain't seen nothin' yet. Wait 'til you see the one in *Shina no Yoru*! Boy, she's really nice."

"You know something? If they got babes like that over there, maybe it won't be so bad, huh?"

"I been thinkin'. Maybe we ought to learn the lingo a little bit so we can talk to some of them babes."

After a few weeks most sections were split into factions—a smaller one that made some effort to study and the others, who slept, read magazines and newspapers, or talked among themselves. Especially in the higher sections some began to display their knowledge of Japanese; a few even began showing off. As time went on the cleavage deepened. Those who were loafing labeled the others "eager beavers" and began to torment them. Instructors in the higher sections gave the few who were trying every encouragement, and some hinted that those who performed well enough might be transferred at some later date to the nine-month course. A rumor soon developed that students in the first three sections would be transferred to Company H. Interest in this rumor reflected the growing bifurcation. The eager beavers were obviously more interested; they repeated it to others and engaged in animated discussions of the possibility. Those who did not care made some remark as, "Yeah, you don't say," and then went on with their tasks.

The eager beavers were continuously harassed by those who seemed determined to learn nothing. The men who felt the brunt of the pressure were those believed to be "bucking" for a transfer. Those who knew very little Japanese and were struggling to learn enough to get by were not tormented in the same manner.

Even when some men began to study openly, the instructors were still unable to maintain control over their classes. Although it had become obvious that many students were capable of speaking fluent Japanese, most discussions were conducted in English, and instructors who were Kibei remained at a disadvantage. Since they could not keep on talking about brothels indefinitely, the sergeants left the classrooms with increasing frequency. Most students continued to amuse themselves. Some members of the more advanced sections devoted considerable effort to translating American military argot into Japanese. Some of the expressions turned out to be quite humorous and were adopted throughout the company. "Sad sack" became *kanashii fukuro*, literally a "melancholy bag." "Goldbrick" was translated into *kin renga*. When "chicken shit" was translated as *niwatori no kuso*, someone noticed that *niwatori* [chicken] literally meant "yard bird." Expressions of another kind that became popular were those that caricatured stereotyped American conceptions of the Japanese people. When inconveniencing someone, for example, a man would bow and hiss, "So solli, honolabulu mistaki." But in time increasing numbers became bored and found it difficult to remain awake. From time to time someone would fall fast asleep, which led to the refinement of techniques for giving "hot foots" and "hot seats."

Each section contained many smokers, and matches were plentiful. "Hot foots" were administered in the usual manner—forcing a match head into the inseam of a shoe and then lighting the other end. Although Army shoes were sufficiently heavy to prevent serious injury, the burns were painful enough to startle even a sound sleeper. Giving "hot seats" was more complicated. Each man sat on a steel folding chair containing airholes. If a piece of paper folded to fit into one of these holes were lighted, it made the metal seat quite hot and sometimes singed the trousers of anyone who slept too soundly. Since everyone expected to be victimized when off guard, periodically a dozing man would jump even when nothing had been done, thinking he was on fire. Those who had experienced hot seats knew that a brief warming of the seat preceded the flames. Thus, anyone who thought his seat was getting warm jumped to get out of the way.

Anyone in deep concentration or sound asleep was a potential target, but the men cooperated most enthusiastically to "get" the eager beavers. One person would stuff the paper into position while talking innocently to the unsuspecting victim. Another would place a match in the tip of his shoe. A third man would light that match with a lighter—to avoid the noise of striking a match—and the second man would move his foot under the paper. In almost every section during virtually every class hour, one would hear a scream followed by howls of raucous laughter. Those who had been labeled "eager beavers" had to sit with their backs to the wall, for others sometimes tried to burn them even when they were awake. Even then some ingenious pranksters managed to start a fire under them.

Since the hot foots and hot seats were directed primarily at eager beavers by those who never studied, tempers occasionally flared. Since the studious were men who were working desperately and sleeping more from exhaustion than boredom, they sometimes became infuriated. Occasionally, men squared off to fight and had to be pulled apart. Thus, one important service provided by the instructors was to keep the rival factions apart. Their mere presence was enough to prevent fires, and one could sleep with reasonable safety. By standing in the classroom, chatting quietly with a few men, the sergeants were able to maintain the appearance of order and to keep down the smell of fire.

As the practice became more commonplace, both the sleepy and the eager beavers became hypersensitive. Those who felt drowsy or wished to concentrate placed themselves in positions in which it would be difficult for anyone to get behind them. Many protected themselves at once on smelling smoke, even when they felt no heat. Thus, the hot seat technique had to be improved. The problem was to get the paper to burn faster—so that it would take effect before the victim could smell the smoke. Experiments were conducted on various ways of folding the paper, and general agreement arose that a conical shape with the point upward was the most effective. One morning in October a further improvement was made. It was discovered that the flames shot up faster and did not even give the victim a preliminary warning if airholes were cut into the paper. Such airholes were designated as "boosters"—a term used by ordnance men to refer to mechanisms in various explosives that facilitate the movement of a spark to the powder magazine. It was also found that the spacing of such holes made some difference, and after some experimentation an ideal pattern of spacing was developed. The booster technique was invented in the morning by a man in section F-1; by the end of the day someone in virtually every section had been victimized with it. One man attempted to increase burning speed even more by using gauze soaked with lighter fluid. This worked very efficiently, but it was immediately outlawed as too dangerous. By common consent the technique was never used again.

Despite such pranks and continued teacher baiting, common accord never developed in the company that no one should study. Indeed, most of the men had no objections to learning some Japanese and were quite willing to make a modest effort. As long as they were not pushed too hard, they were willing to be exposed to the language. But they did not wish to be identified as "eager beavers," and they kept their study time down to a respectable minimum. Nor did they resent the eager beavers as strongly as the handful who appeared determined that no one should learn anything. But most men found it difficult to speak in favor of those who were trying. Resentment over being double-crossed by those who had passed the qualification examination made it easier for those who opposed all serious effort to win the day. The eager beavers were viewed as traitors. Thus, regardless of how the various individuals felt, most students complied overtly with the company norm—to "take it easy." Most men participated from time to time in the horseplay and studied periodically. A number who had considerable linguistic ability were held back by their desire to be a "decent guy."

Those who were deeply involved in the Nisei social world found themselves facing acute pressures. Some of them wanted badly to earn stripes, but they were trapped between their personal ambition and the fear of being condemned within the company. The question of ratings had come up at an early company meeting, and Lieutenant Catalano had replied, "From what I understand you men will get ratings just like the other fellows. You're all Pfc's now. You became Pfc's when you were assigned to this company. I think you'll get at least T/5 when you finish, but I don't know about the academic side. You ask the guys at school about that." Colonel Andersen, commandant of MISLS, settled the question at a subsequent orientation meeting. He indicated that Division F graduates would be used as interpreters only, since they could neither read nor write Japanese. They would *not* be promoted until they were assigned to some unit; he promised solemnly, however, that they would become T/5 upon assignment. Many were upset. Past graduates had gone on their overseas furloughs with a rank of at least T/5, and they were ashamed to go home as privates. Some became so preoccupied with this that they found it increasingly difficult to conceal their desire to get out of Company K. It was now obvious that no one in Division F would learn enough to become a competent linguist. How could they possibly compete with graduates of the "regular" school? These men wanted desperately to get out of the "dummy outfit."

Two avenues appeared to be open to those who wanted out. One possibility arose from the abrupt announcement by the WRA that all relocation centers would be closed, creating serious problems for thousands of families. The MISLS administration, recognizing that many of its personnel would be involved, announced that special emergency furloughs would be granted any man who had to help resettle immediate relatives. Some had to apply for a leave; there was no one else to find housing and jobs for their parents. But others saw this as a way of getting out of the company. Since they would miss fifteen days of classes, there was a possibility of their being reassigned to another division. One man, confident of being transferred, declared openly, "Sure, I'm takin' a furlough. I already put in my papers. When I get it, I'm gonna go down to Chicago and have a good time. Hell, my folks don't need me. I'd just get in the way. I figure by the time I get back, I'll be too far behind so I'll go back to the Turkey Farm and start over again. Might as well start over, huh? I figure it's better that way. I was talkin' to a guy who just got back from Okinawa, and he said it was waste time over there. No fun. He said it was really the shits. I figure the later the better."

Considerable resentment mounted against such men. There was an undercurrent of grumbling about "rats deserting a sinking ship." That the ship was sinking was apparently taken for granted. But a few who were actually needed at home refused to apply for fear of being separated from their friends.

A second possible avenue of escape consisted of volunteering for the Regular Army. Anyone could volunteer for RA at any time and would be rewarded by a substantial furlough financed through separation pay. Several rumors emerged. One was that volunteers would be relieved of duties immediately and be given a ninety-day furlough; another was that any MISLS man who volunteered for two years would receive a minimum rank of staff sergeant. But another rumor was that all volunteers had to enlist for a minimum of three years. Those who were desperate to escape showed considerable interest, but during this period no one from Company K volunteered. Those who took emergency furloughs were surprised to learn on their return that they were still assigned to Division F. Furthermore, they had missed nothing in school.

During the latter part of September persistent rumors that the better qualified students would be transferred to the "regular" school led to some relaxation of pressures against studying. There was word that everyone in F-1 would be drafted for the nine-month course, that about a hundred (the first five sections) would be transferred, that a selected group from all sections would go to the "regular" school. Many eager beavers became preoccupied with grades. They haggled over grades, quibbling over answers until the instructors gave in. Cheating in examinations became so commonplace that the teachers rarely took the trouble to stop it. This intensified the resentment of the others who still did not care. As the "eager beavers" and the "fuck-ups" became more polarized, harsh words were exchanged, and men on both sides had to be restrained.

Changes in study habits were noticeable. A few even had the temerity to study in class. When the lights went out at 23:00, there was a marked increase in the use of Japanese in conversations. The singing of Japanese songs was also more common after "lights out." Some even crammed for examinations on their "own time." Private Shindo was caught studying under his blanket with a flashlight and became the target of considerable harassment. When asked about it, he repeatedly gave the same explanation: "Hell, I thought I might as well because I couldn't sleep anyway." The following incident took place in the small latrine in the Company A area, sometimes used by Company K men when their own latrine was too crowded.

"What'a ya say, Johnny? Hey! What the hell you doin'?"

"Takin' a shit. What's it look like?"

"Naw, naw. What the hell's that little black book you got?"

"Oh! I just forgot it in my pocket."

"Buckin', huh? Goddam! What's the matter? Ain't we good enough for ya?"

"Naw. That ain't it. This stuff's hard. I don't get it. I never studied Japanese before, and I don't catch on."

Toward the end of September a comprehensive examination was

given to check each student's progress. Most men did not extend themselves preparing for it, and with few exceptions they realized that they did not perform as well as they might have under other circumstances. About two weeks later, after the papers had been evaluated, the instructors announced that several changes in section assignments would be made. They explained that promotions and demotions would not be made solely on the basis of test scores; evaluations by instructors of the potentialities of each student would also be taken into account. Those in F-1 were told that 90 percent of them should prepare to move: "Most of you guys are being demoted because you got low test grades. We don't care if it's because you didn't care or because you didn't cheat. Your grades were too low to stay up here. There are a lot of smart guys down there, and they are going to move up. You have to make room." On the appointed day more than half the members in each section were either promoted or demoted. Ironically, in F-1 every "eager beaver" was demoted; of the six who remained there, four were generally regarded as "fuck-ups." Some of the men who had studied very hard, even at the cost of losing several friends, were embittered. Private Shindo was almost in tears; he had been demoted from F-1 to F-6.

Analytic Summary and Discussion

During the five weeks that the newcomers were together on the Turkey Farm, practices that had become habitual to those who had once been in A-9-3—yelling from ranks when they were at attention, threatening NCOs with violence, evading irksome tasks by missing formations—spread throughout the contingent. Recent IRTC graduates as well as other Fort Meade transfers were initially incredulous but quickly adopted them. Those who went to the "regular" school had to mend their ways abruptly, but all of these patterns were perpetuated in Company K.

Given the rare opportunity of affecting the course of their military careers, these men had to make decisions in an atmosphere of confusion—amidst a welter of speculation, rumor, and deliberate falsehood. People do not always say what they believe, but many in Company K learned this too late—after some of their most trusted comrades had lied to them. They had been caught between cross pressures. Like other infantry replacements many of these men placed excessive emphasis on their virility. Not having experienced combat, they had not yet developed the quiet humility and understanding of veterans (U.S. Department of the Army, 1954:383). The braggadocio of such men took its toll, but they also faced the almost unanimous opinion of the Nisei outside their unit. As the smoke of mendacity cleared, it became obvious that Company K was the repository of superfluous personnel: competent infantrymen who could no longer be used and those deemed incapable of learning enough Japanese to become useful linguists. Even before school started, many saw themselves as outcasts. They felt degraded. They were the least competent in the School Battalion; they had the worst barracks, the worst facilities for study, the worst commanding officer, and the worst first sergeant. Not only could they no longer win glory in combat; there appeared to be nothing they could possibly contribute to the war effort or to the Nisei cause. They were on the bottom rung of the status ladder at Fort Snelling; they were last—for many Nisei the ultimate disgrace.

Like so many other rear-echelon troops many became preoccupied with the question of rank (Schneider, 1946), and it was clear that graduates of the "dummy course" were not likely to get too far. Those deeply involved in the Nisei social world found it difficult to maintain their self-respect. Not only were they doing nothing; the were accused of undoing the work of others. Even assimilated mainlanders and Hawaiians, who were less responsive to the demands of outside Nisei, felt useless. What could they do? They could not hide, and self-deluding dreams were implausible. Many reacted with a sense of shame. They became detached; they adopted an I-don't-give-a-damn attitude. They refused to make a commitment to their work; in fact, they went out of their way to make it obvious that they were unconcerned. Seriousness is not welcome in a world given to protective frivolity. They did not wish to appear even more ridiculous by committing themselves to direct competition in the face of hopeless odds. Some became reckless and carefree. They had nothing to lose; they were already at the bottom. Since their only claim to recognition was their virility, some emphasized this to the hilt. They were still infantrymen, and they were against anything prized in the effeminate MISLS.

6. «» impetuous reactions to degradation

Toward the middle of August, as the termination of hostilities appeared imminent, men huddled around radios whenever they could, anxiously awaiting definite word. For more than a week a succession of false alarms had touched off premature celebrations. Official confirmation of Japan's surrender finally came at about 18:00, not long after supper on Tuesday, August 14. It was greeted with whoops and yells of joy; the men pranced about, some embracing each other. Those who were still loitering about in the latrine, the mess hall, or on the street were quickly informed.

"The Japs quit! It's all over! It's all over!"

"No lie? It's on the level this time, huh?"

"Hurray! The bastards quit! We go home!"

"Whee! The war's over! Let's go over the fuckin' hill!"

Academic work in the School Battalion came to a standstill. Married men with wives living in the Twin Cities area were excused to go home, but all others were restricted to the fort. On the following morning the troops were not aroused until 07:15, and there was no call to formation until 10:00. After participating in a parade and memorial service that afternoon, all students were issued passes until 07:30 Friday morning to go anywhere within 150 miles of the fort. Not only was the fighting over; Company K finally got a "break."

Most students realized that they would still have to go overseas; they did not have enough points for a discharge. They speculated over whether the Japanese people would accept the peace terms, how long the occupation would last, and how Nisei interpreters would be treated. Some wondered whether occupation duty would be more hazardous for Nisei than it would be for other Americans. Many were inclined to think so. Many of the instructors were Kibei, and they had indicated that Nisei were not especially popular in Japan before the war. Initial expectations were on the whole pessimistic.

"I think I'd just as soon fight as patrol that fuckin' place. You can't trust them guys. They're liable to sneak up on you at night from behind. Some babes and kids might come after you with grenades. You can't tell who to kill and who to pass."

"Yeah! Me, I no like dat kind. If we gonna fight, we fight. Yeah. Dat way you know where everybody stand, no? De other guy, he try to keel you, and you out to keel heem; so the first guy ween. But dat kind in Japan—no good. You no can tell who to shoot. You t'eenk you get friend, and he shoot you in the back."

A few were genuinely disappointed that they had missed the opportunity to do their share as infantrymen. Private Suehiro confessed, "I know I sound crazy, but I was kinda countin' on goin' over there and seein' some action. My brother got his Combat Infantryman's Badge, you know. What the fuck am I gonna say when we both get home? He's a combat man, and I'm just a fuckin' chair-borne asshole. I used to think about what my girl'd say if I came back with a Silver Star. I used to think about fightin' some Jap snipers or somethin'. I figured we might go right up to the front and get a crack at it. Even if I didn't get a Silver Star or Purple Heart or somethin', at least I'd be a combat infantryman. I don't want to go home with just a fuckin' Good Conduct Ribbon. I don't want to go home and tell my friends that all I did in the war was sit on my ass and learn Japanese."

Questions inevitably arose concerning the fate of incompetent linguists. Combat intelligence troops were no longer needed. Would this mean that everyone would have to take the nine-month course to train for occupation duty? If so, how could Division F personnel possibly keep up? With victory over Japan all military duties appeared pointless, and Company K men in particular began to feel that they were on a treadmill leading nowhere. Their presence at Fort Snelling was a waste of the taxpayers' money and of their own time. This made them hypersensitive and irritable, and soon their resentments focused on a limited number of objects. The discontent mounted until it finally exploded in violence—

first against one another, then against an unpopular cadreman, and then against the all-Caucasian OCS.

Friction: Hawaiian vs. Mainland Nisei

As the prevailing atmosphere became one of irritation and ill will, cultural differences between mainland Nisei and Hawaiian AJAs became more noticeable. On the surface men from the two areas got along fairly well. Open friction was kept to a minimum, mostly by restraint on the part of mainlanders. Several close friendships had developed between individuals across the line, friendships that became strained only in periods of crisis when the Hawaiians were pressured by their fellows to band together. Most mainlanders were fearful of the Hawaiians. Only close friends could criticize or make fun of them without risking a beating, and only such intimates called them "Kanakas" openly. Many Hawaiian privates took orders only from Hawaiian NCOs or from Caucasians. If an order was issued by a mainland Nisei, especially if it involved something unpleasant, they either refused to obey or did so reluctantly and with threatening glares. Mainland NCOs disturbed the Hawaiian privates as little as possible and usually either appointed a Hawaiian as assistant squad leader or made requests through those who were regarded as reasonable. Many Hawaiians were more concerned with stopping quarrels with mainlanders than with participating in them, and their services were needed frequently. Most men on both sides managed to remain on amicable terms and worked together, but there were a handful on each side who had difficulty in concealing their hatred.

A few Hawaiians referred contemptuously to all mainlanders as "Kotonks." Among themselves they frequently reiterated the charges that made up their stereotype of the mainland Nisei. Some complained that the reason the company got involved in so many difficulties was that the Kotonks were *manini*, that they lacked the courage to stand up for what they believed, that they were too cowardly to fight, that they were too egoistical to be considerate of their comrades. They also criticized the Kotonks as being untrustworthy because they would always try to save themselves, even at the expense of getting the rest of the company into trouble. Another common charge was hypocrisy; Kotonks said one thing and did something else.

Some mainlanders contended that Hawaiians were uncouth, and they pointed to the conspicuous few who kept alive this stereotype. They were thought to be sensitive to an unreasonable extent. Sometimes a glance was interpreted as a hostile stare. When continental Nisei stopped talking upon the approach of a Hawaiian, the latter would become incensed, even when assured the conversation had nothing to do with islanders. Among themselves mainlanders often pointed to the recklessness of Kanakas in spending money or in gambling it away, to their sloppiness in dress, and to their don't-give-a-damn attitude that jeopardized everyone in the company. A few Hawaiians openly taunted mainlanders, challenging them to fights at the slightest provocation. Whenever the entire company was punished, each side tended to blame the other. Some mainlanders insisted that Company K had acquired its reputation only because of the disorderly conduct of a handful of drunken Hawaiians.

Only a few Hawaiians were able to speak English without an accent. Although a number of pidgin expressions had been adopted by mainlanders and incorporated into the company vocabulary, language remained a sore point. Hawaiians on the whole were highly sensitive about it and became offended whenever they thought that a Kotonk was making fun of their speech. Some were so sensitive that they became resentful whenever they could not make themselves understood. When a mainlander was unable to understand and asked them to repeat a statement, they became incensed. Mainland Nisei, cognizant of this hypersensitivity, never criticized Hawaiian speech, unless they were with a Hawaiian friend and the whole matter was in jest. Whenever they failed to understand, they smiled politely. Too often requests for clarification led to further misunderstandings.

One widespread belief among the mainlanders was that all Hawaiians were bound by a code of honor to stick together whenever any of their number got into trouble, regardless of the merit of the man's case. This seemed unfair, and this practice more than anything else was resented. It was assumed that Hawaiians would not fight as individuals, that they always fought in gangs. An offended person would first gather enough support and then challenge the transgressor. Hawaiians could always count on the support of fellow Hawaiians. If a small number could not win, they could always recruit more help. Most mainlanders felt that this was a cowardly way to fight, that individuals should settle their differences in man-to-man confrontations. When a Hawaiian was involved in a fight, it was understood that only other Hawaiians could stop him. Mainlanders did not dare interfere for fear of becoming victimized themselves. Another common belief was that all Hawaiians carried switchblades—for emergencies that could not be handled through gang beatings.

IMPETUOUS REACTIONS TO DEGRADATION

Although many threats had been aired, the first actual violence attributed to Company K personnel did not occur until August 19. The incident erupted during a Sunday "beer bust" at Company F, a party to which a number of outsiders had been invited. When some of the visitors became too boisterous, the first sergeant admonished them. In the argument that ensued the sergeant was set upon by a gang of AJAs and beaten. Rumors developed on the following day about the fight. Although the men alleged to have been there were evasive, details were soon elaborated.

On the following day a rumor developed that the company was scheduled to run through an obstacle course during the physical training period—an exercise that had come to be regarded as the least desirable—although only a few would admit that they were getting out of condition and could no longer "take it." Apparently the rumor was taken seriously, for less than half the company fell out for physical training; the others hid in latrines, in the recreation rooms of other companies, in kitchens, or made appointments at the dental clinic or some other part of the hospital. The NCOs did not take names, and nothing was said about the matter, but most men anticipated trouble; too many were absent from the formation. When the company returned to the barracks at 16:30, they warned each other to be "on the ball" for the rest of the day. Thus alerted for trouble, every man was present for retreat formation. As anticipated, a roll call was held and carefully checked by the company commander. After retreat the lieutenant ordered the company to "open ranks." Then, starting with the 1st Platoon, he required each person to hold out the back of his hands as he passed before them. Everyone was puzzled. "What the hell's going on?" One rumor was that a fingernail inspection was being conducted.

After all knuckles had been examined, the lieutenant faced the group and said, "All right, men. You know what we're doing this for."

"Why?"

"The man who said that come up here. I'll tell ya what it's all about. Last night, some men from this company went over to F Company as guests to a beer bust. They repaid the hospitality by beating up their first sergeant. Now I'm going to give it to you straight from the shoulder because I don't believe in pulling sneak punches. That kind of stuff's not gonna go around here. I try not to pass on any chicken shit to you, and I'll back you up to the limit as long as you play square. But if I catch any of you men getting into gang fights, believe me, I'm gonna throw the book at you. Don't expect any sympathy from me if you get into anything like this. Whoever was in that fight last night did a pretty good job on his knuckles. Just watch yourself."

During the evening classes considerable discussion arose over the knuckle inspection. Most members of the 5th and 6th Platoons knew who had been involved and wondered how they had managed to pass the inspection. But nothing more was said about that.

The nightly drunkenness of a small circle of young Hawaiians soon became the source of considerable inconvenience. At the core of the group were three from the 5th Platoon—Takeda, Tajima, and Tsuboi. Others joined their beer drinking bouts from time to time. Mainlanders who were disturbed by the noise did not dare ask for quiet, and those who were accosted by the intoxicated AJAs had to humor them rather than risk a beating.

The first intramural violence occurred on the evening of September 7, when an argument in the recreation room led to a gang beating in the 6th Platoon barracks. The three troublemakers, already primed with beer, were gambling. A tiny mainlander named Yasuda, widely regarded as "cocky" and "always talking out of turn," made a slighting remark to one of the trio; before anyone could stop it the three started beating him. At this point another mainlander named Yamanaka, a former football star from Arizona who had had little previous contact with Japanese, tried to stop the altercation by yelling, "Why don't you pick on somebody your own size?" The three immediately turned on Yamanaka; since he made a formidable adversary, several other Hawaiians joined in the melee. No other mainlander stepped in to help. Yamanaka managed somehow to get away and returned to his bunk in the 6th Platoon. As he stood there, five men charged at him simultaneously and resumed the beating.

The only NCO present was the platoon leader. Since he was a mainlander, he was placed in a very difficult position. The sergeant pulled the men apart and pleaded, "If you guys want to fight, go outside and fight. But don't fight like this. If you two guys want to have it out, O.K. But not five guys on one."

Although he was a Kotonk, the Hawaiians balked momentarily at the idea of striking an NCO, especially one who knew all of them by name. As a discussion ensued, Tajima suddenly lurched at the sergeant, but other Hawaiians jumped on him and pulled him back. He continued to struggle so hard that his T-shirt was ripped off his back. One small Hawaiian known affectionately as "Punchy," who was reputed to be insensitive to pain, and a tall, muscular man named Yonemura, who commanded the respect of his fellow Hawaiians, then stepped in and pointed out to Tajima the folly of striking a noncommissioned officer. Tajima would not listen, and additional Hawaiians joined in dragging him outside. At this point Yasuda, who had started the trouble and had run away, appeared in the barracks, and several Hawaiians gave chase.

After a long, heated argument Tajima was finally persuaded to return to the 5th Platoon barracks. Instead of retiring, however, he went to Teshima, a prizefighter, and Tomoda, one of the most popular AJAs in the company, and told them that the sergeant had challenged all Hawaiians to a fight. Within a few minutes Tajima was back in the 6th Platoon with Teshima, Tomoda, and several others. No further difficulties occurred, however, for the sergeant and Yonemura explained the situation to Teshima and Tomoda, both of whom had many mainland friends and were reputed to be "cool headed." By 23:00 order had been restored, and the sergeant promptly turned off the lights.

In the darkness several Hawaiians continued to argue whether "da sahjint" should have been allowed to stop the fight. The mainlanders went to bed and said little; those sleeping in areas where there were no Hawaiians asked one another in whispers what had happened, but no clearly audible discussions took place. They lay quietly, listening to the Hawaiians. Several were against any fighting within the company, and one of them announced loudly, "Too bad t'eengs like dees have to happen. We all Japanese. Japanese shouldn't fight among demselves. If it's Haole, O.K. But us guys—all da same, yeah?"

It was apparent from the murmurs, the whispers, and the tossing about that neither mainlanders nor Hawaiians in the 6th Platoon slept well that night. Occasionally someone would get up and go to the latrine. Now and then someone would ask his neighbors what time it was. It was not until several hours had passed that Yasuda sneaked back into the barracks. He was assured by some Hawaiians near the door that it was safe for him to return to his bunk.

Most mainlanders later admitted that they had had trouble getting to sleep. They had heard of Hawaiian gang fighting before, but for most this was their first actual experience with it. They found it an ugly experience. They also saw demonstrated the contention that only other Hawaiians could stop a fight involving Hawaiians. It was true that the sergeant was a mainlander, but even he would have been beaten had it not been for the intervention of Yonemura and Punchy. Several subsequently admitted that they imagined themselves being picked on by some Hawaiians. They felt helpless, knowing that their mainland friends could not come to their aid. They wondered what they would do were they trapped in such a situation. It did not take much to antagonize an inebriated Hawaiian. Even innocent remarks could be misinterpreted as a slur. As one man put it, "Goddam, I couldn't sleep at all last night. I don't know why, but I kept thinking of what I would do if I was in Yamanaka's shoes. Jesus Christ! He had a lot of guts stopping that fight. But what the hell can a guy do? Is he supposed to sit on his ass and watch five guys beat up a little fart like Yasuda? I know Yasuda's a prick and always asks for it, but it ain't fair. I kept thinking over and over of what I'd have done if I was in Dave's shoes."

On the following morning the 6th Platoon floor had to be scrubbed. Hawaiians and mainlanders were initially apprehensive of one another, but before long they were working together—sharing brooms, helping one another move beds, carrying buckets of water, and yelling for others to get out of the way. If anything, members of the opposing factions were more considerate toward each other than they had been in the past. Such "GI parties" called for close cooperation; the men had to work together to get the task done quickly. By inspection time most of them were joking with each other in a friendly manner. Little was said about the fight.

At the YWCA dance in Minneapolis on Saturday night, September 22, some Hawaiians—generally believed to be from Company K—attacked a Nisei who worked in the personnel office of MISLS. This man was not just another mainlander outside the company. He had been in A-9-3 and had lived with them on the Turkey Farm; he had been transferred to headquarters when school started. The following account of the beating was given by the victim: "I really didn't know what it was all about. I was standing there talking to a girl, and a Hawaiian guy came up to me and asked me what I was laughing at him for. I wasn't laughing at him; in fact, I wasn't even looking at him. I told him that, and he said he wanted to fight. I told him I didn't do anything and didn't want trouble, but before I could do anything else about five guys jumped on me and started beating me up."

On the following Monday word of the beating spread rapidly, especially among mainlanders: some Kanakas beat up Matsuda Saturday night; Matsuda is in the hospital with a broken nose; six Hawaiians jumped Matsuda, kicking him when he was down. Much resentment arose, not only because the victim was well liked but because he was not the type of person likely to provoke trouble.

"That's dirty! I hear six guys beat him up. I hear they kicked him when he was down and everything. That's really low, though. He's a nice guy. Hell, he wouldn't hurt a flea. He never starts any trouble; he always just minds his own business. Jesus! He's a nice guy."

"Yeah! That guy never bothers anybody. He never had anything against the Kanakas. Lots of his best friends are Kanakas. What do they want to pick on him for?"

This beating was significant in that it made the mainlanders feel absolutely helpless. They knew the victim; he was a friendly person who did not antagonize people. Some had come to feel that if they made friends with Hawaiians and did nothing to offend them, they would be safe from attack. This beating proved that this was not so. One might be beaten anyway, for no reason other than that some Hawaiian did not like his "looks." Furthermore, if other mainlanders would not defend a man as popular as Matsuda, no one else could expect help either. This made each mainlander feel completely vulnerable to the whims of drunken thugs.

Only a few were relatively safe. NCOs who did not "pull rank" were popular; in addition, their close friends included Hawaiian NCOs who could protect them. Those with reputations of being vicious fighters were not disturbed. Utsumi was known to be a boxer of considerable skill; he subsequently placed second in a national AAU tournament. Others had fought on college boxing teams and in amateur ranks and were believed to be "hard as rocks." These men were not only capable of fighting several Hawaiians at once, but they had added strength in that they were close friends. Furthermore, they were on friendly terms with the Hawaiian boxers, most of whom got along quite well with mainlanders. Therefore, men like Utsumi insulted the Hawaiian trio with immunity. A third group that was safe from attack, of whom there were only a few, consisted of sycophants who catered to the troublemakers. They were drinking companions who ran errands for them and served as jesters. These men were scorned by mainlanders and Hawaiians alike.

The beatings reinforced the stereotype of Hawaiians as barbarians, even though only a tiny minority was involved. During the middle of the week a rumor developed, mostly among mainlanders, that a few months before a Hawaiian had slapped a Caucasian woman at a YWCA dance, a woman who had worked hard in behalf of resettlers. According to one account, she had only asked the man to take his whiskey bottle outside; as a result of the incident she now hated all Japanese. The thought of any soldier striking an elderly lady was repulsive; that a Nisei soldier would do such a thing was incomprehensible. Most mainlanders were inclined to believe the story; they had seen enough to make it plausible. The woman had been identified as a "white angel." Mainlanders pointed out that Hawaiians had not gone through the evacuation and resettlement and could not appreciate the sacrifices friendly white people had made to help them; they had run the risk of being condemned as "Jap-lovers." That a woman who had gone through that had been slapped was unforgivable. The warmth of most Hawaiian AJAs was forgotten temporarily as clusters of mainlanders stood about condemning the deed.

Several incidents centering around a colorful mainlander known to most of the company only as "Saipan" intensified animosity against the trio. Saipan was reputed to be retarded; he tried very hard but apparently could not comprehend many of the things that went on around him. He had bushy eyebrows that slanted at a steep angle, and he resembled some of the cartoons appearing in the mass media caricaturing Japanese warlords. Although only a few knew his real name, he was the subject of much discussion. He was believed to exercise certain unknown powers over Caucasian women. He himself boasted of the number of girls he knew, and others who had seen how popular he was at the local USO confirmed his claims. There were also stories of his physical prowess. His physique was impressive, and it was believed that he held a high rank in *kendo* (Japanese fencing). Some contended that he was a disgrace to the company and occasionally played little tricks on him. Since he was good natured and harmless, others felt that it was unfair to take advantage of him. Those who were sympathetic did whatever they could to protect him and to make him feel that he was one of them. Whenever Saipan had been made the butt of some joke, his protectors censured the perpetrators for "taking advantage of a guy who don't know any better."

About 23:30 on the night of September 25, Takeda, Tajima, Tsuboi, and another intoxicated Hawaiian awakened Saipan. Shining a flashlight in his face, they identified themselves as the CQ and his assistants and ordered him to report to the company commander at once. While Saipan was dressing, Takeda asked him repeatedly, "Why you get such a funny face?" Most of the men nearby were outraged, but they felt helpless. Saipan reported to the orderly room; no one was there except the real CQ who knew nothing about the order. When Saipan returned to his bunk and indicated that the lieutenant was not in his office, the Hawaiians told him to report early the next morning.

Saipan reported to Lieutenant Catalano on the following morning and disclosed what the "CQ" had done. The CQ had not reported the incident; when questioned by the lieutenant, he refused to divulge the names of the men involved. A special company meeting was called, and the lieutenant revealed in detail what had happened. Since the CQ would not provide the names, he restricted the entire company indefinitely until those responsible gave themselves up. Only a few blamed Saipan for his part, realizing that he was only doing his best. Indignation focused on the trio, for many found it difficult to believe that anyone could be so depraved as to torment in this manner someone who was so helpless. Even those who had themselves participated in minor pranks on Saipan felt that this was going too far. The Hawaiians got together among themselves, confronted the three, and forced them to act. It was decided that Tsuboi should take the blame, and within a few minutes he reported to the company commander. The restriction was lifted. Most mainlanders felt that it was unfair for one man to "take the rap" for everything; the others involved, especially Takeda, were condemned as being "yellow."

Two weeks later the escalating violence led to the restriction of the entire 5th Platoon. Late Friday evening the same trio began drinking heavily with two women they had smuggled into the recreation room. When it was almost time for "lights out," the three went into the 3rd Platoon barracks, looking for a man they had marked for a beating. Before the evening was over they touched off an orgy of violence.

One of the victims, an assimilated man who knew little of Nisei or Hawaiian codes, long afterward gave the following account of the incident:

I came in from town that night, and I saw a gang of Hawaiians picking on one Nisei. He was in bed, and you could tell he was really scared. They were telling him to come outside and fight. I didn't think it was right so I went up and told them, "Leave him alone. Can't you see he's scared?" Then they came after me. I didn't pay any attention to 'em and started undressing. They told me to come outside. I told them I was sleepy and for them to go to bed, but they wouldn't go. So I finally told Takeda that I'd meet him tomorrow. He was kind of surprised when I said that, but when I told him again, he left. Then, a little while later, that Tajima came running in with a scream and jumped on top of me. My blankets were tucked under so I couldn't move my hands or anything. Jesus Christ! They started pounding me. The only bad one was where Tajima hit me in the head with his flashlight. It must have sounded bad in the dark, and so Nish said from across the way, "That's enough!" Then they turned around and went after him. They tried to push him around, but he's kind of big. They were arguing when Oyama came in. He didn't know what the hell was going on so he just yelled, "What the hell's going on here?" Boy, they really went after him and started pounding the shit out of him. He was in the corner and was yelling "Help! Help!" Then the lights went on, and somebody said "Let's go!" Everybody got up, and just then more of the gang came in. They went up and down, looking for the guy who said that. I guess they couldn't find him, and they left.

After about a half hour Tajima came back again with his flashlight and asked me if I got hurt. He put a plaster on my head and told me to hit him. I guess he figured that if I hit him that made it even. He apologized. I couldn't figure it out though. I guess he thought that would make it right.

I admit that I've hated Hawaiians ever since then. I couldn't stand the sight of 'em for awhile. Now I'm gradually getting over it. But the guys who really got me were the Nisei guys who'd stay in bed and watch their buddies get the shit beat out of 'em. That's something I can't understand at all. Boy, some of those guys are really hard to understand.

On the following morning rumors elaborating details of the incident spread rapidly throughout the company. Accounts of Private Shindo's reaction lightened the tension somewhat. According to his neighbors, Shindo started shaking violently even before the trio reached his bed. When the flashlight was turned on him, he peeked out from under his blanket and whimpered, "Hi, Tajima." This only infuriated his tormentor, who roared back, "No 'hi' me, you Kotonk!" Despite the rumors many questions remained unanswered. Who had antagonized the Hawaiians? Why had they gone into the 3rd Platoon barracks? No one knew. When questioned, a 3rd Platoon man replied, "I don't know why they came in, but they came and beat up almost everybody in the place. Hell, when they made their *banzai* charge, the barracks were rocking. You could hear the thuds as they hit the guys and kicked them and everything. Boy, was I scared!"

When the company commander learned of what had happened, the entire 5th Platoon was restricted over the weekend. He was unable to learn the identity of the culprits other than that they were from the 5th Platoon. The men were charged with being drunk, leaving beer bottles in the recreation room, bringing women into the company area, and engaging in gang fighting. Although the lieutenant said nothing to the rest of the company, he told those in the 5th Platoon in no uncertain terms what he thought of the matter. He did not expect anyone to "squeal," but he hoped that members of the platoon would take care of the matter themselves. Mainlanders were bitter. One of them said, "I don't see why the CO doesn't court martial them bastards. It's always the same guys. Every time it's the same guys. I wish they'd lock 'em up and get 'em out of my sight."

Lack of solidarity among the mainlanders made the reign of terror possible. Many insisted that they would be willing to fight most Hawaiians man to man, and some of them were strong enough to have no reason for fearing anyone on that basis. Although they outnumbered the Hawaiians five to one, they knew that they could not count on support. They were split into too many factions, and they did not trust one another. They had had their differences at school. Having been double-crossed at the qualification examination, they were not sure that they could count even on close friends in a dangerous situation. Mainland privates, therefore, were not expected to fight back when tormented by a drunken Hawaiian. Although a man might consider himself cowardly for taking such abuse, others did not blame him. They admitted that they themselves would not fight back in a similar situation. Yamanaka was both admired and regarded as foolhardy. They realized that, if they tolerated the abuse long enough, some Hawaiian would come along to put a stop to it. Most Hawaiians were known to be opposed to the fighting; it was also assumed, however, that if the mainlanders forced a showdown by banding together, the Hawaiians would unite regardless of their personal feelings.

Those who had close friends in the 5th Platoon were especially troubled; they did not know what to do if their buddy were attacked. On the night of October 16 the trio tried to pick a fight with one of the most popular mainlanders there. Many were afraid that the man might accept the challenge. He was strong enough to handle one or two of his tormentors, and he had several friends who might have fought for him. However, he refused to fight. One of his friends said later, "Jesus Christ! I didn't know what the hell to do. When them bastards picked on Joe, I had to do something. He's one of my best buddies, and I can't just sit by and watch him get licked. But then I didn't feel like getting the shit beat out of me either. I wasn't sure if the other guys would help me if I jumped in. If I was sure that a couple of guys would help, I would have got in there. I was shittin' in my pants for a while, hey. I thought Joe might take 'em on."

Another man confessed:

> I dread every fuckin' night. No shit, hey. I'm almost afraid to go back to my bed at night. I stay away as long as I can to keep out of trouble. They get drunk and beat up anybody they got a grudge against. I want to transfer to any other platoon. I wouldn't care about them guys if I didn't live there, but goddammit I can't keep away from 'em. What can you do if they decide to beat up a friend of yours? No use helping. You'd just get beat up yourself.

But what are you gonna say when you see your friend the next day? He won't be your friend anymore. He won't understand.

Each night several mainlanders were confronted by taunts, insults, and beatings. Since each disorder led to the punishment of entire platoons, if not the whole company, considerable hostility mounted against the Hawaiians, hostility that had to be suppressed. The restrictions did not inconvenience Hawaiians too much, since most of them had few friends in the Twin Cities anyway, but they were felt by mainlanders with friends and relatives there. By this time it was common knowledge that most of the violence was initiated by the three in the 5th Platoon. Some mainlanders developed a lifelong hatred of all Hawaiians. Among themselves they confessed their feelings:

Boy, I'm gettin' to hate them three guys. They're really first class bastards to pick on a swell guy like Joe. They cause all the trouble around here. If it wasn't for them, K Company wouldn't have the bad reputation. Those guys are just unreasonable. They're mean, and they can't hold their liquor. Punks who get drunk when they just smell the cork shouldn't drink. I hope them guys get court martialed. Every night they cause trouble. Every fuckin' night you gotta go to bed wondering if they're gonna come around and pick a fight with you. Christ! I'm almost scared to go to the can at night. I just go to bed early. If I gotta go, I just step outside and take a leak out in the street. I don't mind fighting one of them guys, but you haven't got a chance against all of 'em. The other night I had a dream that I beat the shit out of Tajima. Sometimes I dream of killing them. No shit. One night I dreamt that I killed one of 'em.

Yeah, I been thinking of all the times and places where you could kill them bastards and get away with it. I hope some guy swears out charges against 'em and gets 'em locked up at Leavenworth. Some guys around here'll just jump at the chance to go overseas now—just to get out of this night by night wonderin' if you're next. It's the shits, hey. Every night you wonder if your number's up. I been making some Kanaka friends. You can't depend on the Nisei around here. They won't back you up.

Another man added:

I was up until about 03:00 that morning. Goddamn! I just couldn't sleep good. First I thought of what I would do if them guys came after me. Then I turned over and tried to sleep, but I couldn't get it out of my mind. They kept coming. So then I thought of getting all the Nisei together and forming a gang. I thought of all the tough guys we had. We got enough guys to beat the shit out of all of 'em. But then I figured even if we did it in this company, they could get guys from other companies and we couldn't. That's the trouble with the mainlanders; they won't stick together. You can't trust a Kotonk. I kept thinking of getting together some guys, and then I thought of some bastard telling the Kanakas about me trying to get some guys together. We got guys like that, too. They want to kiss-ass them guys. I lost fight.

One man in the 3rd Platoon admitted to intimate friends that he went to bed each night with a rock in his stocking so that he could defend himself more effectively. Another confessed that he too was improvising defenses: "Funny but that night I kept thinking that somebody was gonna come in and beat me up. Every time I heard somebody around my bed I felt kind of nervous. I thought some Kanaka would yell 'Kotonk!' and start beatin' me up. What the hell could I do? Once I got out of bed to loosen the strings on my shoes so I could get into 'em fast just in case something happened. I wondered if I could reach up and grab my steel helmet in time to sock somebody. I guess I was pretty scared."

The following story, perhaps apocryphal, was related with considerable glee among mainlanders. Takeda, Tajima, and Tsuboi were in the latrine talking about all the cowardly Kotonks. Utsumi, who was taking a shower, became incensed. He told Takeda that if he wanted to fight, he would be glad to oblige him right there. Takeda turned pale and left without taking a shower. No mainlander would have accepted a similar challenge from Utsumi either, but this was beside the point.

Since most of the disorder centered around the 5th Platoon, it was restricted more often than the others, and intense resentment developed there. One man gave an account of the difficulties:

We were gambling and that fuckin' Takeda came in and took a dollar of mine and two bits of Joe's. He just took it. We let 'em have it because there's no sense in gettin' into a fight just for chicken feed like that. Then, some other Hawaiian made him give it back.

Only three guys cause all the trouble. I sleep above Takeda. It's really rough, boy. I hate the bastard. I sweep and mop every day. Then, when I finish he flicks ashes and butts on the floor. He even spits on the floor, and I have to clean up. I've seen a lot of filthy Nisei, but I've never seen a good-for-nothing bum like that. He enjoys seeing me work, you know. He never cleans the fuckin' place. I have to clean up because otherwise I'll get gigged and I won't be able to go into town.

They never fuck around Utsumi. Utsumi is a boxer, and they leave him alone. He tells those guys to kiss his ass, and they take it with a smile. The platoon sergeant is a prick too. He's so damn scared of those guys that he's right in with 'em. They drink with him. He just wants to save his own ass.

When Sergeant Imamoto was around, it was different. He didn't take any shit from any of them guys.

Feelings in the 5th Platoon became so strained that the men had difficulty in doing their routine work. During the weekly "GI parties," when the floor was scrubbed, the trio openly declared their contempt for Kotonks and refused to work. Even the other Hawaiians were unable to prevail on them to do their share.

No open break developed between Hawaiians and mainlanders because the trio commanded so little respect among the Hawaiians. In addition, there were too many close friendships across the line. Among themselves most Hawaiians condemned the troublemakers and did their best to control them. Whenever a mainlander was picked on, someone generally intervened. Although virtually all the trouble had been started by only three men, they realized that Hawaiians as a group were being blamed. Whenever the subject came up, most other Hawaiians spoke apologetically. They approached mainlanders cautiously and expressed their sympathy.

One older man explained, "It's too bad some of dese guys don't have better sense. But dey always like dat. All Kanakas no like dat. Dose guys like dat even when dey back home. Nobody like dem back home. Dat guy Takeda, him smart guy. But he always get in trouble. Nobody like him."

Another added, "A lot of dese guys young keeds yet. Dey no grow up yet. Dey jealous of you guys because you get *wahine*, home, and everyt'eeng. We get naut'ing. What we lose if we get restricted? If we go to town, we just get drunk, maybe go see Wahoo, and get piece of ass. Dat's all. Only time we meet nice *wahine* is when we go with one of you guys, yeah? Mainland *wahines* no like Kanaka boys anyway. No lose naut'ing. Dat's why if CO like restrict our ass, we no care."

Whenever the Hawaiians took a united stand against violence, it was stopped. After the beating of Yamanaka, Yonemura and several of his friends told the three troublemakers firmly that they were not to enter the 6th Platoon barracks again. If they did, the Hawaiians there would not side with them. That platoon was not bothered for the rest of the year.

But more trouble was brewing. On September 28, Company L was activated. It was housed in a row of tar-papered barracks behind the Company K area. These men were also taking a "conversational" course. As they began moving in, difficulties arose at once. The newcomers borrowed several brooms and mops and failed to return them. The delegation that went to retrieve these items reported that the company was about 90 percent Hawaiian. Official records confirm this impression; 202 of the 288 men on the initial roster were inducted in Hawaii. Word spread quickly, bringing joy to the Hawaiians and gloom to the mainlanders. Some of the latter felt that Fort Snelling would now become unbearable, and subsequent events confirmed their premonition. Company L's record of violence eventually surpassed that of Company K.

The Beating of the First Sergeant

The intramural disorders perpetrated by a few troublemakers were condemned by most mainlanders and Hawaiians alike. When violence was directed at the most unpopular member of the company's cadre, however, it won widespread approval. The men identified closely with their commanding officer, even though he was a Caucasian and was not expected to live up to Nisei norms; and they included both the company clerk and the supply sergeant in the fraternity of Nisei soldiers. But the first sergeant, a Nisei from Texas and a Regular Army man, had no familiarity with the culture of coastal or Hawaiian Nisei. He was genuinely puzzled by many of the demands the Nisei made on him; he understood only his duty as a soldier. He was regarded as peculiar.

Difficulties arose in the first orientation meeting of the company. The sergeant began addressing the group in a low voice, and before he had finished his first sentence cries arose from the rear: "Can't hear! Can't hear!" "Open your mouth!"

"If you guys'll keep quiet maybe you can hear something," he retorted angrily.

An undercurrent of mumbling became more audible:

"Fuck that bastard! Who in the hell does he think he is?"

"He's a Buddhahead just like us, even if he was born in Texas."

"Just 'cause he's first sergeant, he thinks his shit don't stink. Somebody ought'a beat the shit out of him."

In addition, the men found the sergeant's personal style offensive; he met almost every situation with threats of punishment. For example, on July 30, the first day of school, physical training was canceled because of rain. About 15:45, soon after the troops had returned to the barracks, the whistle was blown for a company formation. No one in the 6th Platoon building, the farthest away from the orderly room, heard the call. By the time the platoon was finally mustered, Sergeant Endo had become infuriated and threatened to restrict it. Even when Sergeant

Kimura, the platoon sergeant, explained that the whistle could not be heard, he remained adamant. He then went on to declare that the "rain schedule" was not intended for everyone to sleep and that if they did not clean their barracks properly, they would be "gigged" and restricted. Since there was little for them to do, the men could see no reason why they should not sleep, if they so desired. Most of them stayed off their bunks, however, feeling that the sergeant was the kind of man who might drop in for an unannounced inspection. On another occasion he told the men that they were to march to and from school at "attention"; otherwise the entire company would be restricted.

There was widespread accord with the view of a man who commented resentfully, "Jesus Christ! Is that all that bastard can do? All he does every time we have a fuckin' formation is to threaten us. Threats! He thinks he's tough as rocks. 'If you don't do this, I'll restrict ya; if you don't do that, I'll restrict ya.' Goddam! Why don't he restrict our ass for good and get it over with? The bastard!"

The sergeant added to his unpopularity when he attempted to enforce strict uniform regulations. Soon after all privates in the company had been promoted and had been issued Pfc. stripes, the men began arguing over the desirability of wearing them.

"Gee! One great big stripe! If we had three of 'em, we could be proud to wear 'em."

"We get a raise though—four bucks a month."

"Christ! It'll cost about four bucks to get 'em sewed on."

Most men agreed that a single chevron looked lonely, even though it was a symbol of service beyond basic training. Toward the end of the week a rumor developed that the stripe would have to be worn for inspection on Saturday. Several started to sew on the emblem, but others refused, declaring that they would rather have no stripe than only one. A rumor in the 5th and 6th Platoons that Company K men would not be required to wear stripes led to more confusion, and the NCOs were asked for a ruling. No one knew. Those who did not want to sew them on snorted contemptuously about "one measly stripe" and insisted that they would not be required for inspection; those who actually wanted to wear them insisted that they were doing it only for the inspection.

Although many objected to wearing a single chevron, a number began to sport shoulder patches. Several Hawaiians had sewed on the green-and-red patch of the Hawaiian Division, a deactivated unit, while some mainlanders wore the patch of the 7th Service Command, even though they were not part of the cadre. Some sported the shoulder patches of the Southwest Pacific Command or of various divisions in the Pacific Theater. Many Nisei who had served with these divisions had returned to Fort Snelling, and it became difficult to distinguish between students and the veterans who had earned the right to wear such insignia. Most felt that it was cheap to wear unearned emblems, but others argued that they did not wish to be confused with the "rookies" at the induction center.

At a company formation the first sergeant announced that all unauthorized shoulder patches were to be removed and threatened to revoke the pass privilege of anyone who did not comply with the order. He also declared that all first class privates would have to wear their Pfc. stripe. This led to howls of protest.

"What the hell's the difference? You know we're all Pfc's."

"Yeah! You think we're proud of one fuckin' stripe?"

"What a chicken shit place!"

"They can take the stripe and shove it up their ass."

The sergeant refused to budge and intimated that anyone who failed to comply with his order would be restricted. Although many supported the order against unauthorized shoulder patches, most men felt that the wearing of a single stripe should be left to the discretion of each individual.

Not realizing that he was violating Nisei norms, the sergeant continued to enforce petty military regulations. On Tuesday morning, August 7, he made himself even more unpopular. During reveille formation he declared that too many students had been "goofing off." He announced that thereafter roll call would be held at every formation, that each squad leader would report how many were absent, whether or not they had been excused, and that he or some officer would count the number present in each platoon to be sure that the figures were accurate. Everyone knew that Nisei NCOs were reluctant to report absentees, but his open admission that he distrusted them was resented. He then went on to say that too many men had been eating in the mess hall on the Turkey Farm. Since food at the Consolidated Mess Hall was unpalatable, many were remaining in the school area at noon to eat there. The mess sergeant was encountering supply problems, and unless this practice stopped at once, Sergeant Endo warned, roll call would be held at each mess formation. Finally, he announced that henceforth the company would fall out for retreat in Class A uniform—including neckties. This elicited another storm of protest. It meant that everyone would have to change immediately after returning from physical training. Irritation mounted, and by the end of August the sergeant was virtually isolated.

The contempt with which he was regarded was soon expressed in an unflattering sobriquet. On a cold Wednesday morning, August 22, the NCOs in the 6th Platoon were encountering difficulties in getting their subordinates out of bed in time for reveille. Suddenly, two bats began fluttering through the barracks. A few were frightened, but most of the men gleefully gave chase with brooms, firewood, rolled newspapers, and shoes. They ducked when the bats swooped down but continued to throw things at them. In a few minutes there was bedlam; men charged back and forth, laughing and screaming: "Kamikaze! Kamikaze!" "Get them bastards!"

Sergeant Kimura tried to restore order by yelling, "All right, men! Leave Sergeant Endo alone!"

"Is that Endo? I thought the thing was ugly!"

"Goddam! A guy can't even sleep in this fuckin' place. Get the hell off my bed!"

"Where's Endo?"

"There he goes!"

At this point, as Sergeant Endo's voice growled over the loudspeaker for them to fall out for reveille, the men exploded in gales of laughter. News of what had happened spread quickly throughout the company, and thereafter the sergeant was called the "Batman."

Resentment reached a new high on September 5, when Sergeant Endo threatened two sergeants who were standing up for their men. He started a company orientation meeting by threatening to restrict the 6th Platoon over the following weekend if its lockers were not arranged according to the standard operating procedure. Sergeant Kimura objected that this was impossible because of the peculiar manner in which the building had been constructed. He indicated that the squad leaders had attempted to arrange the lockers according to regulations but had found that the clothes racks were so placed on the walls that strict compliance would not leave enough space for all the beds. He argued that the present arrangement was the best that could be done and asked that the unit not be punished for doing its best. When Sergeant Endo still insisted on compliance, the platoon sergeant became incensed and asked for a demonstration of how it could be done.

Sergeant Endo responded coldly, "I don't have to demonstrate anything. I said that lockers are going to be arranged according to SOP or else the platoon's going to be restricted."

"But I tell you it can't be done."

"Don't talk back to me! You heard what I said!"

Before Sergeant Kimura could say any more, Lieutenant Catalano got up and said that he would not tolerate any "sassing back" of his first sergeant; Sergeant Endo had his complete backing, and his orders would be obeyed. He also threatened to "bust" Sergeant Kimura, if he were found guilty of insubordination again.

After the first sergeant had harangued the company for another ten minutes the platoon sergeant of the 4th Platoon asked if the men could be allowed to hang up their OD clothing instead of keeping them folded in their barracks bags. Since the season for woolens would arrive in a few weeks, many had had their winter clothing cleaned and pressed and preferred not to fold them up again. He asked therefore that permission be granted for both khakis and ODs to be placed on hangers just for the next few weeks. Sergeant Endo became very angry: "How many times do I have to tell you guys? We've got an SOP, and the SOP will be followed."

The reaction of the company was almost unanimous; seething anger focused on the first sergeant. Some were so angry that they did not say anything until they got back to their bunks. Chafing under the constant intimidation, the men agreed that the sergeant was "mad with power"; no Nisei officer would be so stubborn in enforcing petty regulations. The sergeant was going too far in asserting his authority.

"Who da fuck he t'eenk he is? Just becaus' he get four stripes he t'eenk he king sheet. He drunk with power. Bust up, dat kind!"

"Ooo da *manini* bastard, yeah? Bust up! Bust up!"

"Jesus Christ! I've seen a lot of pricks in the army, but he takes the cake—and a Buddhahead at that too. We gotta beat the shit out of him."

There was only one topic for conversation. Even those going out on pass took their time dressing and stood about talking with their friends. When one of the eager beavers began complaining about ratings, he was cut off by a curt observation: "Goddam! I never saw so many cheap, lowdown, double-crossing bastards until I got in a Boochie outfit down at Blanding. I never met guys like that before. I didn't think Boochies could be like that. Goddam! Talking about ratings at a time like this!"

The men agreed that something would have to be done, and the most frequently proposed solution was violence. After the beating of Yamanaka in the 6th Platoon, one common comment was, "Too bad Endo wasn't the guy who tried to stop the fight." Many mainlanders indicated that they wished the Hawaiians would beat Sergeant Endo, and some even stated that they would be willing to join them. They even condoned the ganging up tactics, if used against the sergeant.

When the company returned to the barracks at 15:30 on the following day, everyone was ordered to stand by for a locker inspection. Sergeant Endo emphasized that it would be a strict inspection. Friends of those who had decided to disappear went after them, and all but a few returned in time. No purpose was stated, and it was assumed that the first sergeant was on a personal vendetta. It soon became apparent, however, that this was not the case. The company commander, the first sergeant, and the company clerk went through, looking casually at the lockers but asking each man to display his stationery. From time to time the lieutenant stopped to ask a man if he did any sketching. Most men were accustomed to "chicken shit," but they were puzzled; it was not like Lieutenant Catalano. Why was the inspection being held? What were they looking for? They were obviously searching for some particular item, and it had something to do with writing paper.

During the study period after 19:00 several rumors about the inspection spread through the classrooms. The first was that someone had drawn an uncomplimentary picture of the first sergeant.

"They were looking for the guy who drew a cartoon of Endo like a devil. He looks like a fuckin' bat anyway."

"I'll be a son of a bitch! That's what they were looking for, huh?"

"So some guy drew a picture of that prick with ears, huh? I wish I could've seen it."

The rumor seemed plausible. Most of them saw the sergeant as a tyrant who would not hesitate to use his power to get even with personal enemies. Before long, however, the instructors revealed that the inspection had not been confined to Company K. During the second hour another rumor developed that someone had drawn an obscene picture and had sent it to one of the Nisei WACs. When she reported the incident, the administration had ordered the inspection for the entire battalion.

"I heard that somebody drew a picture of a WAC on a stool with a GI peeking through a window, and they were checking up the texture of the paper to see if they could find the guy that did it."

"Haw! Haw! No shit! Is that all? You mean to tell me they went to all that trouble just for that?"

Tension was broken temporarily by the outcome of the inspection.

In their common hatred of "Batman" the men closed ranks to protect one another. On the morning after the inspection the platoon leader of the 6th Platoon called an informal meeting before reveille and warned his men that the first sergeant was out to "get" Sergeant Kimura. He told them that accurate roll calls would have to be taken; if the first sergeant found a discrepancy, he might charge the platoon sergeant with dereliction of duty. He told them bluntly that he expected every man to be in all formations—and on time. No protests arose. Sergeant Kimura, who was from New Jersey, was just as baffled by the Nisei code as the first sergeant, but he was more flexible and went along with the others. He had a keen sense of humor and was well liked. He had covered many times for their being AWOL. He had lied for them, and he had fought for them. Knowing that their sergeant was in trouble, the men suddenly became obedient. The lethargic few were pushed by their neighbors before the squad leaders had a chance to say anything. As the 6th Platoon began to report for all formations, others in the company were surprised. They could not understand until a rumor spread that Endo was trying to "bust" Kimura.

By this time it became apparent that all the noncommissioned officers in the company, student and cadre, were against the first sergeant. The usually reticent NCOs criticized the sergeant openly as unreasonable and unfair. Word soon spread that the NCOs had had a meeting with Lieutenant Catalano at which they had told him what they thought of the first sergeant; the rumor reflected the faith the men had in their company commander's sense of fair play. Another rumor was that Sergeant Kuroda had challenged Endo to a fight. Kuroda, platoon sergeant of the 1st Platoon, had known Endo for years. His turning against the first sergeant was taken as evidence that Endo did not have a single friend left in the company.

During the weekend of September 8 and 9 the mounting rage against Sergeant Endo erupted in another beating. The 5th and 6th Platoons were restricted, and most of these men decided against going AWOL when it was rumored that Lieutenant Catalano would be the duty officer for the weekend. It was assumed that the lieutenant was being punished for the misconduct of his company. Since he knew every member of his organization by sight, those who were restricted did not wish to risk being caught. Even members of other platoons, who had passes, would have to be careful, especially at the YWCA dance. As a result, more men than usual were on hand when the beating they had hoped for finally took place.

At about 03:00 on Sunday morning Sergeant Endo was awakened roughly by Private Fukuhara. The diminutive Hawaiian challenged him to step outside to fight. The sergeant declined, telling Fukuhara that he was drunk and should go to bed. Fukuhara continued to yell at the sergeant, and those waiting outside joined in jeering. When the sergeant still refused to get up, Fukuhara proceeded to pummel him as he lay in bed. The sergeant made little effort to defend himself other than to cover his face with his arms. No one came to his aid. After the gang had left, the sergeant walked to the latrine to wash the blood off his face. He was still alone.

Only the eyewitnesses knew of the incident until breakfast time, when rumors developed quickly. In the latrine, in the recreation room, in the barracks, the beating was discussed avidly. Several woke up their friends to tell them the "good news." Hawaiians and mainlanders alike rejoiced. A common comment was, "It's about time!" At the moment no one considered the possible consequences to the company and to Fukuhara. Rumors about every detail of the beating were shared and compared. A lone Hawaiian had aroused the sergeant. Endo looked outside, and when he saw a number of others waiting for him at the door, he decided to remain in bed and let the little man beat him. Another rumor provided a motive: the "little guy" was infuriated because the sergeant had "talked out of turn" about cleaning the latrine windows. The company commander had ordered him to clean the latrine only, but the sergeant had stepped in to remind the lieutenant about the windows, thus making extra work for Fukuhara.

The sergeant received no sympathy whatever from anyone in the company, including mainland Nisei who condemned Hawaiians for their unfair fighting tactics. Most of the men received the news with joy. No one dared to speak in favor of the sergeant. Hawaiians who felt badly about the gang beatings were quite proud of this affair. "Dees time O.K., yeah? Fair fight—one to one." Although many felt in principle that it was unwise to assault noncommissioned officers, they insisted that this was an exception. As one man put it, "Sooner or later he would have got it anyway. If these guys didn't do it, somebody else would have. It was just a matter of time. Anyway, he had it coming to him." The major disappointment was that the sergeant had not been badly hurt. "Shit, they should have done a better job." He had suffered only a few bruises, and some spoke of a repeat performance to make the injuries more substantial.

Fukuhara was arrested on Monday morning, and several rumors developed about him. One was that he had been "bucking" for a discharge; another was that he weighed only 110 pounds; still another was that Fukuhara had yelled at Endo as the MPs took him away, "I'll get you!" One question that arose was: would he "squeal" on the others who had been present? As a rumor that enlisted men who struck NCOs and officers had to go to Leavenworth was considered, groups of Hawaiians got together to discuss the situation. Tajima looked worried, and other Hawaiians spoke sympathetically to him. Most mainlanders sympathized with Fukuhara. After all, he had only done what everyone else in the company wanted to do. Fukuhara was a hero. They knew that he would face a court martial, but they felt that he was being punished unjustly.[1]

Everyone expected the company commander to be angry. Would there be another restriction? They assumed that Fukuhara would be punished, but what else would happen? At the 15:30 formation a strict roll call was held, and all men were present. No one was surprised when physical training was replaced by a company meeting.

Lt. Catalano addressed them:

> Men, if you want to fight among yourselves, that's up to you. As long as I don't catch you myself, it's O.K. with me. But for God's sake, men, let's fight like men. This gang fighting is going to stop. I think it's the most cowardly thing I ever heard of. If a man is too yellow to fight his own battles, he's got no business fighting. Now, I didn't know about the mess in the dayroom the other night until last night. Your noncoms are with you, men. No one told Sergeant Endo or myself about it. I happened to hear about it indirectly last night when I was over at Company D. I was kind of embarrassed because they knew all the details of a fight in my company that I didn't even know about. They thought I knew all about it. That's O.K., men, I don't want you to be a bunch of squealers. But this cowardly fighting is going to stop. One man against a half dozen! Is that your idea of a fair fight? If it is, I don't want to have anything to do with ya.
>
> Now I want to tell you about this incident last night. I said last week that I would back up all my noncoms, and I meant it. I don't believe in pulling any underhanded stuff; so I'm going to have to give it to you straight from the shoulder. This kind of stuff isn't gonna go. I'm gonna throw the book at

1. Fukuhara was tried on September 19 and was sentenced to four months of hard labor, reduced in grade to private, and ordered to forfeit $18.00 a month from his pay for four months.

ya, if you do anything like this again. What kind of man are you to go pick a fight with one man—with a big gang outside to back you up? Is that your idea of being brave?

This man Fukuhara was supposed to be restricted over the weekend. But he wanted to go into town so I gave him a pass. What does he do about it? He comes back and beats up Sergeant Endo. He beat him up while the sergeant was in bed. Now, men, wasn't that a cowardly thing to do? He had a great big gang waiting outside just in case he couldn't handle the fight alone. What kind of guts is that? I can't think of anything more cowardly than gang fighting.

If some of you men want to fight so much, I'll make you a standing offer. I'll fight any man in this company that wants to fight. I'll take off my bars and fight him man to man. Just come to the orderly room, and tell me you want to fight. We'll have it out. If you win, O.K. We'll forget about it. But if you lose, you'll take a court martial. I'll give you the thrashing of your life and then throw the book at you. But that offer is good, men. If any of you want to fight, just come in and let me know. I'll fight ya.

I don't have to tell you that Fukuhara is now in jail. I don't have to tell you he won't get any sympathy from me. I did him a favor, and look what he did. I trusted him once and gave him a break. He's going to get the book, men, and so will anybody else that tries the same thing.

The lieutenant again asked if there was anyone who wanted to fight him. In the absence of any challenges he asked if the men who had engaged in the brawl in the 6th Platoon were brave enough to accept his offer. Others glanced at the three who had started it, but no one spoke. The lieutenant then ordered all NCOs to report to his office and dismissed the company.

On Thursday, September 13, Sergeant Kimura was transferred out of the company. This was widely interpreted as a repercussion of the beating. Since Sergeant Kimura had had nothing to do with the incident, many objected that he was being punished unfairly. The sergeant himself insisted that he had gone to the chairman of Division F to request a transfer; since he was from the East Coast and had only a meager background in Japanese, he could not possibly learn enough to qualify. Although it was known that he was also a high-point man soon to be discharged, most men assumed that he had been sent away at the insistence of Sergeant Endo.

The Growth of Interethnic Tension

In September 1945 the student personnel of Company A—the OCS company—consisted of 239 enlisted men. Of this number 199 held the rank of T/5; the rest were in higher grades. At that time no Nisei were included, and Company A men became the target of hostility. The policy of giving all Caucasian graduates commissions, regardless of their linguistic ability, while sending Nisei graduates overseas as NCOs had long been a source of resentment. Although there were a number of Nisei officers at the fort, none had received their commissions so easily. Many had earned battlefield commissions; others were "ninety-day wonders" who had gone through the regular OCS training at Fort Benning after superior performance elsewhere, in many cases overseas. That Company A men were sleeping in more comfortable quarters—attractive white barracks ordinarily found only in permanent military posts—was also resented; most Company K men overlooked the fact that the majority of Nisei students had even better quarters—the brick buildings. Furthermore, although well over 90 percent of the troop strength at Fort Snelling consisted of Nisei, passes to various musical and stage attractions in the Twin Cities were distributed mainly to Caucasian personnel. Only a few in Company K actually wanted such tickets, but they were displeased in principle. All this was regarded as unjust discrimination, and the tension that had been developing within Company K soon became directed against the OCS men—the beneficiaries of these policies.

For some reason two Caucasians had been assigned to Company K, and the snobbish attitude of one of them reinforced stereotyped beliefs about white people. Private Janssen was a squad leader, took part in the activities of the group, and was considered a "regular guy." But Private Soboloff, who was in the 6th Platoon, could not get along with the others from the beginning. He made it clear that he did not want to be in the company and spent all his free time across the street in Company A. Since other Caucasians were being trained to be officers, he could not see why he was singled out to be with the Nisei. During the first month he ignored those around him and was generally absent from formations. The others also disregarded him, even when he subsequently tried to be friendly, charging that he considered himself too good to stand in line with them. On one occasion, when he was called out of formation by the company commander, he marched briskly up to the lieutenant in a military manner and saluted. Although he was following regulations, even the company commander was surprised and fumbled in returning the salute. This led to his being labeled as "GI" and "chicken shit"; he was a man who could not be trusted. When Sergeant Kimura was in trouble with the first sergeant, special pressure was placed on Soboloff. Several became concerned about his continued absence from formations, and finally one man told him bluntly to fall out with the others or suffer the consequences. On September 5 he was transferred to Company A.

On September 3 an OCS man committed suicide. Rumors spread quickly, and at that time the reaction was still mixed. Some Nisei were quite sympathetic: "Poor guy. I don't blame him. I can understand why a guy would commit suicide in a place like this." On the other hand, others were resentful enough to yell, "Good riddance! The dirty son of a bitches!" During the evening meal Company A students were asked for confirmation of the rumor, and it was at this time that a few began to realize how demoralized some of the OCS men were. They substantiated the rumor and then went on to discuss ways of getting a discharge or transfer; they wanted to get out of MISLS as badly as some of the Nisei. One of them commented, "Yep, he had one meal of baloney too much."

Bitter animosity between members of the two companies developed out of consistent violations by the OCS men of an agreement concerning priority at the mess hall. An understanding had been worked out between the two commanding officers that the duty platoons of both companies would eat first; then, the various platoons of each company were to take turns being first in line. Both companies were supposed to march to the mess hall in the proper order, but the OCS men did not honor their part of the agreement. When it was Company A's turn to eat first, this did not matter; but tempers flared when it was Company K's turn. While the Nisei were marching, the OCS men simply ran ahead and got into line in front of them. The Nisei could not run because they were in formation; all they could do was to curse the "Haole bastards" as they ran by. The Nisei NCOs objected, but their protests were of no avail. Formal complaints were lodged repeatedly through the orderly room, but the commanding officer of Company A only gave vague assurances, and the conduct of his troops did not change. On August 21, when Company K was scheduled to eat first, the Nisei for the first time were given permission to break formation and to "run for it." Although there was considerable shoving, the racing was initially in good humor. Most accepted the principle of "first come, first serve," and those who had been slow relinquished their places without argument. But a few Nisei tried to push Caucasians out of line, reminding them loudly that they were not supposed to be there. All arguments were stopped before any punches were thrown, however, when others in both companies stepped in to separate the irate disputants.

Being first in line was important not because the food was so tempting but because it was so frequently in short supply. Since the Consolidated Mess Hall served the induction and separation centers as well as the two student companies, the number to be fed varied considerably from day to day. On some days those who were toward the end of the line were given nothing but cold cuts or portions so small that they were still hungry. During warm weather losing one's place usually entailed nothing more than standing in line for an additional half hour. With the coming of cold weather other discomforts were added, and tempers became more volatile. Hawaiians in particular were not accustomed to the cold; they put on more clothing than the others but were still miserable as they stood about shivering. More and more began to argue that if arrangements could not be made satisfactorily through channels, they would have to take matters into their own hands.

Getting one's rightful position in the "chow line" soon became a daily preoccupation. Shoving became commonplace; as irritations mounted on both sides, tempers flared from time to time. During lunch on Saturday, September 15, an open clash was barely averted. Again several OCS men ran ahead of the Company K platoons and got into line in front of them. Two Nisei technical sergeants went up and ordered them out, but they refused to obey. A few tried to make clever jokes, and one man contended that officer candidates had the right to eat first. Several Nisei wanted to fight; one even removed his glasses in anticipation of jumping into the fray. At this point Lieutenant Catalano appeared on the scene. He told the OCS men firmly that they were to get out and that he would eject them personally if he ever caught them in the Company K line again. The lieutenant was enraged, and the castigated soldiers left quietly. As they walked toward the end of the line, they were greeted by jeers as well as challenges to fight. The Nisei knew that their company commander was with them in getting what was rightfully theirs. They chuckled among themselves about how the "old man chewed out the Haoles," and in their merriment the tension was relieved.

At lunch time on Monday, September 17, resentment against Caucasians was heightened in another clash. When the Company K formations reached the mess hall, the OCS men were already there. The line was long and congested. A delegation of Nisei went up to ask them to move, and the following conversation took place:

"Hey, how many times do you guys have to be told? You ain't supposed to be up here. It's Company K's turn to eat first."

"Is it? I didn't even know we were supposed to take turns."

"Aw, don't hand us that shit. We know fuckin' well you know, and so do you. So how about gettin' the fuck out of line?"

"Why should we? We were here first."

"I notice you guys moved when Lieutenant Catalano told you to get out."

"Yeah? Well, we ain't movin' now."

"You guys want trouble, huh?"

"Hell, no! We're not looking for trouble. You're the guys who always start the trouble. We're not bothering anybody. We're just standing in line to eat chow. We got just as much a right to be here as you have, *maybe more*."

"What do you mean by that last crack?"

"I didn't know it was Company K's turn to eat first. Nobody told us anything about it."

"O.K. We don't want no trouble either. What's your name? We'll turn it in to your CO and see what happens."

"Sure, go ahead. Here's my dogtag. Take my name. Our CO won't do a damn thing. He doesn't give a shit what we do."

"Hell, no. Our CO doesn't give a shit. You guys might as well get wise on this deal. I know your CO gets pissed off about it, but he's wasting time every time he comes to see our old man. Our old man just doesn't give a shit if we eat before you guys or not."

"What the hell you think we made arrangements between companies to take turns eating first for?"

"We don't know anything about it."

"How come you guys are so fuckin' smug when it's your turn to eat first? You throw our guys out of line. Now you say you never heard about the rule. How about that? How would you like it if we cut in on your chow line?"

"Go ahead. Try it."

"O.K. If that's the way you guys feel about it, that's O.K. with us too. Don't cry if you get hurt though."

"Listen. Why start trouble? We got as much right to be here as you have."

Most members of the Nisei delegation came away with the impression that the Company A men were saying that they had the right to eat first because they were officer candidates. Some went so far as to say that these men felt that they should have priority because they were white. In fact, most Nisei interpreted the remarks in that light and strongly resented the implication of superiority. As one man muttered, "The bastards! They think they're better than we are just because their skin is white. We're just as good as they are. We're just as American as they are. We're in this shit just as deep as they are. We're all in it together, and we don't have to take any more shit from anybody."

After a long wait each Nisei was served one wiener, some beans, some celery leaves (no more stalks were left), and crackers. One man threw his helmet liner on the floor and stomped out without eating. Those who ate found that it was not enough; as unappetizing as it was, some went back for "seconds" only to discover that there was nothing left, not even celery leaves. Several irate men went to the kitchen to protest, but the mess sergeant insisted that there was no food left. Resentment was then directed at the OCS men. By eating out of turn they had upset the mess schedule and had deprived them of food that was rightfully theirs.

The bitterness was intensified during the next two days. On Tuesday the Company K duty platoon marched to the mess hall at 06:20, the appointed time. They found a line over seventy-five yards long at both wings of the mess hall, a line made up of recruits who had just been inducted the day before and a large number of OCS men. There was no objection to the inductees' being in line, but it had been understood that Company A personnel were to eat after the duty platoons. After standing in the driving rain for a half hour each man was served one strip of burned bacon and one half-cooked pancake. The pancakes had turned out so badly that the cooks apologized as they handed them out. No butter or syrup was left on the tables. The fruits were all gone. Again the OCS men had eaten all the food. The cooks prepared meals in accordance with the number expected on each shift; any irregularity led to those at the end getting little to eat.

The few Nisei who were acquainted with some of the A Company personnel tried to indicate that the OCS men were so demoralized that they simply did not care, but their explanations were of no avail. The argument that the OCS men themselves had nothing to do with setting up the discriminatory system was accepted, but it did not matter. The *Keto* were getting out of hand and asserting their superiority. The general hostility against Caucasians became so pronounced that diverse events were viewed in the light of their propensity to discriminate. For example, when the pianist Hazel Scott performed in Minneapolis, there was considerable discussion of the recent refusal of the Daughters of the American Revolution to allow her to use their auditorium. Even those who disliked Negroes complained that it was undemocratic. One man commented, "Well, I know one thing. After all this, when I get back home again, I won't take no shit from any of them Haole bastards. What the hell are we fighting for, anyway?" One afternoon, when the company was being marched to school by a Nisei lieutenant the men had never seen before, the formation was suddenly halted. The lieutenant ordered two students who were marching without headgear to run back to the barracks to get their fatigue caps. When it became apparent that the column had been stopped by a Caucasian captain, wrath turned against the Haole: "GI!" "Chicken shit bastard!" "*Urusai yatsu da na*!" [What an obnoxious guy!] It became increasingly more difficult for those without ill will to speak up in defense of the OCS men. A rumor that officer candidates received T/5 pay while attending school aroused additional resentment.

As hostility mounted, there was increasing talk of the uselessness of Caucasian language officers overseas. Most Nisei veterans who had returned to Fort Snelling for special duties were contemptuous of them. They indicated that, while a handful of exceptional men were competent linguists and officers, most of them only got in the way. One sergeant, who had just returned from the China-Burma-India theater, dropped by Company K to visit a friend. When asked about MISLS officers, he replied:

You guys have to take a lot of shit from them guys now, but you don't have to worry. Wait 'til you get overseas. Over there the *Keto* might be lieutenants, but they ain't worth a damn. We used to use 'em for typists. We did the translating, and they just typed it up. You don't have to take any shit from 'em overseas. If they get tough, just tell 'em off. They ain't worth a shit. They can't even interview prisoners 'cause the prisoners can't understand their Japanese, and they can't understand the prisoners. The Nisei do the work overseas. Don't worry. Just wait 'til you get over.

Many had received letters from friends and relatives overseas, supporting what the sergeant said. That such soldiers were getting commissions solely on the basis of ancestry was discussed over and over, and their getting better food in the same mess hall only intensified the antagonism.

Them bastards are gonna be our officers! Goddam, if that ain't the last straw! Fuck them bastards! They're all buckin' for their bars. They think their shit don't stink. They think just because they're *Keto* and their skin is white they got a right to eat ahead of us all the time. How would you like to get pricks like that for your officer, huh? In basic the officers used to eat last. They always took care of their men first. Look at these bastards! And do they know any Boochie? Shit, the Japanese won't even know what the hell they're talking about. They're worse than us, and that's pretty bad. We're the worst of the Nisei, and those guys are worse than us. They're good for nothing, but they get bars because they're Haoles.

Caucasian language officers were occasionally assigned to the company during the physical training period, and their obvious incompetence as soldiers reaffirmed once more the belief that their rank was undeserved. Nisei officers knew how to issue commands in close order drill, in marching, and in running them through obstacle courses or speed marches. They had gone through regular OCS training, but the Caucasian MISLS graduates had had virtually no military experience. Although each company was led by its commanding officer in dress parades on Thursdays, junior officers were often assigned this task in the Tuesday practice parades—enabling all the COs to go to the reviewing stand to watch the performance of their units. Since those in Company K were proud of their marching ability, they were further irritated by the ineptitude of inexperienced officers. Most of the language officers simply did not know what to do; one of them even had difficulty keeping in step. They were obviously nervous, and some confessed to those in the front rank that they did not know what commands to issue. During one practice parade, when an order was given for "company mass right," the officer only turned around helplessly, and the men had to use their own judgment, moving to the proper places. The formation became uneven, however, for some squad leaders waited longer than others for an order before turning off on their own initiative. This aroused many resentful comments in the ranks, comments that were made loudly enough for the benefit of the officers.

"Look at these fuckin' Haole desk officers! They don't even know how to give commands in a simple formation like this! They can't even keep in step! How come guys like that get bars? They get 'em 'cause they're white."

"Jesus Christ! Dese bastard's gonna fuck up da company. All da guys know what to do except dem. What for we get officers like dat?"

"Fuck dese goddam Haoles! Dey don't know naut'eeng, but dey get bars. Why dey no give Buddhaheads bars? Maybe dey afraid Buddhaheads too good, yeah?"

"Look at that! The guy can't even keep in step!"

"More better with no officer, yeah? We no need dat kind. Waste time!"

When the matter came up at a company meeting, Lieutenant Catalano agreed that the "chair-borne" officers of MISLS were "no good." Most men felt that any officer in the U.S. Army should know at least the fundamentals of close order drill; they found it difficult to respect one who could not even lead a group in simple formations or call privates to attention without making them snicker. The only reason these men wore bars was that they knew a few Japanese words and happened to be Caucasian; they certainly would not have qualified as infantry officers.

On September 20 the lieutenant announced that the parade that day was to be in honor of the 3rd Infantry Division and that movies would be made of the formations. "I don't have to tell you men to be on the ball. You know what to do."

"Sir, are we going to have our own officers?"

"You mean Nisei officers?"

"Yes, sir. It's kind of hard to march good with them other guys."

"Yes, I know. We'll have our own boys."

A daydream divulged by one of the men reveals how the frustration and resentment were eliciting aggressive impulses:

We get in a big gang fight with the OCS guys. The whole thing starts when one of them calls me a "Jap." I take a lot of shit from him without saying anything until he calls me that; then I let him have it. All the boys are there, and they back me up. They let me fight this guy alone, and they take on the other Haoles. I beat the shit out of the guy. The next day they find out at the hospital that the guy's got two broken ribs, a broken nose, and a broken jaw. They got twenty other OCS guys in the hospital. All I get is a skinned knuckle. I go in to see the CO and tell him all about it, and the CO says I got a right to hit any guy who calls me a "Jap."

Hostility was especially noticeable among some of the Hawaiians, who expressed hatred of such intensity that it surprised most mainlanders. When they became inebriated, some of them would whip out a switchblade and growl menacingly, "Me, I like cut up Haole!" Mainlanders had heard that the Hawaiians enjoyed "racial equality"; furthermore, the Japanese there had not been evacuated. Why were they so bitter? When asked about it, one otherwise quiet Hawaiian replied:

Where you get dat racial equality sheet? Dat's a crock of sheet! We no get racial equality een Hawaii. Yeah, we get equality among da Kanakas, Chinese, Japanese, Filipinos. Yeah, we get dat kind. But da Haole, dey on top. Dey always on top. You get no chance if you no Haole. You go work on plantation, yeah? You do work of bookkeeper. Dere's one Haole dere too. He your boss. You do all da work, and he sit on his ass all day. You get 120 bucks a month, and he get 400. Dat's what dey mean when dey talk about racial equality. Sure, more better dan over here, but dat ain't much.

Another man concurred:

Why you t'eenk we talk funny English, huh? Maybe you wondah, huh? I tell you. In Honolulu dey only get one good school. Before you can go een da first grade, you gotta pass a language examination. Us guys, we grow up weeth Buddhaheads, yeah? We no know no English, yeah? We no can get een. Nobody can get een. Only da Haole kids get een. Sure, dey no get rule saying Buddhaheads no can get een. A couple of guys get een O.K., if dey got Nisei folks who speak English. But most of us no can get een. We never learn how to talk right.

You remember da Massie case? Remember when dey had a big mess when some Haole navy guy killed a Kanaka? Ever since dat time all the colored people een Hawaii hate da Haole. Dey hate 'specially da navy Haole. A lot of Haole t'eenk we forget, but we no forget. We no can. We always remember what dey deed to us.[2]

Although antagonism between the rival companies continued to mount, on the morning of September 25 the clashing groups were momentarily united. A wild race for a place in line led to several heated arguments. Members of both companies readily yielded places to others in their own group, but they were unwilling to yield to anyone else. The OCS men were not nearly as belligerent as the Nisei. "Bust up! Bust up! Dey like fight!" "Take their name tags!" "Pull off their buttons!" "Bust up! *Nagure*! *Nagure*!" [Strike! Strike!] After all the shoving and struggling the men found themselves staring at a breakfast consisting of two half slices of burnt bacon and a piece of toast. It was the first time that bread had been toasted, and only one slice was allowed per person. Since they sat down in the order in which they had entered the mess hall, most tables contained mixtures of men from both companies. Some continued to glare at one another, but most of them were so astonished at the fare that they joined forces in venting their fury on the kitchen staff. For once members of the two companies agreed; they were outraged.

At one table a Nisei reflected, "Goddam! They must be afraid we might get constipated from overeating! The bastards!"

2. In 1931 a naval officer named Thomas Massie was tried for the murder of a Hawaiian who allegedly attacked his wife. After a long drawn-out trial he was convicted, only to be pardoned by the governor of Hawaii. The pardon was interpreted as an assertion of the principle of white supremacy (Packer and Thomas, 1966).

An OCS man concurred, "Boy, you ain't a kiddin', Mac. These guys are so tight, if they squeezed the food any tighter, they wouldn't even get juice."

They were very angry. Together they shouted insults at the cooks as they left, and most of them temporarily forgot their grievances against each other.

Most Company K men were so enraged that they decided to send a formal protest to the administration. A committee of college graduates was asked to draw it up, and it was signed by everyone in the barracks at the time it was circulated:

> We, the undersigned men of Company K, School Battalion, wish to call the attention of the Commandant to the following conditions existing at the Consolidated Mess Hall at which we have been eating since the activation of the company:
>
> (1) Food is generally poorly cooked and unpalatable. The seasoning is poor, and many items are far too greasy. There is little variation in the menu, French toast being served for every other breakfast and boiled potatoes being served with all other meals.
>
> (2) Food is rarely served in sufficient quantity, making it necessary for men to go through the cumbersome procedure of cutting into the long mess lines for "seconds." Upon several occasions, one slice of baloney, some dessert, a scoop of boiled potatoes, and bread had to suffice for a "meal."
>
> (3) Frequently the men serving the food have disagreeable attitudes, sometimes apparently taking delight in giving out small portions.
>
> (4) The utensils are generally not sufficiently washed and unsanitary. Diarrhea in Company K is such a common occurrence that men are constantly being excused from classes to relieve themselves.
>
> (5) Because of poor organization it takes much too long to be served. Most of the time is taken in standing in long queues waiting one's turn. With the coming of more inductees, the activation of a new company, and the arrival of cold weather, the perpetuation of this practice is hardly conducive to raising the morale of any of the men.
>
> Reports of the above conditions have been placed in the "suggestion box" on several occasions since the activation of this company on 22 August 1945, but thus far no action has been taken. We should therefore like to know whether steps have been taken toward setting up a mess hall of our own or toward making changes in the above conditions.

Despite the momentary agreement between members of the two companies in condemning the mess hall staff, that very night interethnic violence erupted for the first time, when an OCS man was beaten. The trouble had begun in fact at supper, as the men from both companies raced for places in line. Private Takeda and a Caucasian reached the line at about the same time. They pushed each other; neither would yield, but others separated them before they could start swinging. Takeda swore vengeance. Later in the evening Takeda and his friends caught the same man in the PX. The argument was resumed, and within minutes the OCS man was set upon by several Hawaiians who beat him mercilessly. It was the same night on which the trio had picked on Saipan. Rumors of both incidents spread quickly in the company. Most men expressed displeasure over the tormenting of Saipan, but they were delighted that someone had finally thrashed an OCS man. Even those who deplored violence felt that the *Keto* had finally gotten what he deserved.

When they learned of the initial beating, many OCS men were enraged. They regarded the attack as entirely unjustified, for they could not understand why the Nisei were so bitter toward them. The race for the mess hall was viewed only as a contest to see who would eat and who would get the scraps. Since they had studied Japanese culture, they thought they knew all about the Nisei, and many were puzzled. The assaulting of one man by a half dozen was resented as cowardly, and about twenty started out for Company K to seek out the assailants and to retaliate. Those bent on vengeance, labeled by others in Company A as the "Southerners," were restrained. They were reminded that within the fort they were outnumbered by Nisei by a considerable margin and that some other solution should be found. After some deliberation it was decided that the matter should be forgotten. The victim went to the dispensary for treatment but refused to divulge the source of his injuries. Had any OCS men invaded Company K that night, they would undoubtedly have received a violent reception.

Two days later more difficulties erupted. Since the Nisei now realized that they would only get scraps if they were toward the end of the line and because they could no longer eat at the mess hall on the Turkey Farm, many had begun to run directly to the Consolidated Mess Hall as soon as classes were dismissed at noon instead of marching back in formation. On this particular day some forty or fifty ran for the mess hall, only to find their way blocked by a Caucasian officer who started taking their names. Before they could escape Lieutenant Catalano appeared and ordered them to surrender. Those who were too close to him to get away did so, but the others continued to run. Some ran all the way back to school and marched in with the rest of the company.

Immediately after lunch an announcement was made over the public address system that unless all those who had broken formation gave themselves up in the orderly room, the entire company would be restricted over the coming weekend. This led to much bitterness. While realizing that several regulations had been violated, many felt the action was justified; for it was the only way in which one could get enough to eat. A rumor that the officer taking names was from Company A evoked screams of protest. It was the Company A officers who were at fault; they did not hold their subordinates to the intercompany agreement. Even Lieutenant Catalano's wisdom and fairness was questioned.

After most of the men had returned to the barracks, the platoon sergeants told the guilty ones to give themselves up; they were "fucking up the whole company." One by one they went. Finally, a few minutes before school formation, someone yelled over the "bitch box" that "the CO wants fifteen more guys." Those who had been holding back got up, and so many reported to the orderly room that latecomers were told that there were already enough names on the list and that they should return to their barracks and "forget it." Those who confessed were given a three-day restriction during which they were required to sign in hourly at the orderly room.

At an orientation meeting on the following day the company commander warned the men not to break formation again. When they objected that the OCS men were doing it, he promised that he would see to it that they were adequately fed—"whether the CO of A Company likes it or not." At supper he kept his promise. Although he was free to leave at 16:30, he remained in the company area and joined the mess formation. As the platoons lined up, he went forward and ordered all OCS men out of the line. He then went inside, stood behind the KPs serving food, and inspected each portion. He asked each man if he had enough; if there were any complaints, he ordered the KP to give the man more food. The mess sergeant was helpless. He had ordered the KPs to "go easy on the chow," but with the lieutenant standing there he could not enforce the order. The lieutenant remained there until everyone in Company K had been served; he then walked up and down the aisle, telling them to go after "seconds" if they were still hungry. There were no further difficulties that night, and the men chuckled among themselves about the forlorn look on the mess sergeant's face.

On the evening of October 1 there was no jostling with the OCS men; they were all away on furlough. However, other difficulties arose with the kitchen staff. Dinner consisted of boiled carrots, beans, and celery leaves! Nothing else was out on the trays! The first few men to be served complained bitterly, but the mess sergeant refused to listen. One of them dropped his tray and ran to summon Lieutenant Catalano. He had already left the company area, but the CQ was able to reach him by telephone. As the line went through, almost everyone shouted some invective. Some demanded seconds and thirds to get enough to eat; others refused to touch the food and went to the PX to buy some candy. In less than fifteen minutes Lieutenant Catalano was in the mess hall—and with him was a major. As the two walked up and down the dining hall, the men, one by one, showed them what they were being fed. Both officers were visibly shaken. As the men filed out of the mess hall, they saw the mess sergeant standing rigidly at attention, being lectured by the two infuriated officers. One observer noted confidently, "This kind of shit's gonna stop. The old man won't let 'em do it."

Hatred of the MISLS Administration

The high command of the MISLS, which included a number of officers and civilians of Japanese ancestry, had always included itself among the "friends" of the Nisei. In pleading the case within the army for the persecuted minority, these officials had to overcome the skepticism and suspicions of their superiors. It had been a hard struggle, but the exemplary conduct of Nisei troops at Camp Savage and Fort Snelling as well as their performance overseas vindicated the confidence they had shown. Most of the high-ranking Nisei were college graduates, men who were assimilated to American ways and had little difficulty getting along with Caucasians. Although some of these leaders were scornful of the "average Nisei," now that MISLS was firmly established, they were proud of the unit's achievements. These officials were stunned by reports of widespread insubordination and by the outbreak of violence. Their initial reaction was one of anger, much like that of a schoolmaster at an ungrateful imp's refusal to cooperate. The "bad apples" had to be rooted out. With vengeance they turned on Company K, but their tactics only led to more difficulties.

The manner in which these officials were viewed by most Nisei enlisted men was revealed by the widespread suspicion that they were doing everything they could to retain linguists as long as possible. A radio newscast on August 13 revealed that some congressmen were considering terminating the Selective Service System. This brought a sudden chill to low-point men; if the draft were stopped, no replacements would be available for them. As the number of points needed for discharge continued to drop, many draftees at Fort Snelling became preoccupied with the question of whether those in G-2, especially linguists, would be released on the same basis as other soldiers. Or would they be classified as "essential" and be forced to remain? The prevailing view was that the MISLS administrators would do anything to further their personal interests, even at the expense of their troops. One man's comment reflected the general suspicion of most of his comrades:

The reason why them bastards are gonna declare all Asiatic language guys as "essential" is that the guys on the post here got jobs that depend on the place stayin' open. They can't be big shots unless things stay like this. Under the table of organization you gotta have so many guys under your command to keep a high rank. If they close up this place, them guys'll be out of work, and they'll get demoted to their permanent ratings. Sure, there are a lot of Boochies in Japan who know English, and they can be used for civil affairs. You can't trust guys like that in combat, but they're O.K. for civil affairs. Everybody knows that. Then, you gotta figure that these instructors here are big shots in the army. They got high ratings. But you know damn well that if they went back to civilian life, they wouldn't be worth a shit. A lot of these sergeants and even the officers would have to go back to their jobs as clerks in grocery stores. That's all they were before, and that's all they're worth outside the army. They know it too. They gotta keep their jobs. They don't give a shit for us. They don't give a shit for Uncle Sam or his money. All they want to do is save their own ass.

Although official announcements did not list Asiatic language specialists as "essential," published accounts of a speech by General Jonathan Wilson, assistant chief of staff for G-2, seemed to contradict them. On August 23 the War Department announced that all servicemen over 38 years of age would be discharged on application *regardless of military necessity*. A man in this category could be held by his commanding officer for only ninety days until a replacement arrived. Although this announcement affected no one in the company, the phrase "regardless of military necessity" was welcomed; G-2 personnel were not singled out as exceptions. The men assumed that this order had come from Washington, D.C., and that local officials would not be able to countermand it. One man commented, "I'll bet Andersen and Kawashita and them bastards are trying like hell to talk the big shots into lettin' them keep all the Boochies. I hope them generals hold their ground." When a summary of General Wilson's speech was published in the *Fort Snelling Bulletin* of August 25, however, he was quoted as describing Nisei interpreters as essential:

The General began by outlining the vital role that Fort Snelling trained Nisei had played in the successful prosecution of the war in the Pacific theater, and then stated, "The Nisei of this school will be absolutely essential to the successful occupation of Japan and to the winning of the peace. Just as the former graduates served as the vital connecting link between Allied soldiers and the Japanese in combat, the Nisei will now serve as the language bridge between the Allied occupation forces and the 80,000,000 people of Japan." He added that the school will be continued for the need for its graduates was "greater than ever." "We must have replacements so the men that have been over there two and three years and who have the necessary 85 points can come back. And we must have men to take your place after you have been over there."

The General declared that Fort Snelling graduates would be used as linguists in tasks such as censorship, screening the Japanese press, watching communications of all kinds, aiding in the preparation of psychological warfare "with peace aims," and that the Nisei would be the ones to find out what "has gone on in Japan these many years." He added a word of caution, "The American soldiers will be intensely hated. . . . The Japs are going to hate Americans as they have never hated anyone before."

A month later the issue was still confused. On September 17 a War Department announcement listed three categories of servicemen considered essential, and Asiatic language specialists were not included. The news spread quickly, and those who had not read it ran out to the orderly room and flocked around the bulletin board.

"Well, that means we get out when our times comes."

"Yeah. I guess Andersen and his boys didn't get nowhere in D.C. The guys in Washington are gonna follow a fair policy."

"Now they can't hold us. It's a War Department order; so they gotta let us out."

"Not bad! I figure we ought to be out in a year or year and a half."

"Yeah. It won't be too long. None of that five year shit anyway."

Three days later abandonment of the point system of discharge was announced. After 1 January 1946 all enlisted men would be released when they had served for a period of two years. Everyone was relieved to learn that a definite period of service had been specified. But many remained worried. Would Colonel Andersen succeed in getting Nisei linguists declared essential and thus "freeze" them in their jobs?

One of the Division F instructors reinforced the prevailing view when he complained in a class: "All the instructors are pissed off and bucking for a discharge or trying to go overseas, but they won't release us. I'm over 38; so I applied for a discharge. But they appeal to us and say they have to have us. I told them I didn't give a shit how much they needed us; I've been in the army for a long time, and I'm going home. They had to let me go, but they're holding me for the full ninety days."

As hostility against the top officials of MISLS crystallized, resentment focused on two high-ranking Nisei. Major Kawashita, director of academic training, and Mr. Kishi, the assistant director, were hated more than the Caucasian administrators. Major Kawashita was widely reputed to be a "bastard" and was probably the most unpopular man in the fort. Colonel Murphy was the only officer among the "brass" respected in Company K; although he commanded the School Battalion, he had nothing to do with academic training.

It was known that these officials were very concerned with the performance of Division F, which contrasted so sharply with the rest of the school. Several incidents added to the resentment. Company K's first encounter with Major Kawashita came on August 13, just when everyone was looking forward to an announcement that the war was over. The major made a sudden, unannounced inspection that caught both instructors and students off guard. The classroom of F-1 was in frightful condition. Chalk that the students had been throwing at one another was splattered on the walls or ground into the concrete floor; books were piled in a sloppy manner on desks, benches, and the floor. About a fourth of the men were asleep; three were reading newspapers; two were studying; and the rest were chatting. The major was shocked. After regaining his composure he harangued the men for several minutes, declaring repeatedly that the top section should set an example for the entire division. Some of the students, however, had difficulty in understanding him; they were not accustomed to hearing such polished, sophisticated English. The first question on his departure was, "What the fuck's he talkin' about?" The reaction was uniform: "That chicken shit bastard."

Ill will was intensified two weeks later when Mr. Kishi, after a similar unannounced inspection, threatened to "bust" all the instructors in Division F. By this time the men had become attached to their instructors. Several with high points had been complaining of being forced to stay, even when they were eligible for discharge, and the students sympathized with them. Most of them regarded the sergeants as "good Joes." Thus, a threat to take away their hard-earned stripes, especially when it came from a mere civilian, infuriated almost everyone.

"Who in the hell does that little prick think he is? These guys are overseas guys—a lot of 'em. The rest of 'em are doing the best they can. What do they expect? You can't teach nothin' to a bunch of hard heads like the guys in Company K. That's expectin' too much. We oughta take that little fart out, and beat the shit out of him."

"Yeah! The little cocksucker! A poor fuckin' civilian at that too! The guy thinks his shit don't stink. He don't know what it is to be in the army. He's probably been a big shot right from the first. I'll bet he goes out to Minneapolis and tells everybody he knows all about the army too. He don't know how hard it is for the instructors here."

"What? Dat leetle civeelian! He say he gonna bust da instructors? No sheet? I like bust up! Who 'n a fuck he t'eenk he ees? Bust up, dat kind!"

"No sheet? Dey gonna bust da instructors? Da *manini* bastards!"

"Who dat guy? I t'eenk to myself he look like preek when he walk in. Me, I no like."

"That's Kishi. He's Kawashita's little boy."

"He's just like Kawashita. They're both pricks."

The reaction of the instructors varied. Some tried to ask the men for more cooperation. Others were defiant. One overseas veteran said, "That bastard can take these stripes and shove 'em up his ass. The only thing I'm worried about is that you fellows might not get your break. We've been bucking for that break for quite a while. Otherwise, fuck him! I don't like to be threatened."

More serious difficulties occurred on the following day when Colonel Andersen dropped in unexpectedly at 11:00. Most of the men were asleep. The colonel awakened those in the first few sections he entered but then merely walked through the rest. In section F-14 he was stunned. Private William Yoneda, generally regarded as the laziest man in the company, was sound asleep. The colonel tried to arouse him, but his efforts were of no avail. He ordered the instructor to awaken him, and the sergeant and several others jostled him. The colonel became very angry and ordered the instructor to send Yoneda into the next room as soon as he was awake. Several minutes later he reported to the colonel. According to eyewitnesses, he sauntered up to the colonel wearing his cap and with his hands in his pockets and asked, "Did you want to see me, sir?" The colonel lectured him for several minutes and asked him to choose between agreeing to remain awake in class or standing court martial. Yoneda protested that he had been forced into G-2 against his will, but the colonel reminded him that he was still in the army. Yoneda replied, "Well, sir, I'd like to think it over," and departed for the barracks. When his platoon returned to the company area at lunch time his neighbors found him on his bunk—sound asleep. Accounts of the encounter between "Willie and the colonel" were repeated with hilarity throughout the afternoon. Assuming that the colonel would not come again that day, many slept through the afternoon classes.

Whenever members of Company K were asked why they behaved as they did, their stock answer was that they had been Shanghaied. The battalion sergeant major, sensitive to this charge, finally called a special meeting of the company on September 4. He said that Company K men were ruining his reputation in the Nisei community by referring to him as "Shanghai." He resented this accusation so much that he personally checked the record of every man in the company and had the following to report: (a) Of all the men in the company at that time 186 had either been interviewed by MISLS recruiters, had formally volunteered for service in G-2, or had sent letters on their own initiative asking to be admitted. In brief, 186 had actually volunteered! (b) Of the remainder, all except two had *not* stated definitely during their interview at Fort Meade that they did *not* wish to come to Fort Snelling. They did not put any negative comments in the "Remarks" column. The sergeant insisted that he had given broad hints to those who did not wish to come to say so in that column; because of his position he could not say it more explicitly. Only *two* were actually forced to come against their will; others had stated that they had no serious objections or that they did not care one way or another. (c) Candidates were selected from the noncommittal Fort Meade Nisei on the following basis: everyone with AGCT scores in groups I and II was included.[3] Then some men in group III who had an exceptionally good background in the Japanese language were added. (d) With the exception of the two individuals who had specialized training that was wanted by the MISLS administration, all those who had stated unequivocally that they did not wish to come to Fort Snelling were left off the transfer order. Sergeant Nishida then reiterated that there were only two in the entire company who were actually on record as being opposed to intelligence duty.

The men accepted the sergeant's charges quietly. As the company marched into the school area, they saw a large group of newcomers in the firebreak, awaiting assignment to huts on the Turkey Farm. Their arrival was welcome news. For one thing, it meant fewer KP duties. It

3. All military personnel were classified on the basis of intelligence tests into one of five groups. Officers were selected only from groups I and II (roughly equivalent to I.Q. scores from 110 to 164); group II was the major source of NCOs (U.S. Department of the Army, 1948:6).

also meant that some replacements would be available after they were overseas. When the newcomers contended that they had been Shanghaied, their claims were met with laughter. That evening an argument erupted in section F-1, when two students insisted that they could not be expected to do good work because they had been forced to come against their will. A man who was generally quite reserved retorted, "Bull shit! You came here because you wanted to get out of the infantry. Why don't you admit it like a man?"

Admission that they had not been Shanghaied, however, did not mollify their resentment against the "brass." Difficulties between Company K and the MISLS administration continued to mount, finally coming to a head over the question of the "academic break." Giving the students a week's vacation at periodic intervals had long been a regular practice. But would the conversational course be treated in the same manner as the rest of the school? When it became widely known that the "regular" school was to receive a break during the first week of October, excitement arose over the possibility of Division F's being included. When the sergeant major was asked about the possibility, he had replied that this was still being negotiated. Since many Nisei worked in the administrative offices, their friends received various "leaks." Rumors developed during the latter part of September that Colonel Murphy was trying to get a seven-day break for Company K; another rumor was that both Major Kawashita and Colonel Andersen were opposed. Both rumors were plausible, and the men concluded that a struggle had developed among the "brass" over the issue. They reasoned that if the battalion commander had approved a week's furlough, only the MISLS commandant or the director of academic training could turn it down. Resentment became focused on these two, especially upon Major Kawashita, since he was a Nisei. Most students were pessimistic. As early as September 10 a rumor developed that Division F would receive only a half week off. This was better than nothing, but everyone kept hoping for the whole week.

On September 17 expectations were confirmed. It was announced officially that Division F would receive a three-day break, beginning on October 4. The instructors pointed out that this could be stretched into a four-and-a-half-day furlough by including Wednesday afternoon and Sunday. But everyone else in the battalion was getting nine days, including two weekends. The OCS men were also getting a full week. Private Shindo seized the opportunity to declare himself openly: "God, I wish we were part of the regular school. I feel like bucking to go up there just to get the seven days. That way you can learn Japanese right, and you get more *kanji* anyway." Others did not bother to argue with him.

To add insult to injury they were to be held until 15:30, after all the express trains for Chicago had left. Students in the "regular" school were to be excused at 11:00. The Milwaukee Road's *Hiawatha* left Minneapolis at 13:15; the Chicago and Northwestern's *400* left at 14:00; and the Burlington Route's *Twin City Zephyr* left St. Paul at 16:10. All other trains to Chicago left late at night and did not arrive until the following morning. Although one or two hours of class time did not matter, it made considerable difference in deriving maximum benefit from limited furlough time. Instead of being in Chicago the first night they would have to waste a night on the train and then would have to sleep part of the following day to recuperate. Could this have been done as a punitive measure? Some of the instructors felt that Major Kawashita was "mean" enough to do it.

The first week of October was a complete waste of time for everyone in Division F. Fort Snelling was virtually deserted. But for a skeleton crew of overhead personnel, those at the Turkey Farm, the few awaiting assignment overseas, and Division F students, everyone was on furlough. Some admitted candidly that they did not deserve a break, but they still wanted to "get even" with the administration. The educational program came to a standstill. The students attended classes, and the instructors appeared in the rooms, but nothing was accomplished. The men were determined to do nothing.

Since many Division F instructors had not had furloughs for over six months, they were given leave at the same time as the "regular" school students. They were replaced by some other instructors who were not yet entitled to a furlough. The visitors had heard about conditions in Division F, but apparently they had thought that their colleagues had been exaggerating. When they took over the classes, therefore, they expected students to stay awake, to stand when they were reciting, and to do their best whenever they were called on. During the first period the new sergeant who faced section F-1 was shocked. He had what was presumably the best section. When he tried to enforce discipline, everyone sat back and laughed at him. When he threatened them, they laughed even harder. Everyone talked at once. Men threw paper wads at one another. When the sergeant realized that he could not control the class, he asked them why they did not want to study. This query elicited immediate responses.

"Why in hell should we work? Everybody else is gone. They're out on furlough. Look at us. Fuck 'em! They can make us come to school, but they can't make us learn anything."

"How can you expect us guys to be like the guys in your division? They get a nice place to live in; they don't have to eat that shit in the Consolidated Mess; they don't have to freeze their ass off in a mess line three times a day; they got nice hot showers in a warm shower room; they got a clean, steam-heated latrine; they get ratings when they graduate. What do we get? They treat us like a bunch of dogs."

"Look at this fuckin' room. Look at how cold it is. We haven't even got a fuckin' stove! We have to sit in here for hours, shivering our ass off. You guys got steam heat. Look at us! We ain't got no heat at all. Look at the P.A. system they were supposed to put in. We're supposed to have it for oral monitoring and Boochie songs. Where's the stuff, huh? They give it all to you guys first, and then they give us the scraps. Fuck that shit! We don't like to get shoved around."

Each hour the teachers changed, and each hour the routine was much the same. During the second period a student in F-1 spilled ink all over one of the large desks. He had not done it purposely; the bottle slipped as he was trying to balance it on the tip of his pencil. As those around the table scrambled to avoid the rivulets of ink, he bowed and hissed with mock humility, "So solli! Honolabulu mistaki!" The instructor exploded and threatened to send the man to see Major Kawashita. At this, everyone broke out in boisterous laughter.

"Hey! You better watch it, Joe. He's gonna send you to see his boy."

"Gee! I'm scared as hell. Ain't you scared? Ha! Ha! Ha!"

"Major Kawashita? Who in the hell's he? Oh, yeah! He's that prick with ears! Yeah! I remember the bastard."

Section F-1 calmed down during the third period, when one of the regular Division F instructors took over. He joked with the men and talked about brothels in Korea. He was as disgusted as the students were, and together they discussed the furlough situation.

"I hear you guys been giving the instructors from the regular school a bad time."

"Hell, Sarge. They tried to give us a bad time so we told 'em off, that's all."

"Well, those guys are used to being King Shit, that's why. Over there the instructor yells, and everybody snaps shit. It's not like this place. Hell, here I yell, and somebody wants to beat me up."

"Well, I figure we get low ratings anyway so we may as well know less than the other guys. But what the hell! They didn't have to fuck us on the furloughs."

The remainder of the period was spent consulting train schedules and trying to find the easiest and quickest way to reach various destinations. All the instructors found it impossible to teach, and by 10:30 several were seen strolling about outside. By the end of the first day all the new teachers had been initiated.

On Tuesday night, the last night before they were to leave on pass, most men reported for the study sessions. They knew that Lieutenant Catalano was the DO, and they did not wish to risk any restrictions. As he approached the classrooms, the alarm went out: "Here comes the old man!" On hearing this, one of the new instructors taunted them, "So! You guys give us a lot of shit, but you're afraid of your CO, huh?"

"Hell, no. We ain't scared of him. But he's a good guy. We just don't want him to think we're fuckin' up. It'll make him feel good."

In one room, where the warning was not heard, the lieutenant walked in just as some students were singing:

"*Shina no yoru. Shina no yo-oru yo—Kimi matsu yoru wa, Obashima no ame ni, Hana mo* . . . Oh! Good evening, sir!"

"Good evening, men. What are you supposed to be doing?"

"We're studying, sir."

"Come again?"

"Well, sir, Japanese songs is one of our subjects, and we're studying."

The lieutenant laughed good naturedly, complimented the man for having the "answers," and warned them not to get caught by someone else. As he walked from room to room, he was greeted cordially with smiles. He addressed several by name and smiled knowingly. He realized his men were putting on a show for him, and they knew that he knew. But everyone was happy.

Just before the troops were dismissed to return to their barracks one of the Division F instructors ran from room to room to warn everyone that Major Kawashita would come on the following day for an inspection. He warned the men to be "on the ball" because the major was a "bastard." One of the other instructors had heard that the major was afraid that the company might be excused at noon and wanted to block any conspiracy between Division F officials and the company commander. Resentment flared.

"That dirty, stinkin' Jap! He don't even trust a first lieutenant like Catalano. Shit, he's a fuckin' major. Why don't he just order Catalano to keep us here until 15:30, if that's so important to him? If it's a direct order, Catalano'll do it. But no, he's gotta come down himself to make sure we stay until 15:30. The dirty son of a bitch!"

"Just to make sure that we gonna miss the train, he's gotta come down. I'd like to kill the bastard. I'd really like to meet him some day after the war. I'd like to beat the shit out of him."

"You won't get a chance to. Before you get 'im there are a lot of other guys who are laying for 'im. Thousands of guys went through here, and there ain't a single one who likes the bastard. They all want to kill 'im. I used to wonder why all the Savage guys hated him so much. Now I know."

"Well, that's what happens when you give a Boochie too much power. It goes to his head. The guy's drunk with power. You can't trust a Boochie."

Wednesday, October 3, was the day on which the break was to begin. For weeks the men had waited anxiously, and at last the day had come. They had planned the use of their clothing and money, scheduled their laundry, and written to their friends and relatives. Because of the shortness of the furlough those whose parents were still in relocation centers would not be able to go "home" like students in the "regular" school, but even these men looked forward to "getting away from it all." The barracks were cleaned thoroughly. No one wanted to take chances. They knew the lieutenant would not restrict them, but some other officer might.

At 09:30 an announcement was made that Major Kawashita would address the company in a half hour. Everyone was surprised, for they had not expected him until the afternoon. One man speculated hopefully, "Maybe we get off at noon." Most were doubtful.

At 10:00 the students were gathered at a firebreak during the rest period; the instructors explained that the major did not wish to "interfere with any of the class work."

He began in his highly polished English, "Men, I wish to take this opportunity to inform you that I am not at all pleased with your conduct. There have been too many derelictions in this division. . . . Many of you have made outrageous charges against the administration. There is not one iota of truth in these charges." He then went on to say that unless such unsatisfactory conduct stopped, there would be no furlough at all—not even for three days. He demanded that the instructors be treated with respect and that the classes be conducted in a military manner. Concerning that afternoon, he made his position clear: "You will remain at your post until 15:20. No one will leave one minute before that. You have a job to do; I expect you to do it. I hold all the officers and noncommissioned officers under me strictly responsible for your conduct."

The few who had been hoping for good news were disappointed. Others were furious at being told what they already knew—on their "own time." Some were so enraged that they said nothing. They sat at the firebreak, glaring at him. One man muttered, "I just want to get a good look at him. I'll never forget him. Some day I might see that guy again."

What the major had done was enough to antagonize most men; the fact that he was a Nisei intensified their bitterness. After the major left, the instructors tried to get the students back into the classrooms. They refused to budge; they insisted that since the major had taken their "break time," they were entitled to a few more minutes to smoke and rest. Some stayed outside for another half hour. Most were angry but fatalistic.

By the time they returned to the classrooms, a few were able to joke. In section F-1 one man ran directly to Webster's dictionary.

" 'Dereliction?' 'Dereliction?' What the fuck's that? I can't even understand what that bastard's tryin' to say. Hey! I can't find it in the dictionary. How in the fuck do you spell it?"

"Don't ask me. He didn't put no *furigana* on that one." [4]

"What was that other word he used? 'Iota' or something like that. What the fuck's that mean? Why don't the bastard use English?"

During the next period someone started referring to him as "Major Cowshit," and before long someone else translated this into Japanese: *Ushi no kuso shosa*. By noon these expressions enjoyed wide usage, and from this point on "Major Cowshit" came to symbolize all the pettiness and anything else obnoxious that took place. He was cursed throughout the day.

Immediately after lunch most men changed into Class A uniforms and packed their belongings. If they could only get away a few minutes early, there was a chance of catching the Burlington *Zephyr* in St. Paul. Arrangements were made at the orderly room to have taxicabs there at 15:20.

Nothing was accomplished in the afternoon classes. About 15:00 Major Kawashita returned to the school area. Someone saw him alighting

4. *Furigana* are the simple alphabetical characters placed beside difficult *kanji* to facilitate reading them, especially in combinations.

from his car, and the alarm spread quickly from room to room. For about twenty minutes everyone pretended to be studying. The major started at section F-15 and proceeded slowly through the classrooms. He found every room in good order, but in each he stopped to say something. As a consequence, when the bugle blew at 15:20, he still had several rooms to inspect. Members of the lower sections ran outside. They waited anxiously for the others to come; after shouting impatiently, those who were in formation finally marched off. The major did not reach F-1 until after 15:30.

The students there were in an ugly mood, but they said nothing. They listened as he castigated them. He reiterated that they were the best section and that he expected them to set an example for the others. Little did the major realize that some of the sullen men before him wanted to murder him. Long afterward one of them confessed:

You remember that time that Kawashita was chewing our ass after all the other guys left? I kept thinking to myself that I wanted to kill the bastard. I kept thinking that I could stab him. I thought they would court martial me, but I figured I'd be doing something for all the boys and for all the other guys who'd have to take this shit from him. Something inside of me made me want to kill him, and I got scared of myself. No shit, I was scared that I might do it because I felt something pushing me from inside. Did you ever have something inside pushing you like that, sort of against your will? I was really scared I would kill him, and I started sweating.

As the F-1 men ran for their barracks, there was nothing but bitterness. They resented the major's not trusting Lieutenant Catalano, but their preoccupation with getting away quickly temporarily focused attention elsewhere. Fewer than thirty in the company managed to catch the *Zephyr*.

The Diffusion of Gang Violence

Jubilance in Company K over the beating of the OCS man on September 25 had provided the Hawaiian trio and their cronies with new targets for their aggression. Although they continued to take advantage of their immunity in tormenting hapless Kotonks, they felt increasing pressure from fellow Hawaiians against the beating of mainland Nisei, especially those within the company. But now they were free to turn on OCS men, for almost everyone was in favor of assaulting them. Even mainlanders frequently expressed a desire to get even with the hated *Keto*.

On October 9, shortly after the academic break, a battalion meeting was held at the Field House. When Major Kawashita rose to speak, boos and catcalls arose throughout the auditorium, but the loudest groans and most inhospitable remarks came from the section seating Company K. Then, Colonel Murphy reluctantly announced some changes in policy. By order of the War Department all ratings of enlisted men had been frozen; students who already held high ranks would retain them, but others would not be promoted on graduation. Most men would go overseas as first class privates and would presumably be promoted by their field commanders. He also announced that all G-2 language personnel with fewer than ninety points had been classified temporarily as essential.

The announcement intensified the resentment against OCS men, for they were not affected.

"Them OCS guys'll get their bars though. That's racial discrimination. Why don't they let the Nisei in the OCS. It ain't fair. The Nisei have to do all the dirty work, and the Haoles get all the ratings. Now they won't even give us promotions, but they keep getting theirs. Why don't they make some rules that apply equal to everybody?"

"They just make the rules for the Boochies."

"Well, we're all Americans, ain't we?"

"Are you an American?"

"What the fuck you think I'm doin' in the army?"

"If our ratings are frozen, the OCS guys shouldn't get their bars."

It was a day of gloom, and the men complained bitterly.

"I wonder if we could sue the government for breach of contract?"

"They wouldn't leave themselves wide open."

"Yeah, but we swore to serve under the Articles of War for six months after the end of the war. Up to that time we ain't got a leg to stand on, but after that we could sue the fuckin' government if they won't let us come home."

"Aw, shit. They won't keep us that long."

"They might. If Andersen and Kawashita had their say, they would."

"Jesus Christ! This is worse than the infantry. In the infantry we get fucked once in a while, but they fuck you from the front—like a man. But here they sneak up from behind when you ain't looking and fuck you from behind. They get you from the back where you can't see."

"I was just thinkin' that maybe I'd better study a little bit, but now I lose fight."

Although no one welcomed the announcement, some eager beavers reacted bitterly. They confessed that they felt depressed, empty. One man, sitting alone in a corner, remarked, "I don't know. I just don't feel like studying or doing anything. I guess I just feel kind of lonely and blue." A few were unconcerned about ratings. One man, generally regarded as a "fuck-up," was unperturbed. He commented wryly, "Shit, I'll be lucky if I leave this place a Pfc."

That night several groups of Hawaiians were drinking heavily, and many mainlanders declared that they wanted to join them. Those who went to the latrine reported that a number of very intoxicated AJAs were looking for a fight. About 22:00 a second OCS man was beaten; this time the incident occurred in the Company A recreation room, where no Nisei belonged. Rumors about the fight spread quickly and once again everyone prepared for a knuckle inspection. Most mainlanders felt uneasy when so many Hawaiians were drinking, but they seemed quite pleased to learn that another Caucasian had been victimized.

By this time the OCS men were finally beginning to appreciate how intensely they were hated by the Nisei in the fort. Since most of them had been sympathetic to the Nisei cause, they found it hard to understand. However, the beatings and even more the gleeful reactions of other Nisei to the incidents made their stance very clear. They had not realized how seriously Company K men took the daily sprints to the mess hall. Their company commander had told them of the intercompany agreement, but they knew he did not care. Even if he did, they disliked him so much that they paid little heed of him. But the violence led them to reappraise the situation, and the jostling in the mess hall line came to an abrupt halt. Some OCS men developed a morbid fear of Company K and especially of drunken Nisei; they did not distinguish between Hawaiians and mainlanders. One of them commented:

It's a horrible feeling when you think about it. What's a man supposed to do when a Nisei picks a fight with him? He doesn't have to start anything; the Nisei comes looking for trouble. Then a man can't even defend himself. He can either take a beating from one man or raise his arm and be dogpiled by a dozen men. We're helpless. I've often wondered what I would do if I were in a situation like that, and I just don't know. What can a man do? It's a heck of a feeling to be in a state like that. Every time you see a Nisei you wonder if he's going to beat you up. Some fellows are afraid to leave the barracks at night. I know darn well a lot of the men are losing sleep thinking about this.

By the middle of October it became apparent that the general pattern of violence that started in Company K was spreading throughout the fort. A number of fights had occurred in Minneapolis as well as within the fort, and Company K had been blamed for all of them. During the weekend of the academic break, however, it became obvious that others must be involved; Company K personnel were all away on furlough. Just as much trouble erupted in the Twin Cities as on previous weekends, and those involved claimed to be from Company K. But Company L and the newcomers on the Turkey Farm were the only large contingents of Nisei left in the fort. By claiming that they were from Company K the culprits had escaped knuckle inspections and thus avoided detection and court martial.

Furthermore, whenever their own men were embroiled in some incident, others in the company soon knew of it; and there was an increasing number of brawls in which no one could recognize the participants. On Wednesday evening, October 10, several clashes involving Nisei soldiers took place in the Twin Cities. Two fights were reported; Company K men who had witnessed them insisted that the antagonists were from somewhere else. On the following Monday a rumor developed that some Hawaiians had thrashed a captain in Minneapolis over the weekend. This attracted considerable attention, for the beating of commissioned officers was uncommon. Throughout the day other rumors elaborated details: The captain was the duty officer; after he had told some drunken Hawaiians to leave a bar they had gathered their friends and given him a *banzai* charge; they knocked him down and then kicked him in the face; the MPs did not arrive in time to help. Even though they knew that no one in Company K was responsible, most men anticipated some kind of punishment.

At a company meeting Lieutenant Catalano confirmed what most men suspected and asked for their cooperation:

I found this morning that almost all the cue sticks in the dayroom have been broken. I also found a lot of beer bottles in there. I know that you men didn't do it. I understand that some L Company men were in there and wrecked everything. I'm going to see to it that this kind of thing stops. At the same time, make sure that none of you do anything like that. . . .

There's another thing I want to tell you about, men. Last night and last weekend we had quite a bit of trouble in town. The ones who caused the trouble said that they're from this company, but they can't fool me because I know every man in this company. I may not know all your names, but I do know all your faces. They've been blaming us for everything, men. Every time there's some trouble, they always call up here first. I've been trying to convince the colonel that we're not the only fuck-ups in this camp. I'm going to keep telling the colonel that, and I don't want any of you getting caught in town for disorderly conduct. If you get caught, then we'll get blamed again for a lot of things we didn't do. I'm finally getting the colonel to see that we're not the worst company in the camp. I want you men to help me convince him of that. . . .

I've also got some good news for you. Remember that night when you had those beans? Well, we started a petition in the company that night. A lot of you fellows signed it; so you know what it's all about. It's against the rules because you can't agitate in the army, but I thought it was justified. So I signed it and sent it through channels to Colonel Murphy. I talked to him about it too. So when it got to his desk, he just signed it and sent it up to the post commandant. Then it got to the post commandant's desk, and all hell broke loose. He called the four mess sergeants of the Consolidated Mess up to his office and really gave 'em a dressing down. Men, things are going to change around here. If they don't, we're going to have some new mess sergeants.

I want you to know this, men. I'm not very popular with the other officers around here because I've stepped on their toes. I'm not popular with the mess personnel either. But if they don't treat you right in the mess hall, you just let me know, and I'll take care of it. Sure, I'm bucking for a promotion too, but I don't want to buck at the expense of my men starving for me. . . . From now on, we're going to have three Nisei cooks in there. We're going to have rice, soy sauce, and all the other food you like. We're going to get that food and have it cooked the way you want it.

Although the welcome news about the mess hall elicited much rejoicing, it was not enough to counteract the general atmosphere of gloom. Heavy drinking by young Hawaiians was becoming more widespread. Furthermore, those most responsible for the disturbances in Company K had established contacts with AJAs of like mind in other companies. As the evenings became more boisterous and disorderly, a number of mainlanders began to wish that the administration would crack down on them. Some openly expressed hope that the assailants would be caught and punished.

"I hope the CO court martials one of them son of a bitches. Then the other guys would be too chicken to fuck around. If a guy wants to goof off, O.K.; but when three guys fuck up the whole company, then they ought to get it."

"It makes everybody lose fight. The guys who are on the ball get restricted on account of a few. What's the use of being on the ball, if you're restricted anyway? May as well fuck off. The CO should crack down on them, but he's too good a guy to do it."

On Wednesday afternoon, October 17, a rumor developed that the company was scheduled to run through the obstacle course during the physical training period. Although most men were reluctant to admit it, they had become soft and flabby from months of inactivity and were beginning to find such exercises strenuous. Substantial numbers disappeared from the 15:30 formation, and only seventy-five appeared at the obstacle course. Although nothing was said about it and Wednesday night passes were issued as usual, the men realized that too many had "goofed off" at the same time and fatalistically prepared for some kind of punishment. Many did not go out. Since many Hawaiians were drinking heavily, several commented that it was going to be a "rough night."

It was indeed a very rough night. In the early hours of the morning three OCS men, who were returning to their barracks from the main part of the fort, were confronted by several Hawaiians who challenged them to a fight. The Caucasians, by this time quite fearful, refused and started walking away. In the brawl that ensued one OCS man was stabbed in the back; a second was kicked repeatedly in the face after he was down. Both required medical care. The incident was reported immediately to the duty officer, who happened to be Lieutenant Catalano.

At 01:00 everyone in Company K was aroused and ordered to stand by for a showdown inspection. The lieutenant, accompanied by the military police, personally examined all knuckles, all items of clothing (for possible bloodstains), and all personal belongings (for the missing knife). After the first two platoons had been inspected with great care, the search turned to Company L. Everyone was told to go back to sleep. The men were puzzled. Why the inspection? If there had been another fight, why didn't the lieutenant go to the 5th Platoon? Did he find what he was looking for?

On Thursday morning everyone learned of the stabbing, and a common reaction to the news was: "Well it finally happened!" Many rumors developed about the episode, and from them a picture of what had happened was reconstructed. There were several sources with which the various rumors could be checked, and grossly inaccurate accounts were gradually weeded out. Some Company K men had seen the fight; a few had contacts in Company A and learned of the rumors there; Nisei who worked in the hospital were helpful, as were the cadremen who could check items with the company commander; several of the Hawaiians had friends in Company L. The fight had started on the "dummy line," when an inebriated Hawaiian picked a fight with an OCS man. Rumors described the manner in which the culprit had been captured in Company L. The men there had been up all night, and their possessions had been searched very thoroughly. The inspection continued until the guilty man was found.[5] Initial reports exaggerated the seriousness of the injuries. It was claimed that the blade had pierced the victim's lung and had just missed a vital artery. The man kicked in the face was said to be in danger of losing his eyesight. At first the stabbing was blamed on some newcomers on the Turkey Farm, and it was contended that some Company K men were among those who had stopped the fight.

A showdown inspection was announced for 15:30. Everyone was surprised. "What the hell! I thought they caught the guy!" The lieutenant went through each of the barracks, but his search was rather casual. At each platoon he asked everyone to turn in their knives: "Don't get caught with a knife, men. You may be searched at any time." All knives were surrendered obediently. Soon thereafter rumors developed. All knives were being collected for blood tests. The tests would be made by the FBI. Knives with blades over three inches would not be returned.

As the atmosphere lightened a bit, men began joking with each other about the incident. For example, a man who was washing his T-shirt in the latrine was asked, "What you do, huh?"

"Aw, I'm puttin' some peroxide on this fuckin' thing to get the bloodstains off."

"Fuckin' up the company again, huh?"

Among mainlanders complaints against the fighting tactics of Hawaiians continued, "That's the trouble with fighting Kanakas. If they can't lick you with their whole fuckin' gang, they pull a knife." Hawaiians on the whole were apologetic, and most of them unequivocably condemned the deed. "Dat's too much. Stabbing is too much. Bust up a leetle beet, O.K.; but no stab." "Buddhahead stab? What a chicken sheet guy!"

One older Hawaiian tried to explain: "A lot of dese guys just young punks. Dey act tough, but dey just yellow inside. Dey no fight man to man. Da reason why dey fight like dat is becaus' dey know dey safe. Dey no get grudge against mainland guys special. Only t'eeng is dat if dey fight another Kanaka, dey have to fight deir own battle. If dey fight mainland guy or a Haole, dey get all da other Kanakas behind dem. Dey just yellow. Dey no can lose dat way. Dat's why dey fight like dat."

Another reiterated the feelings of Hawaiians against Caucasians: "We hate da Haoles becaus' dey always da boss. We do da same work

5. Official records reveal that Private Yukio Hasegawa was arrested that morning and taken to the guardhouse; three days later he was detached from the company.

and get half da pay. Da whole plantation system dat way. Dey get discrimination in da school and everyt'eeng. We get shoved around. Dat's why we no like."

An announcement appeared on the bulletin board ordering all privates to sew on their Pfc. stripes; those who had not already done so began at once. The reason was clarified by the company commander at an orientation meeting:

Men, I want to tell you something that I think you ought to know; so you'll know what we're all up against. In the last couple of weeks a lot of men from Company E and Company L have been raising cain in town. When they get caught, they say they come from Company K. They let the others think they come from Company K to throw suspicion off their own companies. The other night I was duty officer. I got a call that there were some drunken soldiers in that tavern across the bridge and went in to see what it was all about. I found a couple of men, and one of them was especially tough. Finally, I lost patience with him and asked him what company he was from. He said, "Company K!" So I asked him if he knew his company commander.

By this time everyone was laughing. They knew that no outsider could fool him.

Well, that's the way it is, men. I want you to be careful. If you catch any of those guys posing as Company K men, you just bring 'em in here and take care of 'em. When you bring them in, let me know so I can get out of the company area. What I don't see I don't know anything about.

Another example of this blaming everything on us was the stabbing the other night. That was a cowardly thing, men. Don't ever do a thing like that. It's O.K. to scrap once in a while, but don't get the reputation of pulling a knife. . . . Well, we started searching through Company K. If we had stayed here, we wouldn't have caught the man because it would have taken all night. The only reason we went to L Company was that the A Company men who were there said that the men who did the stabbing weren't wearing any stripes. E Company and L Company are the only ones with privates; so we switched over and found the man. Now, men, you saw the announcement on the bulletin board. I want all you men to sew stripes on all your clothes—every damned one. Then, we can't be blamed for anything we didn't do. . . . From now on all men in Company K will wear stripes.

The man in the stabbing was caught because the blood on his clothes was type O. His own blood was type B, and the blood of the victim was type O. It looks bad for him, men; but if he did it, he's got it coming to him.

I want you men to keep away from those men next door. The L Company men are just no good. Last week five or six men were court martialed from that one company alone. Now they're going to have some more. The men who beat up the captain in town the other night are also from that company. I warn you, men. That kind of stuff won't go around here. I don't want you mixing with those fellows. Don't have anything to do with 'em. That knifer is going to get a minimum of five years in the pen and a dishonorable discharge. You know what that means. He won't be able to get a job after. He loses his citizenship. Men, you don't want to ruin your whole life just because you got drunk and lost control of yourself one night. It isn't worth it, men. It's your whole life.

I want to say something about those who started the trouble the other night in the 5th Platoon. Don't worry. I know about that too. I know about some other things you men have done. Don't think that it's all forgotten just because I haven't said anything. I know that a handful of men have been at the root of all the trouble we've been having. I'm not going to pull a sneak punch on them either. I'm warning those men right now. I figure that if I give those men enough rope, they'll just hang themselves. I know that they brought beer in here when it wasn't allowed. But you just wait.

Now about Wednesday. A lot of you fellows goofed off. We checked all the lists and took off all the names of men who had even a half-way decent excuse. But there are forty-four men who were caught flat footed. Now I know that a lot of you men have already bought tickets for the Minnesota-Northwestern game this weekend; so I'll let you go. But next weekend you forty-four men are restricted, and what's more you're going to pull detail. I've made arrangements with the Turkey Farm to supply quite a number of KPs for that weekend, and if that isn't enough work, we can build a new road from Company E to the lake.

The last remark touched off boisterous laughter. Even the guilty forty-four took it good naturedly, for they felt it was fair. After all, they had been warned repeatedly; then they had been caught and had not been able to think quickly enough to provide an adequate excuse. Most men were also pleased with the lieutenant's blunt warning to the trio in the 5th Platoon.

"The CO's O.K.! Boy, it's a good thing we got him."

"No shit, boy. This place would be a mess if it wasn't for him."

"He's a square shooter."

"He's really a good guy. I'll bet he gets chewed out by the colonel for us all the time."

"Yeah. But he never passes the shit down to us."

"Yeah. He's O.K. He's the best fuckin' CO I ever had."

As the pattern of violence that began in Company K soon after Japan's surrender spread to other units in the fort, relations between the OCS men and their immediate neighbors underwent a sudden change.

Remedial measures were attempted as conditions in Company A reached a critical point. On October 19 a second OCS man committed suicide.[6] Thus, the situation was becoming desperate: two suicides, a man stabbed, several others severely beaten, and others constantly facing taunts and threats. An armed guard was placed around the OCS area; although members of Company K were amused on seeing the patrol, it was no joking matter for those within. The initiative was taken by two men, one in each company, who had attended the same university prior to induction. As tension had intensified, the two had been careful not to be seen together in the fort and had even agreed not to recognize one another in contexts in which either felt it would be unwise. The two arranged a secret meeting of representatives of their respective units to seek ways of averting further difficulties. On October 20 two spokesmen for an OCS committee that had been presenting petitions to the MISLS administration, two squad leaders from Company K, and a Nisei from Company I met in a home in St. Paul. As they discussed the situation candidly, all five were struck at once by how much the two companies had in common, especially their distrust and hatred of the MISLS administration.

Both companies were demoralized, the OCS men far more than the Nisei, and they had very similar complaints. The OCS men suspected the intentions of the G-2 administration and were convinced that they would be double-crossed. Not only had two of them committed suicide, but most of the others were considering refusing commissions on graduation. Some were even trying for psychoneurotic discharges—to the extent of studying textbooks in psychiatry to learn to fake some illness. The OCS representatives indicated that much of their talk, aside from their fear of Company K, dealt with maneuvers for getting a discharge or a transfer to some other branch of service. They also disclosed that their committee had proposed a number of reforms, asking for changes in unpopular practices—such as the useless practice parades, discrimination in the selection of officers, the rating system in general, and the substandard food in the Consolidated Mess Hall. The committee had asked repeatedly that Nisei be included in the OCS company, but thus far the administration had been unresponsive.

The OCS representatives also stressed that the greatest resistance

6. According to OCS informants, the two men, while unhappy with their military careers, had additional problems. Both were involved with the same woman at the time of their death, and the woman was subsequently declared "off limits" to the company.

they encountered came from the Nisei "brass" rather than from the Caucasians. One of them commented:

We've asked for a lot of changes all the way from getting better chow in the mess hall to the elimination of those stupid night study hours. It was O.K. as long as the war was on; but now that that kind of thing isn't necessary, we asked that the compulsory study be eliminated or reduced. We've taken that suggestion up there several times, and each time either Kawashita or Kishi would answer, "But the Nisei won't like it if we eliminate your night study." So we agree. Naturally the Nisei won't like it; so we ask them why not let the Nisei have those nights off too. After all, the Nisei need it less than we do; they already know Japanese. But then the answer would always come back, "But the Nisei are irresponsible." It's the Nisei brass in the administration that holds back all these changes on the ground that it won't be good for other Nisei. It's hard to understand, but that's what we've been up against.

Since the two companies had so much in common, the five agreed that the initial step in resolving the conflict was to minimize misunderstanding by familiarizing the men with each other's views. After considering several stratagems, they finally agreed that the best way to accomplish this objective was to hold a joint *kaiwa* [conversation] meeting of the most competent sections of the two divisions. To ensure success it was agreed that the OCS participants in the first session should be selected carefully. Only those with an extensive knowledge of Japanese should be included; furthermore, only those known to be friendly to Nisei were to be included. Since one of the grievances of the Nisei was that Caucasians were incompetent as linguists, actual contact with those who were fluent would dispel this belief more effectively than anything an MISLS official might say. It was also agreed that nothing would be said about the meeting to anyone in Company K; it was to appear as a routine assignment imposed by division officials. Those who were to represent Company A, however, were to be thoroughly prepared. They could talk about any subject on which they had an adequate vocabulary, but they were to be ready to discuss women overseas and their complaints about the MISLS —the two things they had in common with Company K. It was agreed further that the OCS delegation should march to the Nisei area, even though they held higher ranks. This could counteract the widespread belief of many Nisei that Caucasians were snobbish and rank conscious. Some Nisei would possibly be flattered that the Caucasians had come to them. The objective was to show the Nisei (1) that some Caucasians were competent linguists, (2) that the OCS men were just as disgusted with the MISLS regime as the Nisei, and (3) that the two companies faced the same problems and shared much the same outlook.

The plans were drawn up very carefully—to meet in advance any complaints that the Nisei might make—and the suggestions were then taken by the OCS committee to the administration. Again the high-ranking Nisei officials objected, especially to the proposal that officer candidates should go to enlisted men. The committee was adamant, arguing that matters of protocol were unimportant in the current crisis. Since they had no alternative plan, the administrators reluctantly agreed.

At 13:30 on Thursday afternoon, November 1, about twenty OCS men marched into the Division F area, after the Nisei had already been seated in their classrooms. Since they had been told nothing of the plans, many were upset: "What the hell are them Haole bastards doin' in here?" The first *kaiwa* session was held in section F-1. Although the conversations started rather slowly as the Nisei sat about saying nothing, the persistent, friendly efforts of the OCS delegates finally broke some of the reserve. By the end of the hour the men were huddled in several small groups—chatting in English. They stopped using Japanese at the insistence of the Nisei.

All the F-1 students were astonished at the detailed knowledge of *kanji* and vocabulary displayed by the OCS men. Many were embarrassed when they discovered that they did not know enough Japanese to talk to these Caucasians. They did not know how carefully they had been selected, nor did they know that one of them had lived in Japan for thirteen years. The Nisei learned that the OCS men were not plotting with the administration to enslave them; on the contrary, they disliked the administration so much that many of them were planning to refuse commissions. They also learned that OCS men were not arrogant. They learned too that OCS representatives had petitioned the administration to accept Nisei into their ranks on the same basis as all others. This caught most Nisei flat-footed. Many realized for the first time that these men were not responsible for the discrimination; it was an administrative policy. Those in F-1 were impressed. They admitted that they had enjoyed the exchange, and several remarked that the "hour passed fast."

When the bugle blew at the end of the period, several mixed groups walked to the PX together to have refreshments and to continue their conversation—in English. The atmosphere was very friendly.

Since members of the other sections had seen the OCS men marching into the area, they were curious to know what had happened. When they saw their friends walking to the PX with Caucasians, they were even more curious. Word spread quickly to the rest of the company. The OCS men spoke Japanese so well that the F-1 Nisei were embarrassed.

"Jesus Christ! Ya gotta have a dictionary to talk to them guys."

"You ain't a shittin'. Them guys use *kanji* and don't put on no *furigana*."

"Them guys can understand anything we say, but we can't get what they say—too fuckin' hard."

"We lost face."

"Nice bunch of guys. They're really pissed off too."

"They really know their stuff. *Yarareta*! [We were taken!]"

"They ought'a have this *kaiwa* every day. I learned quite a bit."

Other items were disseminated quickly, especially of their bitterness toward the MISLS administration and of their desire to have Nisei in their company. Two OCS men had already killed themselves, and another was contemplating suicide; for the first time sympathy was expressed openly for the two who had died. A rumor soon emerged among the Nisei that about half the OCS men would refuse their commissions; they had been in the service for over two years and wanted a discharge. Some might have gone overseas anyway, but they did not trust the administration. But the Nisei were most impressed by the reaction of the stabbing victim, as it was related to them. He was among those who had come for the session. He had disclosed that if the knife had been a half inch from where it struck, he would have died. News that the victim had come and that he was not bitter against all Nisei spread quickly.

"Yeah? He was here, huh? What's he think of the Boochies now?"

"Just the same as before. He's a swell guy—a good sport."

"Yeah? If some bastard stabbed me, I'd have it in for him for the rest of my life."

"He almost got killed too, huh? Boy, that's really something. I guess some of these Haole guys are all right."

By November 7, when the second intercompany *kaiwa* session was held, tension between the two companies had virtually vanished. Although some haters on both sides kept their distance, most men in the two companies joined forces in condemning the administration. This time the OCS sent its 7th section, made up of students whose Japanese was so poor that the Nisei had difficulty in understanding them. Most conversations took place in English and focused on the rumor that Asiatic language specialists would be frozen indefinitely, a matter of common concern. Furthermore, the promised changes had occurred in the mess hall. Nisei cooks replaced most of the previous kitchen personnel;

the food had improved markedly, and rice and soy sauce had been added; each man was able to get all the food he wanted. Racing for a place in the mess line was no longer necessary. It was not uncommon for Nisei and Caucasians to chat pleasantly in the line, as they made room for each other to break in. Ironically, as the pattern of violence spread to other parts of the fort, the OCS men seemed closer to Company K than they were to anyone else. The plan drawn up in St. Paul turned out to be a complete success.

Analytic Summary and Discussion

Divisiveness has long been recognized as a symptom of demoralization. As men begin to wonder if anything they are doing as part of an organization is worth the effort, individuals and cliques tend to go their separate ways. Once members of primary groups develop suspicions of one another's sincerity, they tend to break up into factions that question one another's loyalty and decency. Those who are suspicious become selectively sensitized to cues that reinforce their beliefs; anything that tends to confirm their suspicions is noted immediately, and many otherwise insignificant acts are misinterpreted as evidence. Once such schisms have developed—between "eager beavers" and "fuck-ups," between Hawaiians and mainlanders, between the assimilated and those still oriented to the Nisei world, between supporters of MISLS objectives and their foes—the men are no longer willing to take risks for one another. Each person tends to look after his own interests and those of his friends.

Company K was blamed for virtually all the violence that broke out at Fort Snelling in the autumn of 1945. Actually, fewer than a half dozen members of the unit were involved. As a general atmosphere of gloom and futility settled over the company, a small band of disgruntled Hawaiians was able to give vent to its pent-up hostility—first within the company and later against outsiders. It was possible for them to do so only as long as their deeds were condoned by the others. When a group becomes demoralized, even informal norms against intramural violence become difficult to enforce; for only a few care enough to risk the trouble that restoring order might entail. The targets chosen by the youthful trio were limited by what their fellow Hawaiians would tolerate. In the 6th Platoon no difficulties occurred after the Yamanaka beating. After the assault on Sergeant Endo, the Hawaiians decided that fighting within the company had to stop; the trio then turned on the OCS men, scapegoats who could be attacked with impunity. Once the others made peace with the OCS, violence in the company stopped altogether. By this time the pattern spread to other units; and even though Company K continued to be blamed, its personnel was no longer involved.

The abrupt termination of the struggle between the OCS men and Company K supports the hypothesis that ethnic segregation facilitates internal conflict by creating divergent perspectives (Mandelbaum, 1952: 93–107). The most important consequence of segregation is differential communication, which makes the correction of erroneous beliefs difficult. As Stouffer (1949: I, 502–07) pointed out, in segregated units many inconveniences that affect all soldiers come to be regarded as the result of deliberate mistreatment. Since the policy of segregation is already seen as an insult, members of minorities become hypersensitive to slights, and many common grievances become magnified and placed on an ethnic basis. The Nisei were convinced that OCS men regarded themselves as entitled to special privileges by virtue of being white, that they cut in ahead of them in the mess line because they felt they had the right to eat first. Even though they were billeted across the street from each other, they had been unable to find out that both units were deeply demoralized and that they shared very similar views about the mess hall, about the policy of discrimination, and especially about the MISLS administration. The rapid resolution of the difficulties, once communication channels were opened, underscores just how much these soldiers had in common.

7. «» fluctuations in company morale

With the advent of the frigid Minnesota winter, life at Fort Snelling became in many respects more disagreeable; but there was a noticeable change in the outlook of most members of Company K. From the miseries they had shared they began to develop a stronger sense of mutual identification. Wherever they went among other Nisei, they found themselves labeled as "fuck-ups" and condemned. They were forced to defend themselves and to explain. As they did so, many began to feel that their grisly reputation was not entirely justified; they were not that bad. A notice that appeared on the bulletin board early in November that three students in Company I had been court martialed for beating their mess sergeant elicited great interest. Discipline was breaking down everywhere.

"Christ, we ain't the only ones. Even I Company guys're doin' it."

"Hell, we ain't the worst company any more. Company L's way worse than us, and now they're startin' to fuck up in the regular school."

Although their reputation in the Nisei social world had not changed, the men no longer looked upon themselves as the worst company in the fort.

Toward the end of October the press began to publicize the exploits of Nisei intelligence troops in the Pacific. Although minor references to their work had appeared before, this was the first widespread revelation of their contributions; and the releases contained many complimentary statements by various Pacific commanders. This led to further discussions of what would happen to the Nisei, now that the war was over. Would the record of Nisei soldiers be forgotten in the same manner as the accomplishments of Negro troops in World War I? Would this favorable publicity be overshadowed by news of rioting in the Tule Lake Segregation Center, where those who had declared themselves disloyal had been confined? Perhaps not, for many felt that even in the already friendly Twin Cities area they could detect greater receptivity on the part of the populace. Some began to feel guilty about not having played a more active part.

"Well, I guess it might not be too bad for Buddhaheads after the war. But Goddam! Look at all the guys who had to die for it! What the hell did *we* do? We just sit on our ass and make it rough for everybody."

"Well, shit. They didn't give us a chance. Why didn't they let us go over right after we finished basic? If us guys got sent to Europe, we would've fought just like them other guys. You know a Buddhahead! Shit, just turn 'em loose, and they go for broke."

According to the announced schedule Division F had only another month of classes, and most students were looking forward to graduation. They were tired of loafing. Qualified or not, most of them wanted to go overseas. In doing so there would still be some service that could be rendered; if nothing else, they could relieve veterans who were entitled to discharges. As the general atmosphere of gloom and futility was gradually replaced by some optimism, the company's performance underwent a transformation that many observers regarded as miraculous. Even bitter critics of the unit were astonished at some of the things that happened.

Reorientation to Peacetime Duties

As attention focused increasingly on occupation duty, conceptions of what to anticipate were constructed from letters, news articles, and rumors. Letters from occupied Japan mentioned the destruction and disorder there. While they pointed to difficulties that might arise, none seemed insurmountable.

"I hear they have food riots over there. I guess they ought'a have some. It'll get worse too. Goddam! Won't it be the shits if we have to guard the fuckin' warehouse?"

"You get gun, yeah? If dey come, shoot 'em."

"Ah, food riot. Sheet! All women and children, yeah? Me, I no like fight like dat. Buddhahead men I fight anytime, but different with keeds."

All indications, however, were that women were plentiful. One widespread rumor, attributed to a monitored Domei News Agency broadcast, was that the Japanese government had recruited 5,000 women to "comfort" the invading troops. The instructors explained that providing for the erotic needs of conquering soldiers was taken lightly in Asia. Classroom discussions continued to center on sexual adventures. Friends had written that the easiest way to finance affairs was through the black market, and some began making plans for taking goods that could be carried easily. What would appeal most to Japanese women? Cosmetics? Jewelry? Food? Would it be more advantageous to carry such items or to have them sent by mail? A letter from a G-2 man stationed in Tokyo attracted much attention. He indicated that in spite of pitiful living conditions, countless places for entertainment were open. He also claimed that Nisei were received just as well as Caucasian troops and that his officers treated him "like a king," since they were so dependent on him.

"No lie! I guess when you get assigned to some line outfit it ain't so bad."

"What da hell! De officers no can do naut'eeng without da interpreter, no?"

"Yoshiwara, here we come! He say da place open, yeah?"[1]

"Jesus Christ! I figured the Boochies wouldn't like the Nisei. I bet they're wondering where all the slant-eyed American soldiers come from, huh?"

"I don't give a shit what they think, if I can get my ass once in a while. I'm gettin' tired of this fuckin' place. I want to get the fuck out of here."

"So da places all open, huh? Dat's good. Now we get *tanoshimi* [a pleasure to which one can look forward]."

Compared with their boring existence at Fort Snelling, life overseas seemed full of adventure, opportunities for romance, and sexual exploitation.

The only source of consistently negative reports was Manila. A replacement pool for Nisei interpreters had been established there, and living conditions on the Philippines were said to be abominable. Everyone had to line up for hours each morning to get water for drinking and shaving. The troops lived in pup tents, and the rain was so heavy that they found themselves sleeping in the mud. Filipinos would sneak into the tent area to steal things from barracks bags. According to one rumor, all MISLS graduates had to take tests in Manila; only those who passed went on to Japan. Another rumor was that Nisei on the Philippines were assigned to KP duty, regardless of rank, while Caucasians drew easier tasks. The most fearful report was that the Filipinos regarded Nisei as Japanese

1. Yoshiwara was a notorious red-light district in Tokyo.

and wanted to kill them. Some contended that Nisei interpreters had to be accompanied by Caucasian guards who protected them from the Filipino partisans. Many recalled the Filipino reign of terror on the Pacific Coast in 1942, and they concluded that Filipinos were barbarians.

"Goddam! That's all the appreciation we get. The Buddhaheads fight like hell to liberate the place, and the fuckin' PIs try to kill them. Those Goddam bastards are too dumb to be independent. I'll bet when the Japs were there, they were kissing their ass and everything."

There was accord that the Philippines would not be a pleasant place for Nisei. Since many assumed that all replacements would be sent initially to Manila, the prospect was not encouraging. Permanent duty there was to be avoided at all costs.

The weekly Japanese movies also played an important part in shaping expectations. *Tsuma no Baai* aroused many favorable comments. The heroine was an Oriental-type beauty, and she did not appeal to most of the men. The film, however, stressed the virtue and devotion of the "true Japanese wife and mother," who patiently endured all the excesses of her husband but still aided and comforted him whenever he failed. Such devotion to the husband, which by American standards seemed to be carried to ridiculous extremes, did not strike the men as implausible, for they themselves had known Issei women, perhaps their own mother, who served their men in this manner. Some thought seriously of looking for a wife with such attributes. The second girl in the picture, a spoiled child who neglected her husband and home, provided a sharp contrast. Several commented that most Nisei girls were much more like her.

"Boy, them Buddhahead babes really take a beating from the old man, huh? No matter what the guy does, they stick along, and they ask for more."

"Yeah. You no see no Nisei gals like that, yeah? Dey just like da other *wahine*. Dey like have good time, yeah? Dey like take all your money and spend 'em."

"Well, all Boochies're like that. I remember my mother never argued with the old man. Anything he said went. But I don't want a wife like that though."

"Ah, you like wife dat push you around, yeah? You catch 'em here in dees country. Plenty *wahine* here who like drag your ass around."

"Shit, I don't want to marry a babe that can't even speak English. I know some Kibei babes that go around like the babe in the picture, but I don't want a wife like that."

Without question the greatest single attraction was *Shina no Yoru*, produced by Toho in 1940 and starring the Manchurian actress Rikoran. When students raved about the actresses in the earlier films they had seen, both the instructors and the more advanced students had commented, "Wait 'til you see *Shina no Yoru!* Then you'll really have something to talk about." When the long-awaited program finally came, the men were not disappointed.

The action in *Shina no Yoru* occurs in the context of the Japanese attack on Shanghai in 1932. Hasegawa, an officer in the Japanese merchant marine, becomes involved with Keiran, a Chinese girl whose home had been destroyed and whose father had been killed by the invaders. In the course of their relationship Hasegawa, through patience and kindness, overcomes her bitter hatred of Japanese and eventually wins her love. The subtle propaganda for Japan's expansionist policy was hardly noticed; for most viewers it was a sentimental story of a beautiful native woman falling in love with a man working with an army of occupation.

The picture was shown in four sessions. On their bus ride back to the barracks after seeing the first part, most students were quite pensive. There was little talking. Almost everyone was obviously impressed by the actress who played Keiran. Although she was unquestionably Asian, she was a beauty by Western standards. Many were also taken by the actress who played Toshiko, a Japanese singer who also loves Hasegawa. While some enjoyed their daydreams privately, others confessed their infatuation candidly to their friends.

"Boy, that Keiran babe is O.K., huh? Goddam! I never seen a babe as pretty as that."

"You ain't shittin', boy. She's O.K."

"How do you like that Toshiko, huh? Boy, she can park her shoes under my bed anytime."

"Boy, that stuff's all right! How'd you like to pick up one of them in the streets of Tokyo, huh?"

Rikoran immediately became the major focus of attention—an object of adoration. (See Plates 5 and 6.) The men talked of her constantly. Since she spoke both Japanese and Chinese fluently, questions arose immediately concerning her ethnic identity. When the instructors were questioned, they disagreed. Some thought she was born in Japan and had been adopted as a child by a Chinese family; others insisted she was Chinese. Still others contended that she was of Japanese ancestry but had been reared in China—a Chinese Nisei! To most students it did not matter; whatever she was, they cherished her.

On the morning on which the fourth part of *Shina no Yoru* was

PLATE 5 *Rikoran*
(Courtesy of the Toho Kabushiki Kaisha)

scheduled, the buses were a few minutes late. When there appeared to be some danger of missing part of the movie, several became enraged.

"Son of a bitch! Where in the fuck's that truck? We're gonna miss the first part, if they don't hurry up."

"Goddam! Seeing Keiran is about all I live for around here."

"Yeah. It's the only *tanoshimi*, huh?"

"Hey, I wonder if we could make it if we ran down there?"

"Don't be an asshole. It's over a mile down there. You can't run no fuckin' mile, and you know it."

"I don't know. I could make it in 10 minutes easy."

"Go call 'em up. Tell 'em to get off their ass and send the fuckin' buses."

PLATE 6 *Scenes from Rikoran movies*
(Courtesy of the Toho Kabushiki Kaisha)

Cheers went up when the buses finally arrived. The drivers were cursed for being tardy, and as the vehicles got under way, they were urged to drive faster. "Hubba hubba!" "Let's get the lead out!"

When the movie ended, the men were desolate, until they learned that they were to see two additional films featuring the same stars—*Byakuran no Uta* and *Nessa no Chikai*. They made no effort to conceal their delight.

Several students, on learning that she was to appear in the next movie, brought their cameras. Some had even purchased extra-sensitive film with the intention of taking time exposures of Rikoran off the theatre screen! At first others were skeptical that this could be done, but they watched with interest and elicited promises that they would be given copies, if they turned out well. In one of the earlier scenes in *Nessa no Chikai*, where Rikoran sang in ^ hotel garden, several close-ups in which she remained virtually motionless for some time provided opportunities for making exposures. Those who missed this chance did the best they could on other scenes. Those who succeeded in getting even a blurred image were envied, as they carried them proudly in their wallets. As one of the older men remarked, "Yep, that babe is really our pin-up. We might as well burn up all the other pictures."

"Goddam! If they got even one babe like that over there, I want to go overseas right now. Fuck this shit around here! Me for overseas."

"Yeah. Waste time around here. Shit, we don't learn nothing sitting around here. We might as well go over. Even if we learned ten times what we know now, it wouldn't make much difference."

"Shit, yes. We just don't know nothing, that's all. We won't learn any more, either. If we go over there, we could learn it from the people, you know."

"Yeah. I know what kind of people you're gonna learn it from."

"Boy! For Keiran I'll learn *any* fuckin' language."

The singing of *Shina no Yoru* became more frequent. It was the only item that enjoyed everyone's interest in the periods devoted to Japanese songs. Some sang it off key; others sang the verses out of order; still others could only hum or whistle it. It could be heard regularly in the shower room.

When civil war erupted in China, many followed the news with concern that was unusual for most enlisted men. One confessed shyly that he had become quite interested in China; in fact, he declared that he wanted to visit China as soon as possible. Those who had previously shown nothing but contempt for Chinese expressed sympathy for the people.

As increasing numbers were beginning to look forward to occupation duty, an abrupt announcement was made that Major Kawashita was departing for Japan, reaffirming the impression that MISLS activities were being expanded overseas. An immediate reaction was, "Aw shit! We get that bastard again. It's better to have him here for four more weeks than to have him for two years over there." Groans of protest arose at noon, when an order was issued for everyone to change to OD uniforms for a parade in his honor. Although no one was surprised, everyone was resentful. Just before the parade began the major spoke from the reviewing stand. Apparently oblivious to the manner in which he was viewed, he closed his farewell address with, "I'll be waiting for you over there."

"That's the trouble."

"Too bad he can't go over with us. We could push the bastard off the boat. Nobody'd say anything. They could restrict us to the fuckin' boat if they want to. Where else could we go? That'd be pretty good."

During the evening classes all the instructors were obviously delighted. Some even participated in the horseplay. Major Kawashita was the most detested man in the administration, and the instructors made no secret of their views. One sergeant contended that things were already better, even though he had been gone for only a few hours. The instructors agreed that most of the "meanness" and "chicken shit" emanated from the major's office; now that he was gone, the school would be run on a peacetime basis of forty-four hours a week.

Just when the troops were becoming convinced of the attractiveness of overseas duty, an electrifying rumor touched off speculation in Company K. At breakfast on Tuesday, October 23, word spread that everyone in Company H had been alerted for overseas shipment. A large part of that company was made up of men who had come from Fort Meade; they had started school at the same time as Division F. They were being sent overseas with only seven weeks training! Considerable excitement mounted, and the shipping orders became the major topic of conversation. Several other rumors developed during the day. One was that General MacArthur needed 1,000 interpreters immediately for occupation duty; 600 were going from Fort Snelling, and 200 more were being transferred from other assignments. Another 200 were still needed. A second rumor was that if enough men were not available from the "regular" school, students from Division F would fill the complement. Members of the more advanced sections were more concerned; if anyone in

Division F was ordered to go, those who were presumably least unqualified would be called. However, others were also concerned. No one knew how many were to go; if 200 more were actually needed, this could include most of Company K.

During the noon hour several men went to the PX and encountered a large number of second lieutenants whom they had never seen before. Since their uniforms were new, many assumed that the OCS men had been commissioned. Unusual events were taking place. A whole company of inadequately trained students was being sent overseas; many new officers were in camp; there was a heavy demand for interpreters overseas. How would this turn of events affect Company K? A general feeling developed that something momentous was about to happen, and everyone was on edge, expecting an announcement to be made at any moment.

Throughout the afternoon more rumors developed. General MacArthur wants 1,000 interpreters, but only 800 are available. About 600 men are shipping out; almost everyone in Companies C, F, G, and H is leaving; some of them have been in school for only two and a half months. Everyone in Company I has been alerted. Companies G and H will graduate this weekend. If enough interpreters are not available to fill the order, F-1 students will be included; section F-1 will graduate with Company H. Instructors were questioned about these rumors, but they knew no more about what was happening than the students.

"I think the F-1 boys'll have a rough time if they go over with them other guys. Shit, they won't get any ratings at all. Them other guys know how to read and write. Even if they only want interpreters, the guys who can read and write can talk better anyway."

"Yeah, but the guys in F-1 are better than the guys in the ass end of H Company. They went to school just as long, and they learned more. They're smarter than the dumbest guys in H Company."

At 15:30 the company was ordered to the dispensary for flu shots; this was interpreted as additional support for the expectation that at least part of the company was leaving. This was believed to be the first of a series of inoculations required for service in Asia. Since compulsory study on Tuesday nights had been terminated, that evening a number went to Company H to find out what had happened. Some visited slight acquaintances in an attempt to get details. Several of the rumors were confirmed. Company H personnel were indeed preparing to leave for furloughs.

On Wednesday morning more rumors emerged. Some men were shipping out from Company F with only three weeks of training. Students who have been in the army for fifteen months do not have to go overseas. Colonel Andersen knew about the mass order before he agreed to eliminate two nights a week of compulsory study. Division F will graduate on Wednesday next week, get a furlough of fifteen days including travel time, beginning November 1, and leave with the others. Another rumor was that Colonel Andersen was meeting that morning with the Division F instructors.

At 08:00 all the instructors were late because of a meeting with the colonel. This seemingly confirmed one of the early morning rumors, and everyone was expectant. "We're going over, boys! They're having the conference about the details." When they returned, the instructors faced a barrage of questions. The following announcement was made in F-1: "The colonel says that 560 guys are shipping out. He told us that he wants you guys to stop fighting and to get to work. He wants no more fights. None of you guys're going yet. He wanted the instructors to volunteer to go over." The tension was suddenly broken. There was a marked letdown. Most men were disappointed. Some argued that they could perform as well as those in the lower sections of the "regular" school and could certainly do better than those who had attended classes for only three weeks. However, excitement over the possibility of going overseas in a few weeks vanished.

It was in the midst of the general preoccupation with overseas service that the rumor of all G-2 linguists being frozen reappeared once more. On October 23 it was contended that Colonel Andersen had told the instructors on the previous day that the MISLS administration was trying to have the Nisei declared "essential" so that they could not be discharged on the same basis as other soldiers. This rumor was reinforced further in the *kaiwa* session with the OCS men who had been fearful of this possibility for some time.

"I hear the fuckin' colonel told the instructors yesterday that they're trying to make the Nisei essential so we can't get a discharge. How d'yuh like that? It's just like him, huh?"

"Yeah, that's what I heard. Them cocksuckers're always figurin' some way to screw us. What the hell! Why can't they leave us alone? Shit, all the other guys get to go home when their time comes. Why can't we?"

"I figure it's something like this. They're probably going to pass a law pretty soon that all guys with more than thirteen months' service don't have to go over now that the war is over. The G-2 guys probably heard about it, and they want to beat the deadline and get us over before the law goes into effect. You can bet there's something like that going on. They're not doin' all this for nothing."

"If they guarantee us we're gonna get out in two years, I'd be willing to work hard for a while, but otherwise fuck them. Christ! If we're too much on the ball, they might classify us as essential, and then we'll get fucked for the next five years. None of that shit for me."

"What d'yuh mean 'might'? I heard they did."

"That's what one of the instructors said."

Since the rest of the army was demobilizing, the men assumed that the General Staff in Washington, D.C., would authorize the demobilization of Asiatic language specialists on the same basis as everyone else. Members of the General Staff would be much too busy to be concerned with such trivia. On the other hand, the need for linguists for occupation duty was recognized. Therefore, the men reasoned that MISLS administrators must be doing everything possible to call to the attention of the General Staff the necessity of retaining all Nisei and language officers. To be sure, the administration had previously promised that all Nisei would be discharged on the same basis as other soldiers. But these same officers had made many promises before, and each person had no difficulty in recalling a number that had been broken. Two other rumors added to the apprehensions. One was that MISLS had been authorized to go on until 1947. The other was that all Nisei in the peacetime army would be sent to Fort Snelling. These rumors created the impression that the G-2 language program was not being curtailed in the same manner as all the other branches of the service; on the contrary, it appeared that it was being expanded.

The OCS men, the instructors, and the Nisei students all had in common a basic distrust of the MISLS administration. They had all heard the rumor before, but this time it appeared more likely to be true. Many instructors had been in the army for several years; some had already served overseas. They were downhearted. Company K men had looked forward to a year overseas and then a discharge. Many had made plans on the expectation of being released in that time, but now they did not know what to do. One of them felt desolate: "Son of a bitch! I don't know. I feel kinda funny. When I heard that rumor, I felt kind of empty. Kinda alone, you know what I mean? I felt sick in the stomach the rest of the day. I don't know. I think everything just looked gloomy, and I felt kinda lost. Funny, huh? I just don't have any pep any more, and things ain't funny anymore."

MISLS administrators continued to deny these rumors, but they were not believed. At a battalion meeting preceding the parade on October 20, Colonel Andersen, after announcing the conversion of the school schedule to a peacetime basis, had commented that he realized that everyone wanted a discharge and had added, "You're crazy if you don't." This statement surprised those who had thought the colonel was trying to hold them as long as possible. On November 8 the long expected news appeared in the Twin Cities press. The announcement read:

> The War Department Wednesday night announced enlisted men who have 50 or more discharge points and are now on furlough in this country will be released as they report back to the army this month. . . .
>
> During the day, the department also . . . put new restrictions on overseas duty for enlisted men and officers—a step affecting approximately 125,000. The department announced that, with a few exceptions, no enlisted man with 21 or more months of honorable service since September 16, 1940, will be sent overseas for permanent duty. The exceptions are those who are graduates of the military intelligence language school, regular army enlisted men, and volunteers for foreign duty.
>
> Officers who have 33 or more months of honorable service, or 30 months in the case of medical department officers, likewise will not be sent overseas for permanent assignment. This will not apply to reserve officers who have chosen to remain on active duty, to regular army officers or to those classed as scarce specialists. Officer graduates of the military intelligence language school at Fort Snelling, Minnesota, also are subject to overseas service if they are not eligible for separation and have less than 39 months of service.

The news struck early in the morning, when the men purchased the *Minneapolis Tribune* to read as they stood waiting in the mess hall line. All fifty point men were eligible for discharge *except* those in MISLS! Now it was official. Large numbers clustered about the newsboys, purchasing papers to see for themselves.

"I lose fight!"

"I hope that fuckin' Andersen's happy. Somebody's got to get something out of this deal. It ain't us so it must be them."

"They did it, huh?"

"We're essential. We get fucked."

"That son of a bitch! That bastard Andersen tells us he knows how we feel. We're crazy if we don't want a discharge, he says. All citizen soldiers want a discharge, he says. Then he turns right around and fucks us when we ain't looking. It's bad enough when they fuck you from the front, but it really gets you when they fuck you from behind."

"They tried to get the teachers to volunteer. Naturally nobody's going to stay here. Fuck this shit! Now we're all stuck!"

"They want to keep up the reputation of this fuckin' school; so they fuck us to do it."

"When I get over there, I'm gonna fuck up by the numbers. Then, they can't say I'm essential. If I can't do my work, that's TS."

The grumbling continued for the remainder of the day. Most men were convinced that they had been double-crossed again. Since the colonel knew how they felt, this act was held to be even more perfidious.

At a company meeting that afternoon Lieutenant Catalano, knowing how his men felt about the announcement, once again denied the rumor and reinterpreted the press report. He advised them not to take too literally what they had read in the newspapers. He explained that there were three categories of scarce personnel: (1) essential, (2) critical, and (3) shortage. He then added, "Those in the first two classifications will be discharged on the same basis as the other men. But if you're in the last classification, well, men, that's just tough." However, many remained skeptical. It was not that they distrusted the lieutenant; they felt that he was not sufficiently high in the administrative hierarchy to know what was really happening. Furthermore, everyone knew that classifications could be changed at any time.

At this point confirmation came in letters from Manila of rumors that MISLS replacements had to take qualification examinations prior to assignment. A replacement pool for Asiatic language specialists had been established in Manila, and all replacements initially went there. If one failed the examination, he would either have to attend school in Manila or be assigned to some menial task. This disturbed the more serious members of Company K, for they realized that Division F students were so incompetent that they could not expect to do well in open competition with graduates of the "regular" school. Another rumor that ratings would be based on the examination added to the gloom. Even those who were willing to serve overseas for a long period as a sergeant found such duty as a private unattractive.

The picture became even more confused on October 29, when it was announced that some Division F students would be selected for OCS. This came as a shock. No one thought that it would be possible for Nisei to win commissions without going to Fort Benning and doing it the "hard way." Furthermore, why Division F? It contained the least competent personnel in the school. During the afternoon classes about a dozen men from the advanced sections were interviewed by a Nisei warrant officer.

Those who had been called were reticent about what took place, and only a few admitted that they would not mind becoming officers. Most of the interviewees claimed that they had told the warrant officer that they were not interested; they wanted to go "with the boys." It was assumed that officers would probably have to serve much longer than enlisted men, and the question was whether even an easy promotion would be worth the cost. A major consideration for those selected was that they had no assurance of being discharged from the service on the same basis as other officers. Again there was concern over being double-crossed by the MISLS. One man explained his refusal: "I figure there's a catch to this deal someplace. It's too easy. I figure they can't do too much to us as EM, but as soon as they get bars on us, they'll have us by the tail. I smell a rat someplace, and I don't trust 'em. Fuck this shit! They can keep their bars."

On November 19 four Company K men were transferred to Company I for OCS training. By this time it was known that only a very small number of Nisei had been chosen for OCS from the rest of the School Battalion. It appeared that when the investigation of the stabbing disclosed the extent and seriousness of Nisei disaffection, administration officials precipitously had ordered some token changes of policy. In making amends Company K was treated far more generously than others with more competent personnel, receiving a disproportionately large share of appointments. Many smirked that the administration "brass" was afraid of Company K.

It was in this context of confusion and suspicion that an increasing number began to consider the feasibility of reenlisting in the Regular Army in order to clarify their ambiguous military status. At 15:30 on October 15 all the men in Companies A, K, and L had been ordered to the firebreak near the Turkey Farm for an explanation of the army's new reenlistment policy. It was very windy and sometimes difficult to hear; furthermore, the policy was stated in such difficult language that many were unable to understand. The gist of what was said was as follows: (1) all men who reenlisted would have a minimum rank of Pfc., if they had six months of service; those with higher ratings could retain their rank if they reenlisted under certain conditions. (2) Men could volunteer for 12, 18, 24, or 36 months. (3) Furloughs of 30, 60, or 90 days with travel pay would be given, the length of the furlough depending on the length of previous service. (4) Mustering out pay of at least $200 would be paid at the time a volunteer was discharged from AUS. (5) Those who volunteered for three years would be allowed to select their branch of service and the theater in which they were to serve. There was considerable confusion over what was said, but many were interested. Still, the overall reaction remained one of distrust.

"What if all the Boochies took three years and volunteered for the air corps in the Caribbean theater? Wouldn't that be something?"

"Fuck, we're gonna be in the army for another year anyway. We may as well take the 30 days and 200 bucks."

"I don't know. I don't trust 'em. Them guys who were reading that paper were G-2 guys. How do you know they read it right? Maybe they got some stuff written down there in small type so they could fuck you. I trust the War Department, but I don't trust these guys."

"That means if a lot of guys volunteer the draftees get out, huh? Will it mean we might get out at the end of a year? What if a few draftees are still left at the end of the year you sign up for? Do we have to stay until all the draftees are out?"

"Does that mean that all the draftees are going to get out in a year? Hell, they gotta let the draftees out before the regulars."

"If I was sure of getting out in one year, I'd reenlist right now and take the 30 days. But it sounds too good. I don't trust them bastards. They promised us a lot of stuff before. They're a bunch of double-crossing pricks. So fuck 'em! I ain't signing nothin'."

"If they get enough volunteers, maybe we won't have to go over. Even if we go, maybe we won't have to stay very long, huh?"

"It sounds fair enough, but there's probably a catch someplace. The wind was blowin' too hard, but it sounded pretty good to me."

"If you get a discharge from AUS, then you're not under oath no more for the duration plus six months. If the new oath is for one year, they have to let you out."

"I don't know. They can always find some way to fuck you if they want to. They never leave themselves wide open. They'll get you one way or another. I don't trust them."

The advantages and disadvantages of volunteering were weighed against each other. One obvious advantage was getting a furlough of 30 days, probably over the Christmas holidays. There was also a remote possibility of getting out of G-2 and being assigned to some other branch of the service. On the other hand, one would lose $100 of mustering out pay, since anyone serving in AUS overseas received $300. Furthermore, draftees might be discharged sooner than expected, and G-2 personnel might be released at the same time. Separation from one's friends was another consideration. But the strongest argument in favor of reenlistment was the definiteness of the termination date; what many feared was being caught indefinitely in occupation duty. When they had been inducted into AUS, they had taken an oath to serve under the Articles of War for the "duration plus six months." Everyone realized that this term could be defined at the army's convenience. By reenlisting they would be released from this oath, and the new commitment would be for a definite period. At least something would be certain.

Several asked Lieutenant Catalano about reenlistment. The lieutenant warned them that after a term of enlistment in the Regular Army one was still subject to reinduction into AUS, although it was unlikely that MISLS officials would go that far. He also pointed out that if a man reenlisted after the furlough following graduation, he would receive another furlough. He explained further that anyone who reenlisted before October 1946 would still be eligible for benefits under the G.I. Bill of Rights. However, after enumerating these advantages, he concluded, "But, men, you can't tell what's going to happen. If I were you, I'd just stay in this man's army." Most heeded his advice. Between November 8, when the lieutenant made these remarks, and December 20, only twenty-nine volunteered from Company K.

As the third semester drew to a close, classroom work became somewhat more orderly, and increasing numbers made a serious effort to learn some Japanese. Some were anxious to do well in overseas examinations, not only to get higher ratings but also to avoid being stuck in the Philippines. Some studied from sheer boredom; they were so tired of doing nothing that they even resorted to studying Japanese. A few even went so far as to bring their textbooks to their barracks; they were studying openly on their "own time." Informal pressures against studying continued, but they were less intense. The term "eager beaver" was replaced by "wristwatch." One of the established practices in MISLS was the awarding at graduation of wristwatches to students voted by the faculty as the most competent man and as the man who had improved the most. As more and more students made an obvious effort to earn high grades, one man in F-1 composed a song for them. He called it "*Tokei no Kyoso*," or the "race of the watches," and sang it to the melody of "Tokyo Rhapsody," which had become quite popular. One afternoon a student who leaned back against the wall accidentally tipped a fire axe that was balanced on two nails. As the axe rocked back and forth, it made a sound resembling the ticking of a large clock. All the next morning the men sang "*Tokei no Kyoso*" to the accompaniment of the swinging axe. Thenceforth, whenever a man was thought to be studying too hard, someone rocked the axe to call the attention of the others to him. At times a man who recited exceptionally well was asked for his Alien Registration Card, but the whole matter was taken in good humor. Although little was accomplished during the final days of school, the instructors appeared grateful that things were relatively quiet.

Mobilization for the Company Commander

Toward the end of October several changes took place in the company. Because of a new War Department policy of not sending high-point men abroad, virtually all NCOs in Company K were transferred to the Turkey Farm. Since most of them had had three or four years of active service and were exempt from overseas duty, there was little point in continuing their training. On November 8 there were only 187 enlisted men left in Company K—1 sergeant, 1 T/5, and 185 Pfc's. The NCOs' places were taken by the former squad leaders, and this placed the burden of maintaining order on men who held the same rank as all others.

By this time everyone knew the "acting gadgets" of their platoon and recognized their limited authority. Acting NCOs were, on the whole, hesitant about giving orders, but the others generally complied with their requests. Squad leaders generally got things done through personal ties and obligations. Whenever anything came up that called for a decision among the men themselves, squad leaders were usually consulted, if not called on to settle the issue. "Acting gadgets" were also sources of information, and frequently rumors were checked with them. A man who wanted to do something unauthorized generally told his squad leader of his intentions. Squad leaders rarely scolded anyone; if they did, it was in fun. Most men sympathized with the acting NCOs. As one man put it, "It's a shit job. No stripes, and they get all the work and responsibility."

Appointed platoon leader of the 6th Platoon was Private Henry "Cyclone" Nakasone. Nakasone was an unusual name, and it was regularly mispronounced in an Anglicized manner. Because of his exceptional ability as an athlete and infantryman he was highly respected; most of the privates rated him a cut above the other "acting gadgets." Those who knew him regarded him with affection, for he was reasonable and good natured. Many joked about his inimitable stentorian voice; they often commented that he was the only soldier they knew who could march an entire battalion of troops unaided by a public address system. Although he was only a first class private, he sometimes bawled out offenders as sergeants were wont to do, but no one seemed to mind. The members of the 6th Platoon—Hawaiians and mainlanders alike—sometimes referred to him jokingly as "Simon Legree."

In contrast, Sergeant Inamasu, platoon leader of the 5th Platoon and the only sergeant left among the students, soon made himself very unpopular. At 15:30 on October 30 a fight almost started on the road by the school. The argument began when members of both the 5th and 6th Platoons expected to return to the barracks as the duty platoon. Since the 6th Platoon had cleaned the latrine that morning, its personnel assumed that it was their turn to be excused from physical training. Sergeant Inamasu and members of the 5th Platoon insisted, however, that on their last turn to return to the barracks the entire company had been excused from physical training and that therefore they deserved a rest. As the argument became more heated, members of the 6th Platoon completely surrounded the sergeant; finally tears began to flow as some became too enraged to control themselves. The sergeant claimed authority by virtue of rank; furthermore, he insisted that he had been ordered to bring in his platoon. When several expressed their doubts, he became angry and shouted that if anyone thought he was a liar he could step up and fight. Punchy accepted the challenge, but the other Hawaiians dragged him back. Private Nakasone was not present, and the squad leaders who were fighting for the 6th Platoon finally agreed under protest to take their group with the rest of the company.

"Dat fuckin' Inamasu! He t'eenk hees sheet no steenk. We bust up!"

"Just 'cause he's the highest rankin' guy, he's gettin' too fuckin' fresh."

"Me, I no like when Boochie pull rank."

"He tell us to come out and fight becaus' he know we no can do naut'eeng. Why he no say dat some night when we all happy? Me, I like fight heem right now."

"No, no, Punchy. No fight now. Next month we go overseas. You fight now, you no can go."

Sergeant Inamasu was discharged on November 26.

Although the men resented Nisei NCOs who "pulled rank," they obeyed those whom they felt had earned their stripes. At a company meeting on October 25 Lieutenant Catalano introduced the first sergeant who had replaced Sergeant Endo. He explained that First Sergeant Kondo was not a MISLS NCO, who had received his stripes by studying Japanese; he was an infantry sergeant who had won his rank in IRTC cadres. Although he said very little, he immediately commanded the respect of the men.

"Any Boochie that can win a first sergeant rating in IRTC I take my hat off to. That's really hard work, and you have to be on the ball. That guy must be O.K. He looks like a good guy."

"Any Buddhahead that can go up that high in an all-Haole outfit must be pretty good. I hope he ain't too chicken shit, though."

A week later, when the sergeant was making an announcement, men complained of not being able to hear. He yelled, "Let's have it at ease!" and to everyone's surprise the order was obeyed. When some Caucasian KPs working nearby continued to make noise, Sergeant Kondo went over and upbraided them severely. This pleased the men immensely.

"Now that's what I like to see once in a while. All the time, ever since I got in this fuckin' army, I see the fuckin' Haole chewing the shit out of the Buddhaheads. I want to see some Buddhaheads chew out some Haoles once in a while."

"Boy, that sergeant don't take no shit from nobody, huh?"

"Shit, he's an IRTC man. He's had a lot of Haoles under him there. He's used to chewing their ass."

The operation of the company suddenly became markedly more efficient. When the men were warned that AWOL would no longer be tolerated, they grumbled about "chicken shit" but reported for all formations. They made various excuses for appearing on time, but they were present. Their "goofing off" days were over. Yet there was little animosity toward the first sergeant. Several had already been in the orderly room concerning personal problems and had found him sympathetic and helpful. Word had gotten around that he was strict but a "good Joe."

With the transfer of most NCOs some mainlanders felt that the company area was unsafe. The tension from the reign of terror had not passed completely, and many anticipated more trouble. About 22:30 one evening considerable commotion developed in the 6th Platoon when Punchy became so inebriated that he urinated all over the floor. Because he was so friendly Punchy was well liked by Hawaiians and mainlanders alike. As he staggered about the building yelling and urinating, mainlanders pretended to be amused and said nothing. Several Hawaiians got out of bed, quieted him down, and after much struggling finally succeeded in undressing him and putting him to bed. Many mainlanders were disgusted, but they talked of it only among themselves in quiet whispers.

The night of October 31 was one of the wildest in the company's history, but somehow everyone managed to survive without any casualties. As might be expected on a payday, by early evening the barracks were a bedlam of confusion. Because of the rain many had not gone out on pass. Some went periodically to the PX for beer and soon became intoxicated. Those who still had money were rolling dice or playing cards, and sometimes tempers flared as the losers tried desperately to save their month's earnings. Several who had gone into town returned drunk, and some brought back bottles of whiskey. The duty officer visited the area several times, and each time he found something wrong. All the lights were on long after 23:00. When he entered the barracks, no one yelled "Attention!" He was ignored, as the men went right on drinking and gambling. When he started to castigate them, no one addressed him as "Sir." When he made a check of those who were restricted to the company area, he found that almost all of them were missing. At bedcheck so many were missing from their bunks that the CQ gave up. So many men were drunk that night that the CQ spent much of his time helping friends of the intoxicated put them to bed. About 02:30 the next morning Tajima became enraged at the CQ and punched a gaping hole through the screen door of the orderly room, which he mistook for the CQ. Several Hawaiians had to drag him to bed.

On the following morning the duty officer returned to the company area during reveille. Because of the hard night the troops were quite sluggish in falling out on the road. One man was still missing; the DO caught another in the latrine during reveille. Since the CQ had reported the company "all present," as usual, he was given a severe reprimand. Everyone expected the worst; some considered the possibility of their passes being withheld for a month. It was taken for granted that Lieutenant Catalano would be punished again. At 15:30 when the company was ordered back to the barracks for an orientation meeting, trouble was anticipated.

"Well, we get our ass chewed today."

"Lucky if we don't get our asses restricted."

"Expect the worst."

The lieutenant, however, merely warned the men not to drink if they could not retain control over themselves. He announced that the man who had punched the hole through the orderly room screen would this time only pay for repairing it; if it ever happened again, however, whoever did it would stand court martial. "I'm gonna give it to ya straight from the shoulder, men. I don't like sneak punches. If it happens again, you'll stand a court martial." The men were again impressed with the lieutenant's attitude. Although a minority in the company had been responsible for the serious trouble, they had expected mass punishment. Instead he had simply warned the few and had done nothing more. When the call came to fall out for retreat, the entire company was on the road in forty-five seconds! The new first sergeant was astonished.

There was some hilarity later that day when the platoon sergeants went from man to man asking each how long he had been in the army and how much of that time had been served in the continental United States. Everyone with a year of continuous service on the mainland was entitled to an American Theater of Operations ribbon. Most mainlanders were qualified but ambivalent. Since it was a noncombat award, many made derogatory comments even though they were secretly pleased.

"Hurray! We finally get a campaign ribbon! How about the one for capturing A.P. Hill? Don't we get a Silver Star for that?"

"Gee. I'm thrilled! I better go out and get my picture taken. Ha! Ha! Ha!"

"I wouldn't want to be seen in that chair-borne stuff. Shit, they can keep it. It wouldn't be bad if you had a lot of other ribbons, but with just one blue one and a Good Conduct ribbon, it looks too fuckin' cheap."

"What makes you think you gonna get a Good Conduct ribbon?"

"I lose fight. We no do naut'eeng yet, and we get two ribbons already. By da time we come home from Japan, we no can carry all the lettuce, yeah?"

Most noticeable during this period was the deep affection and respect that the men had developed for their commanding officer. Even the few who did not like him agreed that he had been their champion at considerable sacrifice to himself. They could not help but noticing the frequency with which he had been assigned as duty officer on weekends. Yet he had never complained about this. They knew that his superiors had often called him on the carpet, for the cadremen told them privately. Although he occasionally alluded to his troubles with the colonel, he never took them out on his subordinates. Whenever he explained army regulations or policies, he was direct and straightforward, and the men appreciated it. When he asked about the length of their last furloughs before going overseas, for example, he responded, "I know that a lot of you are worried about your furloughs. I can't promise you a thing, men, but if I find out anything definite about it, I'll let you know. If there's time, you'll get it. I remember one time when I was at Camp Ritchie I saw a man graduate at 16:00 and on the boat at 06:00 the next morning. It's like this. If there are fifteen days before your boat is scheduled to go, you'll get fifteen days. If they're ten days, you'll get ten days. If there are no days left, that, gentlemen, is tough shit."

By this time most men did not conceal their high opinion of Lieutenant Catalano. They trusted him, and they took their personal problems to him with confidence that he would do whatever he could for them.

Their feelings about him were manifested in several ways. He was a Caucasian, but no one ever referred to him as a "Haole."

"Boy, he really took a lot of shit from us. I think K Company's beginning to get on the ball now because of him. Boy, his patience really held out good, huh?"

"Hey, I hear there's gonna be a concert at the University of Minnesota next week. If we stay on the ball, I wonder if the old man'll give us a pass to go out and hear it?"

"Sure, the old man's reasonable. If you fuck off, it's TS; he'll throw the fuckin' books at ya. But otherwise, you get all the breaks. He's really square from the shoulder. He never fucks you up. You go ask him, and if you got a good reason, he'll give you a pass. If you got a lousy reason, he'll tell you where the fuckin' passes are and turn around the other way. He's O.K."

"Yeah. If you ask for it, he'll give it to you. Even then, he's pretty soft-hearted."

"He's the best fuckin' CO in this goddam camp."

Most Company K men, especially when they were in their own area, were slovenly in appearance. They did not walk with the erect military bearing expected of soldiers; their uniforms were often mixed or incomplete; military courtesy was virtually ignored. But this stance was transformed suddenly whenever Lieutenant Catalano appeared. If they had the time, they hastily buttoned up, tucked in their shirt tails, and straightened their caps. As he approached, they snapped sharply to attention and saluted. They *always* addressed the lieutenant as "sir"; he was the one officer who was respected. No other officer in the fort, Caucasian or Nisei, received such treatment; most of them were viewed with contempt or resentment. Thus, Lieutenant Catalano seldom saw his men as other officers saw them.

A minor incident one afternoon, when physical training consisted of playing softball, revealed the orientation of most of the men toward him. When the hour ended, an order was issued to stop playing and to get into marching formation. In direct opposition to the order, one man picked up the ball and batted it into the air. Anticipating a reprimand, everyone anxiously followed the flight of the ball. It came down directly on the company commander! Several yelled for him to be on guard. The lieutenant caught the ball with a smile and threw it back to the batter, who by this time was quite worried. When the lieutenant continued to smile, there was a sudden burst of applause and cheering. "Nice catch, sir!" The general feeling of fellowship and camaraderie definitely included the lieutenant. "Boy, the CO's really a swell Joe!"

Few means are available for enlisted men to do favors for officers. Whenever an opportunity arose, the men seized it. They knew that their performance in parades pleased him; they could see him smiling broadly whenever he was on the reviewing stand. Although they agreed that the exercise was a stupid waste of time, they worked diligently week after week to make him "look good" in the eyes of the colonel.

Saturday morning inspections by high-ranking headquarters officers also provided such an opportunity, and the men on their own initiative went to extraordinary efforts to transform their old, dilapidated barracks into model quarters. At 16:30 on October 26 it was announced that a headquarters inspection would be conducted the following morning. It was to be a full field equipment inspection. At 08:00 company personnel were to have all their equipment displayed on their bunks and were to fall out for personal inspection.

Griping was good natured, and spirits were high. Although the lieutenant had said nothing specific about what was to be done, most men started working at once. Talking amiably and joking, they helped one another clean their equipment. Some of it had not been used since basic training and needed scrubbing.

"That's what we get for gettin' a fuckin' IRTC sergeant. Goddam! We're back to basic training again. I haven't had one of these full field inspections since I got out of Wheeler."

"Lucky we got no rifles to clean, huh?"

"You ain't a shittin', boy."

"Don't be too damn happy, Sak. They got plenty of rifles in the warehouse, and they might just decide to have some of 'em cleaned."

In the 6th Platoon the squad leaders were asked by the others to take charge. A division of labor was worked out informally. Some washed the windows; others cleaned the stoves. Still others dusted the boards above the beds, the rafters, and the shelves that went from one end of the barracks to the other. When these tasks had been completed, the men individually or in small groups polished all their shoes, gave each other haircuts, and went over their equipment. The cleaning of mess gear that had not been used for several months presented special problems. Someone made a solution with soap that worked unusually well with steel wool, and this was passed from bed to bed until everyone had polished his equipment to a bright sparkle. The few who were lying about became targets of curt comments; before long they found it preferable to leave altogether. Even those who usually "goofed off" participated actively. The men worked well together; each contributed freely whatever he could. Hawaiians and mainlanders forgot their differences and helped one another generously. Very few left the company area that night, even though there was no restriction. They worked until about 21:00. Long before the lights were turned off most men were already in bed.

On the following morning the 6th Platoon was up at 05:30, and the men washed and shaved before the rest of the company got up. The other platoons had scrubbed their floors the night before. The 6th Platoon also needed extra time because it was the duty platoon of the day and had to clean the dayroom and laundry room as well. At 06:00 the 3rd and 4th Squads cleaned the dayroom; guards were then posted to keep everyone out until inspection time. The others stretched a hose from the latrine and give the floor a thorough scrubbing with GI soap and lye, after all beds and footlockers had been moved out of the way. As the floor began to dry, each person began to lay out his equipment. The platoon leader arranged a display of how the items were to be placed, and the others copied it. Squad leaders went back and forth, straightening out the beds and checking the displays. Each man made a final check before changing into his Class A uniform. By 07:45 the platoon was ready. Most of the men were obviously pleased. The barracks had never before been so clean. They joked about asking Lieutenant Catalano to remove his shoes before stepping on the clean floor. There were rumors of what some of the other platoons had done; in one of them the men had gotten down on their hands and knees to scrub the floor—"just like in basic training."

At 08:05 the company was called out for personal inspection. Everyone was ready; all buttons were buttoned, and all shoes were shined. Much to their consternation and disappointment no "brass" arrived. Three young lieutenants represented headquarters. The company passed the inspection easily. The lieutenants warned the men, however, that the "brass" would be coming soon and that they would have to remain "on the ball." There was a letdown. They had worked so hard for three "punk looies."

The next headquarters inspection was scheduled for Saturday, November 3. This time the men were assured that Colonel Andersen himself would come. No restrictions were imposed, but very few left the company area on Friday evening. From 16:30 until late in the night, most of them were in their barracks preparing for the inspection. There was little in the way of formally organized preparation. The men worked quietly on their own. Some climbed on the rafters and shook off all the dust; every exposed piece of wood was washed or carefully dusted. All windows and storm windows were removed and washed. The latrine ran out of toilet paper, as members of all the platoons used the soft tissue for the final wiping of their windows. Although they cursed the "fuckin' colonel," most men were smiling. Whistling, humming, and joking was common. Very few were idle. Those who had finished their own tasks went to help their neighbors. Groups gathered around footlockers, using one another's shoe polish and advising each other on the best ways of getting GI boots to shine.

About 17:30 the men in the 6th Platoon decided to scrub their floor. It was extremely cold, and this created several unexpected problems. They discovered that the company hose was frozen. While several struck it with heavy implements to break the ice inside, others began hauling water from the latrine in buckets. This turned out to be an arduous and painful task. It was freezing weather, but the work could not be done effectively while wearing overcoats; furthermore, the latrine was about fifty yards away. Finally as a desperation measure Nakasone threw the hose on one of the stoves amid gales of laughter. The hose thawed out, but the stream of water that trickled out was so feeble that the bucket brigade had to continue. The skin of those who were working outside turned bright pink from the cold air and water. Windows had to be opened for cleaning, and the barrack also became cold, especially since all the stoves were out. Despite the discomforts, however, the men continued working. They scrubbed the floor on their hands and knees, voluntarily taking turns. When half the floor was finished, the hose began to leak, and even the inadequate trickle stopped. Several went out into the snow to repair it, but by the time they had put on a patch the rest of the hose had again frozen. When the task was completed, the men jubilantly pushed their beds back into line, dumped coal into the shining stoves, and took their first break.

"Take your fuckin' shoes off when you come in here, goddammit! We don't want no fuckin' snow or dirt on the floor!"

"If we don't pass inspection tomorrow, I lose fight. Look at my fuckin' hands—all frostbitten!"

"Jesus Christ! It's really cold out there. I couldn't feel my fingers any more after that third bucket."

"Me, I bring da water one time; den I stay here. Sheet! Too cold, no?"

Delegates from the other platoons visited the 6th Platoon building to see what was happening. After seeing for themselves, they decided not to "GI" their floors. The mops were all frozen stiff. The soapy water that had been swept out had frozen on the steps; the steps then became so dangerously slippery that the ice had to be chopped off with hatchets and fireaxes.

On the following morning the men were sluggish in getting out of bed. Although they were aroused at 06:00, many could not get up. They had to skip breakfast in order to have enough time to get ready. In the 6th Platoon a strange thing happened. The floor could not be mopped. The boards had turned white from the severe scrubbing the night before, and the water from the mops only made them darker. "Only dry mop the place!" "Use clean water, goddammit!" The floor was carefully swept and reswept. By 07:30 the entire platoon was ready with their displays.

At 07:45 the company fell out on the road. Everyone was present. The work had not been done in vain; this time the colonel had come. There was no love for Colonel Andersen, but the men wanted to be inspected by a high-ranking officer, one who could influence decisions regarding Lieutenant Catalano's promotions. The colonel stated immediately that it was too cold to hold an outdoor inspection. As the men ran back to the barracks, each stopped at the door to clean his boots carefully before entering. The floors had changed color. They were almost white, and wet boots would make noticeable tracks. There was continued joking about taking off one's shoes "like in Boochieland" before entering the building. "Stocking feet only, goddammit!" Anyone who inadvertently carried in mud was censured by those near the door and forced to clean it immediately. Those who slept near the doors kept broom and mop in hand to clean up after any traffic. One man was posted outside to warn the platoon when the inspecting officers were approaching. Since the 6th Platoon was the last to be inspected, the men had to wait twenty minutes. They sat on the edge of their footlockers, talking quietly. Occasionally one would get up to realign his shoes with the bedpost or to straighten a wrinkle in his blanket. When the alarm came that the inspecting party was coming, everyone jumped into place. When Private Nakasone called the group to attention, all boots snapped together in one sharp thud.

The colonel was obviously astonished and pleased. As he walked slowly from man to man, he stopped frequently to compliment the lieutenant. The lieutenant walked behind him with an unsuppressed grin of satisfaction. As soon as the inspecting party departed, there were hoots of laughter and joy. The men were jubilant; the colonel had said nothing, but they knew that he was pleased. The barracks were impeccably clean.

Each man had a haircut and shave. All boots were shined. All fingernails were clean. Men with poor vision had put on their GI glasses, even though they did not ordinarily wear them. "Boy, everybody was sure on the ball today, huh?" Given the quality of the facilities the company had been assigned, the transformation was dazzling. Given the reputation of Company K, it was nothing short of a miracle.

Soon after the colonel's departure, the lieutenant spoke to the company over the public address system: "I want to say, men, that the colonel was very pleased, and I'm proud of you. He told me that in all his years with the MISLS he has *never* held a better inspection. I thank you, men. I'm proud of you."

The men were ecstatic. The reaction was, "*Mochiron! Mochiron!* [naturally]." Then, when the lieutenant announced that everyone was free to leave on pass at 09:00, there were more yells of joy. The men had been certain of the colonel's reaction. They had worked very hard, and they were proud of themselves. Although they were usually skeptical of anything the colonel said, this time they repeated over and over, "It's the *best* inspection he ever had in MISLS!"

Affection for the lieutenant was such that even the punishment that he meted out was accepted cheerfully. Those who had been caught rarely received any sympathy from their fellows. Most men agreed that the CO was fair; if anything, they felt that he was too understanding and lenient.

On Tuesday, November 6, a rumor developed that Private Takeda had been arrested and was awaiting court martial. No announcement was made, but members of the 5th Platoon confirmed that he was no longer there. Mainlanders were delighted that the CO had finally decided to take action against him. Hawaiians on the whole were reticent, but they could not help but notice the approval of the others.

"It's about time. He asked for it."

"The old man's been giving him breaks for the last three months. I hope that son of a bitch gets the fuckin' book."

"I hate to see anybody go to jail, but he's the one bastard who belongs there."

"Just as long as he don't ship over with us on the same boat, I don't care. I hope they get them other two guys too while they're at it."

By the following day everyone in the company assumed that Takeda had been arrested, and details of his latest offense circulated in the form of rumors.

"Hey, I hear Takeda got court martialed yesterday. He's gotta do thirty days in the guard house."

"Thirty days! Is that all? Shit, I figured he was good for six months at least!"

"No shit! Is that all they give him? What'd they get him for?"

According to some eyewitnesses, Takeda had been drinking beer in the PX and was becoming obviously intoxicated. After several drinks he went to the counter to demand several bottles all at once—contrary to a recent order of the post commandant that only one bottle could be sold to a man at a time. The clerk refused, and Takeda reached over the counter, seized him by the throat, and attempted to assault him.

Most mainlanders felt that Takeda had gotten off too easily, and even the Hawaiians were surprised at the lightness of the sentence. A man with his long record of creating disturbances should have drawn a much longer term. On November 13, when the men returned from their Armistice Day passes, more rumors developed of the way in which Takeda had gotten off so easily.

"They say Takeda got away with only thirty days because he cried like a baby and put on a show for the review board."

"No shit! That goddam chicken shit bastard! He'd do it too. He's pretty brave when he's got his boys with him, but he's chicken hearted when he's all alone. Remember that night Utsumi wanted to fight him? He just chickened out."

"All them guys're like that. They fight in gangs because they're yellow to fight by themselves. They won't fight man to man."

"Yeah! Remember that night when them bastards picked on Saipan and woke up the 5th Platoon? The CO wanted the guys who were responsible to show up in the orderly room. Remember? Takeda was the bastard who did it all, but Tsuboi was the guy who went up and got hell from the old man. That bastard let Tsuboi take the rap all by himself, and the old man was pretty sore too. That's chicken shit, huh?"

"Oh, dat guy no good. I know dat guy back home in Hawaii, you know. Young punk—no fuckin' good. Him always dat way, no fuckin' good."

On November 14 Takeda unexpectedly returned to the company. He was reticent about where he had been for the past eight days, and the others did not press him. He appeared to be a changed man, and the others assumed that his sentence had been commuted to enable him to graduate with his class. Actually, no report of his being court martialed appeared in the company records. No notice was ever posted on the bulletin board.

About ten days before graduation a move got under way to purchase a gift for Lieutenant Catalano. Almost everyone agreed that something should be done as a token of appreciation. When it was pointed out that army regulations prohibited giving presents to superior officers, the cadre suggested getting a gift for his wife; the regulations said nothing about that. Questions of the legality of that were drowned out by those who had taken basic training in all-Nisei units: "We did it in the other camps. Why can't we do it here?" The vote was overwhelmingly in favor of getting something for the CO through Mrs. Catalano. The platoon leaders took a collection among their men. Everyone who had money gave something; some even borrowed to make their donation. Those who had been absent were assessed as they returned to the barracks.

"What's goin' on?"

"We're chippin' in for something for the old man."

"Yeah? How much you guys kickin' in?"

"About a quarter."

"A quarter! Is that all?"

"Well, put in as much as you want."

The end of the training period was Friday, November 16. Two days before, one of the instructors expressed surprise that Lieutenant Catalano was planning to let the company leave on furlough on Friday night; he had been told that a graduation exercise was to be held on Saturday and that all instructors had to attend. A rumor about the graduation ceremony spread quickly and elicited much resentment.

"What the fuck! That's useless. We can't understand the fuckin' speeches anyway. Fuck the goddam diplomas! I don't want mine. This school ain't worth a shit even if that bastard Kawashita does call it a campus. Fuck the wristwatches! They can pass 'em out anytime. Son of a bitch! We get fucked all the time. Kishi must be the bastard that did this. We know Kawashita didn't do this. They're all the same—a bunch of bastards."

Absolutely nothing could be accomplished by holding graduation exercises. The wristwatches would be of interest only to the two receiving them; furthermore, even they would prefer to receive the award where no one else would see. The speeches were of no interest to anyone; if the valedictorian spoke in Japanese, few men in Company K would be able to understand him. The lieutenant had promised them that they could leave as soon as they were through with their duties on Friday. Had he not consulted the battalion? What had happened? When one of the instructors insisted that it was Lieutenant Catalano who wanted this graduation, most men dismissed it as a "rumor." But it was not entirely implausible. They knew that he was proud of the company; perhaps he wanted the ceremony for sentimental reasons. But most of them were convinced that he would be willing to forgo it for them.

On the following day there was still no definite word about the rumored graduation ceremony. In spite of the rumor of the lieutenant's wanting it, most men put the blame squarely on the administration. It was known that Major Kawashita, Mr. Kishi, and some of the other Nisei "brass" liked to refer to the MISLS as a "campus" and to the troops as its "student body." Since graduation exercises were regarded as a vital part of an academic program, it followed that these officials would insist on one for Company K. Regardless of what Fort Snelling meant to civilians, to the men it was just the army. Those who had attended college were amused at the attempts of some officials to play the role of academic dignitaries and referred to them sarcastically as "small-town big shots."

A shakedown inspection that had been announced for Thursday began at 08:00 and was not over until 10:00. It was more an inventory than an inspection. The inspecting officers were all Nisei lieutenants, some of whom had personal friends in the company. If anyone had excess items, the officers said nothing. If some item was missing, the officers simply marked it down for "exchange" or "salvage" so that the man would not have to pay for it. Thus, records were changed to match what each person actually had in his duffle bag. When it was over, the officers sat down on the bunks and chatted with their friends. Then, each man went in turn to the supply room to get rid of unwanted equipment.

That afternoon all students were ordered to the school area, by platoons, to turn in their books. Several difficulties arose. Some of them had decided to steal useful books. Others had lost track of dictionaries that had been assigned to them. Some books had been assigned to groups, each man signing for a different book. Many had forgotten which books were theirs. When confronted with a list of items with which they were charged, each person simply picked up anyone's book from the desks and returned it. Those who came later found their own books missing and picked up someone else's books. By the time the last group reported, no books were left on any of the desks.

At the orientation meeting at 16:00 the lieutenant announced that passes would be issued that night on the pretext that the post had no ticket agent and that everyone had to go into town to purchase tickets and to make other arrangements for travel. Was this the act of a man who would hold up his men for a ridiculous exercise?

At noon on Friday it was announced that a company iron was missing and that no one would be permitted to leave on furlough until it had been returned. Furthermore, all floors had to be scrubbed and be inspected before 18:00. By this time everyone knew that they could leave that night.

At 13:15 Friday the entire company fell out for graduation exercises in Class A uniforms. The ceremony was held on the lawn in front of the laundry room. Colonel Andersen and all members of the Division F faculty were present. In his speech the colonel said that Company K, after a bad start, had turned out well after all. He went on to add that their graduation from the oral school did not necessarily mean that they would receive undesirable assignments overseas. There was "plenty of work and glory for everybody." He then complimented the men for doing an "excellent job under very trying conditions." He also spoke of the inspection that he had held two weeks before and read an excerpt from a letter of commendation he had written to Lieutenant Catalano: "In all my four years with the Military Intelligence Service never have I held a more satisfactory inspection. The condition of the facilities and the bearing of the men were the finest I have ever seen." When he added, "That speaks well for the company and for the company commander," there were grins and nods of approval. Then, a wristwatch and two letters of commendation were awarded.

After the awards had been made Lieutenant Catalano got up to address the group: "I'm proud of you, men. What the colonel said about the inspection makes me very proud. Also, I want to thank the instructors who are here for their patience. God knows they had to have patience for Company K."

After the colonel's departure the lieutenant spoke more candidly: "You're leaving for your furlough at 18:00. The battalion doesn't know about it yet, and they won't know about it until after you're gone—unless you go and tell them. The MPs have been tipped off; so they'll let you go. Now for God's sakes, men, don't let that missing iron hold you back. If any of you fellows have it, bring it back and don't hold your buddies back. Nobody's going to ask any questions. Just bring it back."

The lieutenant was about to add something more, when one of the acting NCOs came out with a large package. When told that it was a gift from all the men in the company, the lieutenant was speechless. Tears streamed from his eyes. Clutching the package, he rushed into the orderly room, waving his hand toward his men as he ran. They were overjoyed.

"Boy, da old man really happy, yeah?"

"I guess he didn't expect it. Him swell guy."

"He's the best fuckin' CO I ever had. He's O.K."

"Boy did you see the smile on his face? From ear to ear!"

Word was passed from man to man as they remained in formation that the gift was a large silver plate. Most of them approved the choice. With the lieutenant gone the first sergeant took command. Before dismissing the company he added, "I've only been with the company for a few weeks, but I can honestly say that this is the best outfit I've ever been in." Everyone laughed, but they agreed that the company had been "on the ball" during the few weeks that the new sergeant had been there. He reiterated that the iron had to be turned in by 16:00 if the company was to leave on time. He also announced an inspection for 17:00.

There was much joking about the iron. Men smilingly accused each other of stealing it to "fuck up the company—after everything the colonel said." Private Harada, who had been awarded the wristwatch, was the target of catcalls wherever he went. Since he was a very popular Hawaiian, however, he encountered no serious difficulties. At this point the academic format was rudely interrupted by an announcement on the public address system: "All you guys who want your diplomas come to the supply room and get 'em!"

At 16:30 all the gambling stopped, and brooms and mops were again brought out. All the barracks passed inspection. Nothing more was said about the missing iron, and at 18:00 all furlough papers were ready for distribution. The men were delighted. They were getting a head start on their last furlough before going overseas. The CO had "fixed" it with the MPs. For once they had received a real "break."

Disenchantment and Recidivism

Men returned from their furloughs over a period of several days, beginning on December 3. The length of each man's furlough depended on his destination, for travel time had been added to his fifteen days at home. Early returnees were issued daily passes to the Twin Cities, but many remained indoors because of the frigid temperature and because most of them had so little money left. Friday, December 7, was the deadline, even for those with maximum travel time.

Saturday morning began amid some confusion. The CQ sounded the first call at 06:10, but no one paid much attention. At 06:20 he warned that reveille would be held in five minutes. Most of the men got out of bed, but there were several obvious gaps in the company formation. With the exception of the 1st and 6th Platoons, all acting NCOs reported "all present." The new acting platoon leader of the 1st Platoon, disliked as "chicken shit," reported a soldier who had refused to get out of bed. One 6th Platoon man had to be reported absent; he was always twenty-four hours late from long passes and furloughs. Exaggerated claims of erotic exploits were exchanged. After breakfast three rumors emerged. One was that Caucasian inductees had used the barracks during their absence. The second was that Lieutenant Catalano had received his discharge. The third was that shipping orders had been delayed because of lack of transportation to the Pacific Coast.

At 09:00 an orientation meeting was called. Mr. Catalano announced that he was now a civilian and proudly displayed his "lame duck" lapel button. This drew loud cheers; everyone was delighted over his good fortune. He introduced the new company commander, Lieutenant Skinner, and added, "This was, to put it crudely, the most fucked-up outfit in the camp. But when we finished we weren't one of the best; we were *the* best company in the whole goddam battalion. I want to thank you, men, not only for your cooperation but for the gift. You sure pulled a fast one on me. Right after I finished bawling you out you present me with a gift like that. I kept it around here to show it off for about ten days. I'm proud of it."

He also announced that arrangements had been made for a company party on Monday at the Hotel Radisson in Minneapolis. Money from the company fund and other sources had been combined so that it would be possible to rent the ballroom, have a turkey dinner, and hire a ten-piece band—all for only $2.50 per person. He had invited the mayor of Minneapolis as well as the governor of Minnesota to make the party the biggest the fort had ever seen.

Since Lieutenant Skinner wore a Combat Infantryman's Badge, the men's initial impression of their new commander was favorable. He was a combat veteran. He talked like an infantryman, and the men assumed that he would not be "chicken shit" like so many G-2 officers. The lieutenant addressed the group briefly and announced that the company would probably remain at Fort Snelling at least until the first of January. Since there was not much to do, passes would be issued freely, as long as everyone stayed "on the ball." When he indicated that proper uniform regulations would be enforced strictly, someone remarked that that was "chicken shit." The lieutenant replied, "Maybe so, but that's what it'll be." He then ordered that all beds be taken out of the barracks for airing; inductees had been sleeping in them, and they might have brought in bed bugs, lice, and "crabs."

In the last shipment of linguists from Fort Snelling, Hawaiians had been separated from the mainlanders and given a fifteen-day delay-en-route in Hawaii on their way to Japan. A rumor had developed that the same arrangement would be made for Company K, and the Hawaiians were elated at the prospect. Wild cheers erupted when Lieutenant Skinner confirmed the rumor. Hawaiians immediately got together in clusters to consider plans for their second furlough. Several mainlanders joined them, offering their congratulations; they realized that the Hawaiians had not been home since being inducted. A few Hawaiians announced loudly that they were delighted to get away from the Kotonks, and several mainlanders also indicated among themselves how pleased they were not to be sailing together. In general, however, animosities were barely noticeable.

On Monday, December 10, Sergeant Kondo was discharged, and Private Nakasone became the acting first sergeant. Nakasone's appointment won widespread approval. The men accepted his authority, for they respected him. Many welcomed the change, for the orderly room once more became accessible. They could ask questions and check on rumors. But a rumor spread about him: Nakasone would not be going overseas, for he had been appointed drill sergeant of Company G.

It had been apparent for some time that an overwhelming majority of the men were not enthusiastic about the company party. All other companies had had them, but most graduates of Company K did not feel that they had anything to celebrate. Lieutenant Catalano had approached them as early as October 25 and had raised the question several times thereafter. Each time he had received lukewarm reactions. He had also suggested having a group picture taken, but there was no enthusiasm here either. Although a number subsequently confessed that they would have liked such a photograph, they did not speak up. When a show of hands was requested, the picture was voted down. On November 8 the lieutenant had asked how many would bring their wife or sweetheart to a company party; only one man raised his hand. Although most men agreed that the company was improving, it was obvious that few were proud of it, and many were wary of the conduct of their comrades.

When Mr. Catalano heard that a substantial number were not planning to attend the party, he appeared briefly at the monthly payroll formation. He stated that he had remained behind for an extra six days just for the party and then asked, "Who's not coming?" He reiterated that unless each member contributed $2.50 the party could not be held at all. It would cost well over $1,000; and every cent in the company fund, the PX money, the rice money, and every other source had already been pooled for it. The men agreed that it was only fair to pay whether one planned to attend or not. "Just as long as you pay, it's O.K. They'll go in the hole if they don't collect. You don't have to go." To be sure that everyone paid, two squad leaders were posted at the door to collect $2.50 immediately after each man had received his salary. No one objected. It was for the "old man."

Almost everyone attended the party, but only a handful invited guests. Some brought Nisei girls, but most of the men—even those whose wives lived in the Twin Cities area—went alone. The temperature was 10° below zero. The men braced themselves for what they feared would be an ordeal. Although the party was scheduled for 19:30 it did not get under way until 20:15. The turkey dinner was served as promised. None of the dignitaries appeared, but most of them sent congratulatory telegrams. Only a few speeches were made. A resounding cheer went up when Mr. Catalano rose. He said that when he was first appointed company commander, Colonel Murphy had said to him, "Heaven help you!" He reiterated that "we ended up as the best company in the battalion." Then Colonel Murphy was introduced. He asked the men if any of them were interested in fighting in the ring and ended his talk amiably with, "Some of you should fight in the ring. Heaven knows you did plenty of it outside the ring."

The bar was crowded. Many were drinking highballs during dinner. Only a few were too drunk to eat, but by the time dinner was over several had become quite inebriated. When the floor was cleared for dancing, those who had come with Nisei women danced, but the others stood about watching. The Caucasian hostesses from the USO were left standing, and Nakasone and his committee were embarrassed, not being able to find enough men willing to dance with Caucasian women. Many stood by the orchestra admiring the shapely vocalist in a strapless evening gown. At most only about thirty-five couples occupied the floor. Several pointed to the Caucasian girls and urged others to invite them to dance; they acknowledged the breach of etiquette, and one man said loudly, "That's a Boochie for you. Holler when the Haole won't give you a break, and when they come, you won't take 'em." Despite widespread self-condemnation the women remained unescorted.

It was well past midnight when most of the company began to straggle back to the barracks. About 02:00, when Nakasone was making an unofficial bedcheck to be sure that everyone had returned safely, a disturbance erupted in the 6th Platoon. A Hawaiian who was very intoxicated accused another Hawaiian of having called him a bluff. He wanted to fight, and several other Hawaiians jumped out of bed to quiet him down. After a noisy struggle they finally put him to bed. Punchy led the peacemakers, yelling over and over, "Me, I no like see trouble!" Most of them were drunk, but they cooperated to avoid trouble. "We all same from Hawaii, huh? So forget, forget!" At 02:30 the lights were turned off.

On the following morning several complained of hangovers. A formation was called at 09:00, and there was much grumbling when the men learned that they had been aroused from a warm bed only to "police up" the area. Roll was not taken. It was only a nuisance call. When a radio newscaster announced that the high for the day before had been 4° above zero and predicted 10° below zero again for that day, their discontent was compounded. On the whole, the men seemed to have enjoyed themselves. They talked happily of how drunk and happy Mr. Catalano had been, how he had stayed away from his wife for six days just for the party, how Nakasone had stayed up most of the night taking care of others even when he himself was sick from too much drinking, and how some were so drunk that they could not remember what they did with the women who were with them.

After the company party mainlanders became preoccupied with a pressing question: would they be in the country long enough to go home for Christmas? The men had been told that they would have to wait at Fort Snelling until sufficient space was available on trains. They also knew that there might be delays at the port of embarkation until there was enough shipping space. At a company meeting on December 12, Lieutenant Skinner announced that the "ready date" was three days away; after that the company would have to be prepared to leave on 24-hours' notice. All processing would have to be completed before the 15th. Since there would be nothing to do after that, he promised that everyone would be issued a pass every night for twenty-four hours until a definite departure date had been set. He also announced, however, that the wearing of paratrooper boots and the blousing of trousers over combat boots would no longer be permitted. When the men protested that those in other companies were doing it, he replied that he did not care what others were doing; army regulations would be carried out in Company K. He ended his remarks with a threat: "So, men, if you're on the ball, which you're not, there'll be plenty of passes. But if there's one uniform violation, no one in the company will be able to get out, no matter what kind of excuse you may have. So don't fuck up, men. You'll wreck everything for the whole company for the duration of your stay."

For the next two days everyone was restricted to the company area, for processing had to be completed by Saturday. One afternoon the company was assembled for the reading of the Articles of War. The lieutenant assigned the task was obviously bored and apologized profusely to his audience for having to subject them to such a waste of time. The only point of interest was the section stating that company punishment could not exceed one week. This elicited loud laughter; Lieutenant Catalano had given some troublemakers as much as thirty days punishment. There was no sympathy for them; it was just "TS." Once clothing and equipment had been issued and checked, and inoculations and medical examinations had been completed, there was nothing more to do. Lieutenant Skinner kept insisting, however, that there were "thousands of things" still left to be done. Since so many in the company had been processed for overseas shipment at Fort Meade, they were familiar with the procedure. Whenever they asked the CO to specify just what had to be done, he became quite vague. Everyone remained restricted. As they sat about chatting or gambling, they were called out from time to time for various formations. On Friday afternoon the company was ordered out for a speed march. It neither broke the monotony nor served as exercise. The temperature was hovering near zero, and the men were wearing so much clothing that they could not move about freely. Furthermore, the streets were so slippery that no one could trot consistently at a double time cadence. The group shuffled about for a half hour amid complaints of "chicken shit." It was impossible to raise a sweat; within a few moments the mucus in the nostrils froze, making it extremely difficult to breathe.

"Why we have to do dees? Seely, huh?"

"They don't like to see you sittin' on your ass. If your ass gets too big, you can't fit the gangplank."

"Fuck you! No joke, goddammit! Me pissed off."

"This is really chicken shit. They can't think of anything else for us to do."

"I wouldn't give a shit if we got something out of the deal or if Uncle Sam got something out of it, but nobody gets a fuckin' thing. We can't even wear out shoe leather like this."

From time to time someone slipped and fell on the ice, and the catcalls they elicited provided the only levity in the undertaking.

On Saturday morning, December 15, the company was ordered to march to the field house to attend the graduation ceremony for Company A. At the exercises General Wilson condemned severely the OCS men who had refused to accept commissions. Only 43 of the 239 graduates in Company A had agreed to further service! He said that while others were dying overseas these men had stayed home in comfortable barracks, had the "best food when civilians went without," and had the "best teachers" and a good education. These men had taken everything from the government and were now unwilling to render the service for which they had been trained. He added that this criticism applied to some extent to the Nisei instructors who had also refused to volunteer. There was a shortage of teachers, and everyone "must give to the country from which we took so much." His comment about the "best food" brought stifled laughter from those in the Company K section, for they knew what kind of food the OCS men had had to endure. Most of them admitted readily that the general was correct in saying that the students at Fort Snelling, Caucasian and Nisei, had had things much easier than those who had fought overseas. However, they were not impressed with his argument. Most of the Nisei said little, but many in Company K sympathized openly with the OCS men who had insisted on a discharge rather than going overseas as a G-2 officer.

"He talks that way so he can fuck us good. When guys give patriotic speeches, they want something from you and are afraid to come out and say what they really want."

"He just wants us to reenlist to save his own ass. Why don't he just come out and say that he tried to screw the OCS boys but had to let 'em out because the law said so? He couldn't push Congress around like he pushes us around."

"Fuck the general!"

"Where in the hell was *he* when all the boys were sleepin' in the mud and dyin'? He was sittin' on his ass in D.C., eatin' big steaks, and fuckin' his WAC secretary every other night. What the hell was *he* doin'? He just don't know what it's like to be in G-2 on the ass bottom."

By this time considerable resentment was developing against the new company commander. In spite of his repeated claims that he was doing everything he could for them, many were beginning to suspect that he was insincere. His combat ribbons had commanded respect at first, but the men were developing doubts. Whenever questions arose about some inconvenience, he put the blame of Mr. Catalano, and this further alienated them. Some openly questioned his integrity.

Distrust of the new company commander soon crystallized. He had promised many passes, but he restricted the company on every conceivable pretext. He insisted that there was much work to be done; yet everyone was sitting about day after day with nothing to do. Furthermore, in spite of "all the work" the lieutenant himself was seldom in the company area. From time to time emergencies arose; papers had to be signed, and decisions had to be made. But the lieutenant could not be located when he was badly needed. Inspections were announced, and the men made preparations; however, no inspecting officer arrived.

"Promises, promises, but all he does is talk. What do we fuck around so much for, if we're so busy?"

"Yeah, where are them 24-hour passes?"

The cadre and the acting NCOs rebelled. Since they knew that the CO did not arrive in the company area until late morning, they decided to hold reveille in bed. "Waste time having reveille. If we had something to do, O.K.; but we just get up and go back to sleep again. If it's warm you can say, 'This is the army,' but this is no joke at 13 below zero." It was indeed very cold, and there was no point in getting up at 06:00 when there was nothing to do. Each morning all platoons reported "all present." Roll call was never held, and very few went to breakfast. If anyone was unable to get back in time for some formation, he notified one of the cadre, who then took appropriate steps to cover up. When the lieutenant ordered two inspections a day to keep the area clean, his order was simply ignored. Since he was seldom there himself, he did not know that inspections were no longer held at all. The general view of the cadre was: "What the bastard don't know won't hurt him."

The men became sullen and recalcitrant, for they had little to lose. He had promised daily passes, but he refused to issue them and continuously threatened to withhold them altogether. Since occasional passes were all that anyone wanted, once they were taken away, there was nothing more. They did not fear a court martial, for they felt that their sentences would be suspended to enable them to depart with the company. If a man with Takeda's record could be excused, by comparison AWOL was only a minor offense. Thus, resistance developed to all his orders. They laughed at his edict against wearing boots and deliberately walked about with their trousers bloused. Unlike their former CO, Lieutenant Skinner could not recognize them even if he did see them. Even without passes the men went into town whenever they wished. On December 20 one man had been AWOL for two days. Of course, the lieutenant had not missed him, but the cadremen were becoming worried and sent one of his friends out to look for him.

The men were genuinely puzzled. What could the lieutenant possibly gain by acting in this manner? One of the cadremen proposed an explanation: "His wife lives in this area, and he's afraid that if he makes a mistake they might transfer him someplace else. I hear she's pregnant. That's why he's so chicken shit. He wants to stay with her."

On December 12, Private Nakasone, the acting first sergeant, announced that the Hawaiians would leave on Monday, December 17, and that the mainlanders were to depart on December 24. Later that afternoon an announcement was placed on the bulletin board that Company K would not receive a break over the Christmas holidays because of "military necessity." Questions arose at once about passes. The reason for limiting passes to twenty-four hours after the "ready date" presumably was that the company commander could not allow his men to take extended leaves when he did not know just when they would be needed. Now that the shipping dates were known, the men could see no reason why they could not be given three-day passes or short furloughs.

At 09:30 on December 14 Lieutenant Skinner called an orientation meeting. He insisted that there was so much work to do that the entire company had to be restricted. He said that the Hawaiians were leaving on Monday and that their processing had to be completed over the weekend. When asked why the mainlanders had to remain, he pretended that he did not know their departure date. When one of the men asked him to indicate just what had to be done so that everyone could cooperate to finish as quickly as possible, he replied that all salvage had to be turned in and that a showdown inspection would have to be held. Since these tasks required no more than a single afternoon, the man pressed him further; he was given an evasive reply, "and a lot of other things." When pressed further by others, he stated, "There were a lot of things that should have been done before you men left on furlough. But those things weren't done. That's why we have so much left now." Without mentioning his name he again placed the blame on Lieutenant Catalano. But the men knew better. They knew that Lieutenant Catalano had made special arrangements with the infirmary for inoculations before the furloughs. Furthermore, those working in the orderly room contradicted the new CO.

Later that evening Ito, the company clerk, and Nakasone had an argument over weekend passes for the mainlanders. A shakedown inspection could not be held until a clothing check sheet had been prepared for each man. Since the Hawaiians had to be processed at once, it was impossible for the few typists in the orderly room to complete these forms for the mainlanders until after the Hawaiians had gone. Nakasone argued, therefore, that all mainlanders should be issued passes over the weekend. Ito, however, had direct orders from the lieutenant not to let anyone leave. Word of the disagreement spread quickly.

"The new CO is yellow. It's the first time he's ever been a CO; so he's afraid. I hear he's buckin' too."

"Remember Captain Larsen at Meade? He fucked us, and we fucked him right back. The same guys are here, and we can do it again. We ought to really get this bastard so he won't forget us."

"Christ! A pass is the only *tanoshimi* now. If he takes that away, we got nothing to lose. We ought to really fuck up then."

"Christ, yeah. I figure somebody'll fuck up and get us restricted anyway. Then, we got nothing to lose so we can really go to town and get that bastard."

Even the Hawaiians were resentful over the restriction of the mainlanders; it was so obviously unnecessary. The latter were not only not needed; their presence would only complicate matters. With the mainlanders gone the Hawaiians would have additional bunks on which to lay out their equipment.

Anger against Lieutenant Skinner mounted on Saturday morning, when it was learned from the cadre that he had decided not to come to the company area at all over the weekend. He had been absent most of the day before. Those who lived in Chicago wanted to go home just once more before going overseas; they could not even ask him for a pass, for he refused to come to his office. No officer was available to sign the Class A passes required for distances of more than 150 miles. "He's as bad as them Fort Meade officers." Several Chicagoans decided to go home on their Class B passes, on the chance of not being stopped by MPs. Those working in the orderly room were certain that the CO would not miss them and looked elsewhere whenever anyone came in to pick up passes. It was understood that these men were going at their own risk.

Class B passes were good only for the Twin Cities. Those on the Chicago-bound trains, however, encountered no difficulties. The following was a typical encounter with the military police:

"Where's your pass, buddy? Where you going on this pass?"

"Chicago."

"Chicago! On a 150-mile pass! Ha! Ha! Ha! Ha! Ha!"

"Is Chicago further than 150 miles?"

"Don't hand me that shit! Ha! Ha! Ha! Ha! Ha! When you coming back?"

"Tomorrow night."

"O.K. Ha! Ha! Ha! Ha!"

All Hawaiians had been asked by the cadre to be back by noon

Sunday, and all mainlanders had been asked to return by midnight. At 08:00 Monday morning seventeen were AWOL, but the company was reported "all present." Lieutenant Skinner did not arrive until after 10:00 and was so preoccupied with the final processing of Hawaiians that he did not miss the eight soldiers who were still absent at that time. An unannounced inspection was held, and the lieutenant found the 5th Platoon barrack so dirty that he ordered the occupants to scrub the floor. In the 6th Platoon he found one area very dirty and asked one of the men nearby what time he had gotten up that morning. When he replied "07:30," the company commander found out for the first time that reveille had not been held. Some of the men had forgotten that the lieutenant knew nothing about it. The cadre was severely reprimanded, and everyone was ordered henceforth to stand reveille.

Although a large number had gone to Chicago for the weekend, most of them had returned on Sunday afternoon trains and were in by midnight. Eight men, however, took Sunday night trains, expecting to be in by 07:30 Monday morning, confident that the lieutenant would not be there anyway. Four managed to get back by noon, and one of them, not knowing that he had not been missed, went to the company commander to explain why he was late. Then, the lieutenant realized that all his men were actually not in the company area at reveille. When a roll call was taken, four were still missing. That morning a Milwaukee Road freight train had crashed between Wabasha and St. Paul, blocking all the tracks. All passenger trains behind it had to backtrack to La Crosse and proceed to the Twin Cities on Burlington Line tracks. The four who had been trapped on the Milwaukee Road night train were unable to return until 17:00. The lieutenant was already gone.

In the midst of this confusion fifty Hawaiians departed. Mainland friends helped carry their bags. There were fond farewells and exchanges of addresses. Most mainlanders were ambivalent. Many had close friends among the Hawaiians and already missed them. On the other hand, the troublemakers had also departed. On Monday evening, December 17, for the first time in months the barracks were quiet.

On the following morning the four who had been AWOL were called before the company commander. He lectured them on the seriousness of their offense—being AWOL after they had been alerted for overseas duty. He quoted the 128th Article of War on "desertion in the face of hazardous duty," a capital offense. After explaining how generous he was being in not holding them for court martial, he excused them. Technically, it was a serious matter, but the four were unconcerned. They had been reassured by the cadre: "You don't have to worry. He won't turn you over to the battalion. If he does, they'll find out how fucked up this place has been since he took over, and they'll find out how lousy the reveille reports have been. Then, they'll chew his ass too. He won't tell 'em." The cadremen were so disgusted with the lieutenant that they did not even scold the four for getting them into trouble.

One NCO muttered, "Fuck that bastard! I'm not going to call out the company in this cold weather unless the son of a bitch is here. The hell with him!"

Another added, "Do you know what the cocksucker wants? He wants two inspections a day! I'm afraid he might come check up on us tomorrow morning; otherwise, I'd let everybody sleep. He's really chicken shit."

Nor were the others upset. The four were targets of good natured gibes.

"Hi, AWOL!"

"Whadda ya mean? What's eighteen hours among friends?"

"Friend? Do you call that son of a bitch your friend?"

On Tuesday morning, December 18, the company was called out for reveille at 06:10. Because of what had happened the day before, all but two or three in each platoon fell out. It was about 5° below zero. Everyone was heavily bundled, and as soon as the report was taken they rushed for the barracks. The latrine was very cold, and icicles had formed inside the building. There was no hot water for shaving. As the men tried to clean the barracks, water on the mops turned to ice, making them useless. Even when they thawed out the mops by placing them on a stove, the mop water turned to ice on those parts of the floor more than three feet from each stove. All the stoves were full of coal and burning, but everyone was still shivering. It was the first time the company had been up so early in sub-zero weather.

Inspections were announced each morning, and from time to time an officer came to check. Those who passed were issued passes in the afternoon. As the temperature dropped to 15° below zero, however, few were inclined to leave just for a short period. It was now obvious that there was absolutely nothing left to do, but the promised 24-hour passes never materialized; it also became apparent that the daily inspections were being held just to be sure that everyone returned. Only one man, whose wife lived in Iowa, was permitted to stay out until Saturday morning. If he was not needed, obviously no one else was needed either. Why was the lieutenant so reluctant to issue passes to soldiers who had only a few more days left in the United States? By this time most men had concluded that he was very frightened.

On Friday morning, December 21, Lieutenant Skinner announced that the company would leave on Monday, Christmas Eve—something that everyone had known for over a week. He pretended that he had not known, but no one was impressed. Then a long discussion ensued over disposition of a case of prewar whiskey that belonged to the company. Lieutenant Catalano had arranged to have the men supplement their GI insurance with an especially attractive offer made by a private firm; so many had taken advantage of the opportunity that the grateful agent had presented the unit with the gift. Lieutenant Skinner had indicated that he was planning a company party at which it could be consumed; now he insisted that the colonel had disapproved. When asked how the colonel had found out about it, he replied that apparently Mr. Catalano had made the disclosure at the party when he became "too happy." He then suggested giving the whiskey to the cadre, but the men protested. Finally, after considerable discussion he persuaded them to give it to the cooks in the Consolidated Mess Hall, if they would agree to give Company K one wing, serving special food over the weekend and an outstanding breakfast on the morning of their departure.

The men were suspicious. Perhaps Mr. Catalano had told Colonel Murphy about the whiskey, but it seemed unlikely. He was not that drunk. Furthermore, when Lieutenant Skinner mentioned trying to get a mess hall, it became obvious that he was looking for some legal way for enlisted men to consume whiskey on the post; they knew enough about army regulations to know that it could not be done. They concluded that Lieutenant Skinner must have asked Colonel Murphy's permission, and the colonel had to refuse regardless of his personal feelings. Since nothing else could be done, they agreed that some good meals were better than nothing.

"For a whole fuckin' case of whiskey they ought to give us some good chow."

"If you ask me, it's a waste of good whiskey."

"I hear the CO took a couple of bottles home."

"I wouldn't doubt it, hey. And that son of a bitch said it wouldn't go to any officers."

"I wouldn't mind if Catalano got it."

"Hell, no. He took a lot of shit from us. We really need him now."

"Yeah, he left us when we needed him the most."

"They oughta give us some good chow."

"They would if they really get it. That bastard probably took it all himself and is handing us that shit to cover up."

"There's something about that guy. I don't trust him."

Later that morning a roster of men on the shipping order was posted on the bulletin board. They were organized into four platoons of four squads each, arranged alphabetically. Every tenth man was named squad leader, although the platoon leaders had been chosen from the acting NCOs. Private Nakasone was designated acting first sergeant. A rumor developed at this point that Captain Asakura, commanding officer of Company G, would be the train leader. This was greeted with groans. The captain was highly respected as a wounded veteran of the 100th Infantry Battalion, but he was generally feared and disliked by those in his command. The men remembered that Captain Asakura had once made things very unpleasant for them when he was the duty officer. Other rumors followed. Captain Asakura once restricted his company for two weeks, when he found a single cigarette butt on the floor. Lieutenant Skinner had asked permission for the troops to wear fatigue uniforms to the train, since it was too dark to tell the difference anyway, but Captain Asakura refused. Although most men found it difficult to believe that Lieutenant Skinner would ask for anything for them, they were not surprised that Captain Asakura had insisted on regulation Class A uniforms.

At 10:15 on Sunday the men were ordered to fall out for the first time in their newly formed units. They had to wait in the snow in subzero temperature for twenty minutes. Finally, one of the acting NCOs ran to the orderly room to ask the lieutenant to come out. He checked to see that everyone was wearing an overcoat, took the attendance report, and dismissed the group. The men were disgusted: "What the hell! He must think we're really dumb. Who in the fuck's gonna stand out here for a half hour without his overcoat?"

He informed them that everyone was restricted to the company area; they could not even go to the PX or the post theater. The men accepted the restriction, for the lieutenant insisted that the matter was beyond his control; they assumed that the order had come from battalion headquarters. Restrictions of this sort were to be expected before shipping overseas, especially after completing the medico-genital examination.

At 13:15 the company was assembled in the 3rd Platoon barrack, where Captain Asakura briefed them on what to expect during the trip. The manner in which he spoke dispelled the anxieties of most men; in fact, he was not at all like his reputation. As he read the rules, he indicated which he would enforce strictly and on which he would overlook minor infractions. He seemed very reasonable. He told them that they could leave the train whenever there were long stops and promised that he would notify them ahead of time so that they could be ready. Those with relatives in Spokane, Washington, could wire ahead to tell them that the troop train would make a long stop there. The last time he had taken a group to Fort Lawton, in Seattle, the soldiers had been allowed out on pass every night until 08:00. He also stated that everyone would be paid before they boarded the ship and that there was an excellent chance that they would have an opportunity to spend some of the money in Seattle.

Then someone asked him why they could not go to the post theater. The captain was surprised at the question and replied, "I don't see any reason why you can't. That's up to Lieutenant Skinner." One of the acting NCOs ran out to look for him, but the lieutenant was no longer in the company area. When the captain reiterated that he could see no reason why they could not go anywhere on the post, the men realized clearly what they had long suspected—that Lieutenant Skinner himself, and not the battalion, was responsible for all the restrictions. That Captain Asakura was so astonished at the question indicated that he had assumed that they had been given freedom of the post. Inadvertently, the captain had disclosed that the lieutenant was a liar. The men began to doubt other reports, such as the ban on fatigue uniforms that had been blamed on Captain Asakura.

The afternoon was spent gambling, doing laundry, packing, and writing letters. A number went into town without passes. Of course, this surprised no one. "This 'no pass' deal never bothered Company K." When Lieutenant Catalano had orders from the battalion to restrict his men to the company area, he usually modified the order to give them freedom of the post to enable them to go to the movies or to the PX. Consequently, the men rarely left the fort. Seen in this light, attempts by Lieutenant Skinner to blame Mr. Catalano for all inconveniences made them very bitter. Many plotted vengeance.

By 17:30 more than half the company was gone. Beginning about 19:00 each group that had gone into town returned laden with food and drinks for those who had remained behind. In each platoon there was an abundance of soft drinks, beer, whiskey, hamburgers, hot dogs, rice balls, *tsukemono* [pickled vegetables], and sandwiches. So much food accumulated in the 6th Platoon that a man was sent to the other barracks to invite friends to share it; he returned, however, with news that each platoon had just as much food and that no one was interested. One reason for the oversupply was that so few were present to eat it. By 22:00 it became apparent that at least two-thirds of the company had gone either to the post theater or into the Twin Cities. As they returned, they huddled about the stoves—eating, drinking, and singing or humming three currently popular songs: "White Christmas," "I'm Always Chasing Rainbows," and "I Can't Begin to Tell You." With each returning group the food supply became more abundant. Some sat about quietly discussing what they had done on happier Christmases of past years.

Long after 23:00 Lieutenant Skinner made a surprise visit to the company area. All the lights were still on, and at least a fourth of the men were still AWOL. He started going from one barrack to the next to tell them to turn off the lights and to retire. But the men had only their clothing and mattresses. Their blankets had been packed that morning, and it was too cold to sleep. Some tried sleeping between two mattresses, but they found this impracticable.

When the lieutenant reached the 4th Platoon, he found several men drinking heavily. As he entered, only one man at the opposite end of the building saw him and called "Attention!" The drinkers did not hear him. They continued to curse their "chicken shit CO" until he was standing directly behind them. Finally, one of the drinkers stiffened.

"Hey! Hey! There's the CO!"

"Aw, fuck the son of a bitch!"

"Naw, naw. Pour the bastard a drink."

The lieutenant left quietly. As he walked out of the building, a man running from the latrine almost bumped into him. He immediately spread the alarm that the CO was in the area.

"I almost bumped into him by the can."

"You should'a clipped the bastard."

The lieutenant did not disturb the remaining platoons.

On Monday morning, Christmas Eve, the first call came at 03:30. Those who were sleeping were awakened by the others and by 03:35, when they were called to breakfast, the men ran out in anticipation. They had been expecting special food throughout the weekend and thus far had received nothing out of the ordinary. This was their last meal at Fort Snelling, and for their case of whiskey they expected an unusual breakfast. Would it consist of ham and eggs? Hash and eggs? Wheat cakes? To their utter astonishment they were served soggy French toast, one strip of bacon, and half a grapefruit. No one said much. They ate quietly. From time to time someone muttered, "Waste of good whiskey."

"We should'a drank it."

"The CO must'a made a deal and kept half the bottles himself."

Everyone was ready to leave at 05:00, the time posted on the bulletin board. But there was an unexplained delay. Finally, the company commander called over the public address system for the 5th Platoon to "police up" the company area. One man yelled in reply, "Tell the CO to shove it up his ass!" The lieutenant appeared at the platoon in person and supervised the work. The company did not leave until 05:45.

It had been snowing all night, and it had piled up from two to eight inches. It was still snowing as the 182 mainlanders—all but Private Janssen and a handful who had reenlisted—marched away from the company for the last time. Lieutenant Skinner did not leave the company area.

"Catalano would have come down to the train to see us off."

"Shit, if you use Catalano as a standard to judge COs, you'll be pissed off all the rest of the time you're in the army. But Skinner is a cocksucker though."

The group departed from the same siding at which they had detrained 187 days ago, almost to the hour.

Collective Responsibility in Transit

As the snow crunched and grated under the feet of marching troops, cadence that was at first sharp and rhythmical soon degenerated into shuffles. It was still dark, and the fluffy snowflakes continued to drift down lazily. What was left of Company K—182 mainlanders—sloshed over the three-quarter mile route to the same loading zone to which they had come from Fort Meade a half year before. Waiting for them there were pullman cars, five Nisei lieutenants, mess and medical personnel, and two train officers—Captain Asakura and Lieutenant Clinton. Colonel Murphy was also there to see them off.

There was a wait of about twenty minutes before the first group began to board. The cold was biting, and cries erupted from ranks: "Let's go! Let's go!" Inside the trains the men found themselves in pitch blackness; the generators had run down and frozen, and the lights would not work. It was also extremely hot, but most preferred it to the cold. At 06:25 the train started with a jerk, and everyone cheered wildly. Their stay at Fort Snelling had been eventful, but they were delighted to leave.

"I hope I never see this chicken shit place again."

"Fuck you, Snelling!"

"Geisha gals, here we come!"

Their joy was short-lived, however. The lights were expected to go on once the train got underway, but they did not. Except for occasional flicks of cigarette lighters and the flashlights of the porter and the conductor, the cars remained pitch black. They also remained hot and stuffy. Before long the men began to doff their clothing. Some took off everything to replace their long johns with light underwear. By daybreak almost everyone was in fatigue uniforms. After a long stop at the Minneapolis station the train finally got under way at 08:10.

"I hear it costs twenty bucks to stop and start one of these trains."

"Troop train's a losing proposition then."

"That's no shit."

"Especially with K Company. Shit, with all the dough it costs to send us over, they could hire the same number of Boochie interpreters for life."

Since it was so warm, many fell asleep. At 09:00 a 1st Platoon man came through distributing apples; the others then realized that they would have to do KP at least once during the trip.

Captain Asakura gave the troops considerable freedom. When the train stopped at Breckenridge, North Dakota, for example, he ordered everyone out of the cars while the engine took water. He even allowed them to go across the street to a drug store, telling them only to stay away from the bars—for the time being. Although there was not much to buy or to see, the men were delighted; being credited with having enough sense to return in time was refreshing. Nakasone placed one of his assistants in charge of each of the cars. Although there were no clearly drawn lines of authority, no difficulties arose. Communication was efficient. Information went from the train crew to the officers to Nakasone and the platoon leaders. Since the latter were drawn from the ranks, they immediately passed it on to their friends. But there was little levity, for it became apparent as evening approached that the lights were still not working. Supper was served in the dark; the men ate quietly and then sat about in small groups chatting. Very few could sleep; those who had been drowsy had slept all day. The darkness added considerably to the misery of having to spend Christmas Eve on a troop train.

About 20:00 the men began getting accustomed to the darkness.

"Jesus Christ! I can't see a fuckin' thing. Flash your teeth, Joey, so I can see where you are."

"No joke, goddammit! I'm pissed off. The fuckin' army can get a train any time they want it. Why in the hell do they have to pick Christmas? Shit, this is bad for the train company too. They need all the cars they can get for the civilian rush."

"Them bastards at Snelling wanted to get us, and they really did."

"Well, shit, they didn't have to take off the fuckin' batteries."

"Goddam! This is a lousy deal for Christmas Eve, huh?"

"If G-2's gonna treat us like this, I'll be fucked if I'll do any work for them. When I get overseas, I'll get somebody else to interpret for me. Fuck them! I'm gonna get even with them bastards for this."

There was no bitterness toward the train crew nor the officers. They too were away from home on Christmas. Late that evening, when the train stopped at Minot, North Dakota, the men swarmed out, dressing hurriedly as they ran for the canteen and restaurant. The stew they had for supper had not been filling, and many were still hungry. Since each bought enough for all his friends, the cars had an oversupply of ice cream, sandwiches, and fruits. Those who had purchased too much ice cream had difficulty finding enough people to eat it. Those who were already asleep were awakened to join in the feast. As they shared freely, something resembling a holiday spirit developed.

Several admitted they were thinking of home. They talked nostalgically about what they had done on happier Christmases in the past—before the war and evacuation. A number were constantly whistling, humming, or singing "White Christmas." The song had a melancholy effect; it made them homesick.

"I'm dreaming of a white Christmas, just like the ones I used to know . . ."

"Shut up, goddammit!"

"Last Christmas I got drafted. On Christmas Eve I was on a fuckin' troop train for Blanding. Shit, everybody was singing that fuckin' song. I was really homesick though. I felt like bawling. You know; we were rookies yet. We weren't used to this kind of shit."

"I like that song, but it sure makes me sad. The first time I got to know it was when I was in Tule Lake. We sang it in camp in the Christmas of '42, and I thought about how Christmas used to be back home. It was kind of sad then. I never been home since. I haven't been home for Christmas since '41. I don't know. Out in Chicago it don't seem like home; you don't feel it. Get what I mean?"

"Yeah. But this is really the shits though. It's bad enough not being home without having to be on a fuckin' troop train—and a blackout on top of that. Shit!"

"I still remember last Christmas. We always get fucked on Christmas. We were down at Wheeler takin' basic. Goddam! Just on Christmas Eve our platoon pulled battalion guard! Son of a bitch! I went on at 10:00 at night and had to freeze my ass off until 02:00 in the morning."

"You saw Santa Claus come in with a fuckin' rifle on your shoulder, huh? That's tough."

"Shit! All the other guys got to eat good chow, and they got passes to go into town to tear off a piece of ass. And we had to stand guard. Then, all the guys came back drunk, and we had a hell of a time. We always get fucked on Christmas."

"I never saw a real white Christmas at home. We never had no snow back home. The only white Christmas I ever saw was in Chicago, but I wasn't home then. I sure wished I was though."

"After this I'm gonna buck for a discharge no matter what I gotta do. This is enough of this shit for me."

Some sang Christmas carols. The group singing began spontaneously. One man would start humming a Christmas carol, and about five or six others in the vicinity would join him. At times everyone in a car would sing together. From time to time a small group would start singing some Japanese song, but this elicited little interest.

About 19:30 a small group gathered in one of the washrooms to consume the whiskey that they had taken from the orderly room. The cadremen had suspected that Lieutenant Skinner would not make the arrangement with the mess hall cooks as he had promised. When the "special" breakfast turned out to consist of French toast, they allowed one of the men to take as much whiskey as he could carry in his duffle bag. The whiskey belonged to the company, and neither the cadre nor the students wanted Lieutenant Skinner to have it; they agreed that they would rather pour it down a drain. As word was passed that some of the whiskey was aboard the train, the men went back, one by one, into the washroom. Even those who ordinarily did not drink accepted a "swig." The company whiskey, augmented by bottles purchased at various stops, was consumed within a half hour. Seven remained in the washroom, trying to drown their sorrows. They sang Christmas carols at the top of their voices and then turned to sentimental tunes such as "Juanita," "I Can't Begin to Tell You," and "The Old Mill Stream." Soon their songs became more bawdy. At this point those who were trying to sleep in the adjoining car protested that there was too much noise.

"Sad case! Here it is—Christmas Eve—and we spend it sittin' in a shithouse singin' Christmas carols."

"Talk about gettin' fucked!"

Since thirty-six soldiers had been assigned to cars containing only twenty-four sleeping units, those in the lower bunks had to double up. When the porter came to make up the bunks, the men helped him. One held the flashlight for him; others helped put on the sheets and blankets. Few difficulties arose in deciding who would sleep alone in the upper bunk. If the three in a given unit did not know one another well, the two smaller men occupied the lower, and the third slept above.

On Christmas morning most of the passengers were up by 07:00, when the smell of bacon filled the cars. The lights were working again. Late risers were pulled out of bed by their friends, and the porter began breaking up the bunks by 07:15. Breakfast consisted of bacon and eggs, and the men were pleased.

"Better than the Consolidated Mess."

"We should've given the whiskey to the Boochie cooks on the train."

"Don't worry. They got theirs last night, I think."

At 08:30 an inspection was announced. The cars had already been swept, and there was nothing more to do other than refrain from throwing cigarette butts on the floor. At 09:00 Captain Asakura came through for inspection. After approving the cars he announced, "In a little while we stop in Havre for a good ten minutes. You can get off and buy what you want. A half block from the station there are some liquor stores. They may be closed today, but if they're not, you can get a snort if you want to. This is on the QT, but if you want to get a bottle, make sure it's wrapped up so I won't be able to see what you've got. I wish you men a merry Christmas."

The men chorused in response, "Thank you, sir!" "Same to you, sir!"

"And they said he was chicken shit."

"Christ, if it was Skinner, we'd have reveille and bed check. If we had a ten-minute break, he'd double time us around the fuckin' station and then restrict us to our seats. This guy's O.K."

"Sure. He's a combat man. He's seen guys suffer and die. He knows that this chicken shit stuff don't win the war. He's human."

At 10:15 the train pulled into Havre, Montana. Since an eastbound troop train was already standing in front of the station, the men had to go through it. Some soldiers on the other train, combat veterans from the Pacific, had already tried to purchase whiskey at the store the captain had mentioned and had found it closed. But they had had an opportunity to explore other possibilities and had found a black market source. Although a few made purchases, most balked at the prices. The veterans showed no animosity toward Nisei troops; they only expressed regret that the latter were going overseas rather than home. When the whistle for the eastbound train blew, passengers from both trains scrambled wildly to get back. Most Nisei were not interested in purchasing whiskey, but the captain's permissive attitude was appreciated nonetheless.

The evening meal consisted of weiners, and although they were well prepared, somehow they did not seem adequate for Christmas dinner. When a signout sheet for those who wished to meet friends and relatives in Spokane was passed around at 18:00, several signed primarily to get something to eat. But the train did not reach Spokane until well after midnight, and most of them were already asleep.

On Wednesday morning, December 26, the company was awakened as the train was making its way through the Cascades; several commented on the scenic beauty of the area. Everyone was packed and ready to debark two hours before the announced arrival time. As the porter swept the floor, the men on one of the cars decided to take a collection for a tip. There was no obligation to do this, but his cheerfulness and efficiency had made the trip less unpleasant. When those in the other cars heard about this, they too took a collection. The amount collected averaged about $6.00 per car. They had done so much sweeping and mopping of late that what the porter had done was genuinely appreciated.

The train pulled into a sidetrack in Seattle at 10:25 in the midst of a rainstorm. The troops were loaded on trucks and driven to Fort Lawton. No shelter was available at the detrucking point, and they had to stand in the pouring rain, as Nakasone and the officers conferred over details of where the company was to be billeted.

Fort Lawton was located northwest of Seattle, adjacent to the Magnolia Bluff district. It was on a peninsula in the Puget Sound, seven miles from the center of the city. It was a quagmire of mud. The "streets"—the spaces between the buildings where the trucks passed—consisted of a succession of puddles of unknown depth. Walking from one part of the camp to another was a challenge; pedestrians had to pick their route, hopping over one mud puddle after another and periodically getting stuck.

It was not until 11:30 that the group was marched down a hill to some tar-papered barracks. They were poorly constructed; the ground could be seen through the cracks in the floors. The buildings had not been cleaned for months, and no brooms or mops were available. The roofs leaked, and several beds had to be moved. Each of the barracks contained a single stove, and one of them had no grates. Coal was plentiful, but no kindling wood could be found; several search parties returned unsuccessful. The mattresses were so dirty that a cloud of dust arose whenever they were struck. Furthermore, there were not enough mattress covers to go around. Since they were thoroughly drenched, the men hesitated to lay any of their clothing or equipment on the beds. The latrines in the area were filthy and could be located easily by their stench. There difficulties developed quickly. Since no toilet paper was available in any of the latrines, newspapers had to be used, and several of the bowls were soon clogged. The men were utterly disgusted; this was without question the dirtiest army camp they had ever seen.

"What a dirty fuckin' place! Maybe Haoles can sleep in a place like this, but not me."

"Look at the fuckin' dust on the window! It must be a half inch thick."

"Holy shit! I thought the train was dirty. Now look where they put us!"

But the latrines contained brooms, and men ran to all the latrines in the area to collect them. Before long they were hard at work cleaning their barracks.

At the time Fort Lawton was being used primarily as a transshipment point for combat veterans returning from the Pacific. The veterans were housed there temporarily until train space became available to take them to various parts of the country for discharge. An air of informality prevailed. The war was over, and nothing military was regarded as being of any importance. No one cared about uniform regulations, and no one bothered to salute. The camp personnel was preoccupied with Christmas and the processing of large numbers of returning veterans. Hence, Company K was left in Captain Asakura's hands. An orderly room was set up, and Nakasone was placed in charge. Dispensing with the roster of acting squad leaders prepared at Fort Snelling, Nakasone selected several of his friends—most of them squad leaders at Snelling—to help him with various chores.

At 15:00 an orientation meeting was called. Captain Asakura announced that under Fort Lawton rules only 50 percent of a company's personnel was authorized passes at any one time. Since the group's own acting NCOs were to handle the passes, however, he saw no reason why anyone who wanted to leave could not do so. Nothing was scheduled other than an abandon ship drill at 08:30 the following morning and a showdown inspection at 09:00. There was also to be an ammunition check; but since that was up to the officers with the group, the matter could be dropped. "So if somebody asks you if we had an ammo inspection, you tell them we did. You men only have a few days left; so when you get passes, take off. You'll be able to see Seattle once, maybe twice." The men raised several questions. When would their baggage be unloaded from the train? Where were they to eat? Which latrine were they to use, and who was supposed to clean it? The captain promised to find out about these things. One man asked what time the lights had to be out. The captain replied, "Unless you want 'em out, leave 'em on. When the DO comes, just say to him, 'Oh, lights out already?' and turn 'em off."

Since they thought they would be leaving immediately, almost everyone in the company went out on pass that night. Several commented on their good fortune in having Captain Asakura as commanding officer. He understood them and did everything he could for them—even breaking camp regulations. They agreed that no one should go AWOL; the captain would be embarrassed, and Nakasone would get into serious trouble.

At 13:00 on December 27 the company was called out for an "overseas physical." Since it was the final medical examination before going overseas, some expected to spend several hours at the hospital. But it consisted simply of lining up before a medical technician who conducted a "short-arm" inspection. Each man's left arm was painted red to indicate that he had passed. Several commented that the technician was more interested in getting the red paint on them than in finding anything wrong. A man with syphilis could have passed easily, for the room was too dark to see.

"Jesus, the exams are gettin' tougher every time."

"All you have to have to pass is a prick."

"If you didn't have that, they'd transfer you to the WACs and send you over anyway."

"Remember what that first sergeant at Wheeler said about overseas physicals? 'If you've got two legs and an asshole, you'll pass.' Boy, he wasn't shittin'."

Since so many of the men had been at Fort Meade, they knew what to anticipate in the processing for overseas duty. They knew that they would be issued new equipment and that any item with flaws would be replaced. They had been looking forward to this and were pleasantly surprised to find the procedure at Fort Lawton very lax. The showdown inspection was conducted by the Nisei lieutenants. Each man was handed a checklist on which the clothing and equipment with which he was charged were listed. They were told to indicate on this sheet what they already had, what they needed, and what they wished to salvage in exchange for new equipment. This provided opportunities to get extra items; they could mark for salvage items they had no intention of returning. The officers expected them to do this; their only request was that sufficient restraint be exercised so that it would not be too obvious that the requests had been padded. The officers also decided to send in a mass requisition for the entire company. If each man went to the warehouse and drew his own order, he would have to turn in everything he had marked for salvage. But it was unlikely that anyone at the busy warehouse would bother to count the salvage items of a whole company. Furthermore, this enabled the officers to get battle jackets for themselves which they otherwise would have had to purchase.

The clothing issue began at 15:00 on December 28. The men had to stand in line by platoons in the rain. When the transaction was completed at 21:00 there was considerable dissatisfaction. Those who were toward the front of the line received everything they had ordered; the others did not. They received clothing that did not fit, and the supply of popular items was exhausted. Those with misfits and shortages complained that there was something wrong with the procedure; each person should have gone to the warehouse. The difficulties had arisen when members of the first two platoons took more than their share. Some had as many as ten pairs of socks. Some of them had taken the extra items, not for personal use but for black marketing in Japan. They were roundly condemned, for their comrades had to do without. Some charged that Nakasone's friends working in the supply room had helped themselves to more than their share. Since some of the items turned in for salvage were still in excellent condition, the dissatisfied were allowed to pick out whatever they needed. Since the unit had been alerted to leave on the following day, no opportunity remained for further adjustments.

Saturday, December 29, was the day on which the mainlanders were to leave for overseas duty. Most of them had never been outside the country before. All platoons were aroused at 05:30. After an early breakfast everyone had finished packing by 07:00. It was still raining, and the men continued to curse the weather. At 08:30 the company was called into formation. The barracks bags had been jammed with more equipment than the men had ever been issued before. Everyone wore a combat pack and cargo carrier on his back, and those who had special items that they intended to sell in the black market carried hand bags as well. Some of the smaller men staggered when they tried to stand. They assumed that they would not be required to walk too far with such an unreasonable load and stood about waiting for trucks to come. They were surprised when Captain Asakura called them to attention. To their utter astonishment he then ordered them to pick up their barracks bags and started to lead them up a steep hill. It was a march of over 200 yards. The onerous burden would have been challenging enough to carry on level ground, but marching uphill at a steady cadence was too much. Since they took pride in being a "rugged" company, not a single man fell out for the first fifty yards. They started panting heavily, and some were gasping for air. Then, one man collapsed, and the company formation disintegrated. Everyone was waiting for someone else to be the first to drop out. Only about a third of the company managed to keep up with Captain Asakura. When he reached the assembly area at the top of the hill, the officer in charge asked him to take his troops back to their barracks! The captain's order to halt was relayed down the hill. The company was scattered on various points on the hill; some were still closer to the bottom. As the conference above proceeded, the men dragged their bags to the side of the street and sat on them. Those who had reached the top listened to the discussion and passed back word of what was happening. The shipping schedule had been changed, and there would be some delay. When this was announced, howls of protest erupted from the hill.

"Why can't we wait on the ship?"

"Why don't we leave the bags up there, now that we carried 'em this far?"

"Why don't the bags go ahead? We can stay behind."

"Why didn't they send trucks in the first place?"

When the men were ordered to bring their bags all the way up and to load them on a truck, a rumor developed that the bags would be returned to the barracks so that they could be carried up the hill again in the afternoon. Very few took it seriously; that was too stupid even for the army. Those who had reached the top went back to help some of their friends carry their bags.

As the men stood about the assembly area, one of them shouted loudly for the benefit of the Caucasian officer in charge, "Why didn't they tell us before we got here? All they had to do was pick up the phone."

Someone replied in an equally loud voice, "The phone must have been too heavy."

This broke the tension. Despite their anger at what they considered inexcusable inefficiency, most men laughed good naturedly and went back to the barracks. As they were sitting about recovering from the ordeal, they were shocked to learn that the trucks had indeed brought the bags back to the barracks! The men refused to unload them and demanded that the truck driver take them back to the assembly area.

"Aw shit! What the hell are they doin'? This is no joke."

"That's the second dumbest thing they could have done. The dumbest thing was to make us carry 'em up there to begin with."

"Fuck, why couldn't we stack them on that big loading platform?"

"We could leave 'em any fuckin' place. I sure don't feel like carryin' these bastards up that hill again."

"Shit, leave mine out there. If anybody's got enough poop to cart it off, he can have it."

"Why didn't them sons o' bitches phone or something. Why didn't they at least come to the road when they saw us coming? Why did they wait until we were all the way up there and then tell us?"

"Why didn't the trucks come down here? A couple hundred yards is nothing for a truck."

"That's the way the fuckin' army works. Some prick fucks up, and 180 guys get screwed in the ass."

Finally, after considerable discussion, the men agreed to bring their bags back into the barracks. Some were secretly delighted when they saw that those who had been greedy and had brought supplies for black marketing had suffered more than the others.

Much time was spent discussing the weight of the load. Everyone had carried heavy loads before, and none had expected it to be so staggering. Some thought that the duffle bag alone weighed at least a hundred pounds, probably an exaggeration. With combat pack, cargo carrier, handbag, overcoats, ODs, and steel helmets they estimated the total weight to be over 150 pounds. Many took pride in being "rugged," and they confessed that they did not realize the extent to which they had gotten out of condition. At this point a man who had become so exhausted that he had to rest elsewhere staggered into one of the barracks.

"Hey, General, go out there and get your bag. The trucks just brought 'em in."

At first he thought it was a joke; then he saw that all the others had their bags beside their bunks, "They brought 'em back? What the hell was that—a dry run?"

"Yeah. They want to give us plenty of practice carrying them things so we won't be nervous when we carry 'em for real."

It was not until noon that the officers learned that all shipments had been postponed until January 6. After a pay formation Captain Asakura had to return to Fort Snelling, and First Lieutenant Maruyama, a medical administration officer, was placed in command. There was nothing to do for a week, and those who still had money went out on pass. New Year's Day was spent in diverse ways. Many did not even leave camp. Others went into Seattle in small groups, whistled at girls, went to a movie, and then returned to tell imaginative tales of what they wished they had done. Saipan was apparently spending considerable time at a burlesque theater, and others insisted that they went there only to watch him. When one man walked out of a restaurant without paying, he was roundly condemned. Since most Seattle establishments had been cordial to Nisei soldiers, the men were utterly disgusted.

"How can a Boochie stoop that low?"

"Aw, he's just a no good tramp."

The conduct of the combat veterans attracted much interest. Some Company K men had wanted to go into combat; others wanted to be like combat men. They noticed that most of them were informal and friendly. None of them saluted officers, and they paid little heed to uniform regulations. They were on the whole sloppy and practical. When walking about in the mud, they simply rolled up their trousers—contrary to army regulations. If anyone said anything about the regulations, the standard reply was "chicken shit." Several Nisei observed that combat veterans did not polish their boots. Boot polishers were then dubbed "drug store cowboys" and "USO commandos." Some refused to polish their boots and even walked about with their shirts unbuttoned.

Since Fort Lawton was filled with veterans returning from the Pacific, it was easy to get information concerning troopship regulations, conditions in Japan, and black marketing. The veterans mixed easily with the Nisei and were soon advising them on what to do. The most common advice was to take cigarettes. Several of the veterans advised them to throw away all their clothing and equipment and to pack all the bags with cigarettes. It was easy to get more GI equipment overseas. The worst thing that could happen was being forced to sign a statement of charges, and the cost of lost items would constitute but a fraction of what one could earn selling cigarettes.

Since the group did not leave on schedule, another opportunity arose for adjustments of equipment shortages. It was at this time that it became generally known that Lieutenant Clinton was partly responsible for the difficulties. The most common problem was the shortage of 34-short field jackets and of socks smaller than size 12. Captain Asakura had been told by Lieutenant Clinton, who was in charge of the warehouse detail, that these items were not available. However, when the men went individually to the warehouse for exchange and salvage, they learned from the warrant officer in charge that field jackets of all sizes were available. When the warrant officer learned of their plight, he invited them to exchange any jackets that had not been worn. When this information spread, those who had worked on the warehouse detail recalled an exchange between Lieutenant Clinton and one of the clerks. She had indicated that while no more size 10½ socks were on hand, she could send upstairs for more, but the lieutenant had said, "No, just give us 12s." When the woman objected that they would be much too large for anyone who wore 10½, he had replied that it did not matter. The discovery that field jackets had been available all the time led the men to place all the blame on Lieutenant Clinton, the only Caucasian officer in the group. Then someone recalled that Lieutenant Skinner had commended Lieutenant Clinton as a fine officer and had told some of them that they were lucky to have him as a train officer. He was a friend of Lieutenant Skinner! Some went so far as to charge that Lieutenant Clinton had acted deliberately to inconvenience them as much as possible. It was another attempt by the MISLS administration to get back at them!

Wednesday morning, January 9, was embarkation day. A company formation was called at 08:00 to be sure that everyone would be out of bed on time. As the men were turning in their mattress covers, a rumor spread that some Japanese being deported from Tule Lake would be on the same ship. Many were resentful and cursed the disloyal Nisei who "made it tough for the rest of us."

The major preoccupation, however, was with carrying the duffle bags up the hill. Since Captain Asakura was gone and management of the company was in the hands of Nakasone and his friends, the men decided among themselves to do it on their own. They knew where the bags had to be taken. Some waited in hope that trucks would come, but when it became known that they would not, small groups started up the hill with their bags. They wanted to carry them without being burdened by additional packs and a steel helmet. Guards were posted in the assembly area, and arrangements were made to take turns watching the bags.

"It's not bad if you go about ten feet and then take a rest."

"That's what I did the last time."

"Anybody who can climb that fucker with a bag without puffing is O.K."

"Fuck, anybody who can go up that son of a bitch even without a bag is O.K. I thought I was just gettin' soft, but even the guys in shape were puffin'."

"No shit! That's really a hill."

By the time the formation was called at 11:15 everyone in the company had carried up his belongings. At 12:00 the company was assembled at the trucking area. There the men had to stand in line until 12:45. Trucks finally came, and they were loaded by units. A small group of Nisei went first; then a handful of Caucasian soldiers boarded; then went a sizable contingent of Negro troops. Company K was last and the largest single unit in the shipment.

At 13:10 the trucks arrived at the Seattle port of embarkation. After unloading the men again had to stand in line. As they waited, some Red Cross women brought coffee and donuts. Some had eaten neither breakfast nor lunch and were hungry; others were afraid of becoming seasick and declined to eat. Others, arguing that the chances of getting seasick were minimized if their stomachs were full, took all they could. At 14:00 boarding began. Officers checked each man's name and serial number carefully. Again each soldier had to carry all his equipment and duffle bag up a slope, this time a narrow gangplank that was much steeper than the hill. Some joked about dropping their bag over the gangplank; those who were concerned lest they do that failed to see any humor in the remarks. They watched the Negroes and Caucasians ahead of them as they groaned and grunted up the gangplank. Occasionally, a man faltered, slowing down the entire loading operation. But there were no delays once Company K started. Drawing on all his reserve strength, each man went up, determined to carry his belongings without stopping all the way to his bunk two decks below the main deck. Some almost collapsed when they arrived.

Passive Resistance on the Transport

The U.S.S. *General Weigel* was built for the Grace Lines and commissioned for the navy on 6 January 1945. The ship had a displacement of 19,650 tons under full load, a cruising speed of 21 knots, and at that speed a range of 12,400 miles. The compartments, which could hold 5,342 soldiers, were water, air, and fume tight. The ship had four 5-inch guns, four 50 mm antiaircraft guns, twenty 20 mm machine guns, two radar units, four signal searchlights, and a mine sweep. It was manned by a coast guard crew of 400 men, 39 officers, and 30 chief petty officers. Facilities were available for boxing, music, reading, and movies. After a staff had been recruited from the passenger personnel, a newspaper—the *Weigel Echo*—was mimeographed and distributed daily to everyone aboard. Chaplains, both army and navy, were aboard. The latrines—"heads"—were large and generally empty. Mess facilities were excellent. The replacements had little to complain about; the ship was new, large, clean, and fast.

With only a few exceptions Company K men were assigned to a hold two decks below the main deck. Since no specific bunk assignments were made, each was able to sleep near his friends. Once unpacked, they roamed about the ship. They talked to the sailors to learn about its facilities. They were gratified to discover that all the decks were small; there was not enough space for extensive physical training. Furthermore, they were delighted to find that the water fountains contained fresh ice water. Many sighed in relief.

"Well, we're finally on."

"Even Willie Yoneda made the boat. I thought they'd never get him."

"Finally made it! About time!"

The passenger list was small for such a large transport. Army personnel numbered 333—26 officers and 307 enlisted men. Diverse military occupation specialties were represented: 187 Asiatic language interpreters, 75 riflemen, 11 field artillery, 2 Asiatic language translators, 2 coast artillery, 1 cavalry, 1 engineer, 1 military police, 1 ordnance, 1 truck driver, and 25 unspecified. One officer and 75 enlisted men were Negro; 20 officers and 46 enlisted men were Caucasian; 5 officers and 186 enlisted men were Nisei. Company K had been joined by a small detachment of Nisei, who introduced themselves as CIC (Counter-Intelligence Corps) men trained in San Francisco. In addition, there were 88 navy passengers, all Caucasian, of varying rank and special skills. Finally, 182 Japanese prisoners of war—18 officers and 164 enlisted men—were being returned to their homeland. Thus, over a third of all the men on the ship—passenger and crew—were of Japanese ancestry.

All officers were billeted in first class staterooms above the main deck and had their own spacious dining hall and lounge that spanned the entire width of the ship. They were provided with mattresses, linen, pillows, and blankets. All navy, coast guard, and marine enlisted personnel slept on the mess deck. Although they had no access to the officers' lounge or dining hall, they were given linen, pillows, and mattresses and had beds similar to those in the staterooms. Similar quarters were given to the few army enlisted men of the upper three grades and their personal friends. Although similar quarters were unoccupied on one side of the ship, the remaining army enlisted personnel were assigned to the holds below the mess deck. The only enlisted men of the lower grades who managed to be assigned to the mess deck were Private Nakasone and the few Caucasian privates friendly with the NCOs. About twenty Caucasian privates, about a dozen Nisei, and all the Negro enlisted men were assigned bunks in a compartment one deck below the mess deck. The remaining Nisei enlisted men were assigned to the compartment below that. In these compartments the bunks were four tiers high, extending from the floor to the ceiling; they consisted of canvas stretched on a rectangular bar of steel. These passengers received no linen, no pillows, no blankets, no mattresses. They had nothing other than what they themselves carried. Since there was so much extra room, the bottom bunk of each unit was used for the duffle bags of the three who slept above. In this way the narrow aisles could be kept clear. Since two sets of bunks were placed side by side, each man slept only a few inches from another on one side, above, and below.

Although army officers enjoyed many facilities and privileges denied enlisted men, the distinctions were considerably greater in the navy. Meals for enlisted men were served in a huge mess hall that extended from one side of the ship to the other. Navy and coast guard personnel ate on one side, which was equipped with low tables and benches. Army personnel ate on the other side; the tables were high, and meals had to be eaten standing. Officers were served in a dining room that resembled an expensive restaurant. Movies were shown enlisted men on a screen about six feet square stretched in the middle of the mess hall. The audience had to squat, sit, or stand on the floor. Furthermore, whenever someone opened a hatch, the screen fluttered so much that it was difficult to follow the action. Not more than fifty could be accommodated at one time. Although only a handful of officers were aboard, they were able to sit in comfortable stuffed furniture. The screen for their movies was large and stationary. As enlisted men roamed about the decks, they could not help but seeing the officers' quarters through the portholes. The clean white sheets and the wash basin in each room made them envious. While they were given a few cubic feet in a stuffy hold, officers were living in hotel rooms. As they began to realize how uncomfortable it was to sleep without pillows and after they had bumped their heads two or three times, their resentment against officers mounted. They complained bitterly of the attractive staterooms with carpets and drawers. They had heard of the "caste system" in the navy. Now that they could see it, many bristled indignantly at the differential treatment; the contrast was too obvious.

Singled out for particular attention was the difference in shower facilities. Those quartered on the main deck and above had showers with hot fresh water. Since the enlisted men slept in such close quarters, it was important that they kept themselves clean. But their showers were located at the bow of the ship. The water was cold; furthermore, it was saltwater. The passengers discovered quickly that their soap would not melt in saltwater. Although the room was well ventilated, it was cold. Taking a shower when the ship was rocking was challenging; they had to balance themselves on the slippery floor. In addition, the dirty water on the floor splashed all over the room, and clothing left on racks and benches sometimes fell into the puddles. Taking a shower was difficult and unpleasant.

Each day the *Weigel Echo* published reports received over Mackay Radio. News of the mass demonstration of American soldiers in Paris, reported under the headline "NEW CASTE SYSTEM," struck a responsive chord. The demands of the disgruntled veterans there were: abolition of officers' mess, with all rations to be served in a common mess hall on a first come, first served basis; opening of all officers' clubs to enlisted men; abolition of reserved sections for officers at recreational events; abolition of all special officers' quarters, and the requirement that all officers serve at least one year as enlisted men; reform of court martial boards to include enlisted men. In the discussions that followed the passengers agreed unanimously with the demonstrators.

"I think them guys in Paris are on the ball. The American army might be better to the enlisted men than the other ones, but they sure have a lot of shit."

"Yeah. Look at this fuckin' boat. Twenty-six officers and 300 EM. The 26 officers sit in a big drawing room to see a show, and all the EM have to squeeze into one wing of the fuckin' mess hall. What kind of shit is that?"

"They got enough first class bunks on this ship to let all 400 guys on it travel first class, but they put us in them holes just to let us know there's a difference between us and the fuckin' officers. Boy, if that ain't shit for the birds."

"Shit, yeah. There's a lot of room on this tub."

From the very first day all passengers were assigned to some duty. The enlisted men were assigned as guards, latrine cleaners, table waiters, and other such tasks. Three Company K men were assigned as runners; two were sent to the ship newspaper staff. Those whose names started with letters from A to H were ordered to clean the mess deck daily. Those whose names started with S and T were assigned to the latrine detail. Others went into the galley. The rest, over half the company, were assigned to guard duty. The guards complained bitterly, for they were on duty for four hours and then off for eight hours. Those whom Nakasone had selected as corporal of the guards were responsible for waking up the men on their shift at odd hours and assigning them to their posts. The irregular sleeping made the task disagreeable, even though there was little for them to do.

Although the assignments had been intended for the duration of the voyage, so many complaints of inequities arose that changes had to be made after eight days. Everyone hoped to be transferred to an easier detail. Those who were on guard did not care what their new duties were to be; they wanted a change. A large Negro crew was taken out of the galleys and sent to the latrine-cleaning unit, but the Nisei who had been in charge was left in command. The Nisei who had been in the latrine detail were sent to the galleys. After a single day they were enraged, charging that it was worse than KP, most of which was performed by the Japanese POWs. Those who had been on deck-cleaning crews traded places with the guards, but the whining continued.

"Shit, this detail crap gets me pissed off. We get the same old shit every day. Goddam! It might be a long trip too. I expected to lay around and take life easy. Son of a bitch! What do we get? More details than before."

"Yeah. Me, I expected a vacation. *Kembutsu* [sight-seeing], you know. Shit, I'm gonna get sick. I ain't gonna clean the shithouse every day."

What made most of them unhappy, however, was navy discipline. The Nisei, who had become accustomed to lax enforcement of regulations, were most vocal in their complaints; but the Caucasians and Negroes were just as displeased.

Most passengers were outside on one of the decks at 11:25, when the ship began to move. When the transport was out of Puget Sound, she started to rock and sway. The waves striking the bow could be heard in loud thumps, and the constant creaking annoyed those unaccustomed to sea travel. Many became seasick. Others complained of being dizzy and of having a headache. Discussions arose of how to avoid seasickness. Some said that the susceptible should not eat; others argued that those who were ill should go outside for fresh air; still others recommended lying down.

"It's all in the mind, boys."

"Oh, fuck you! Why don't you shut up?"

Many on night duty were too sick to report for work, and several guards were missing from their posts. They were so miserable that they did not care about being court martialed or anything else. They just wanted to be left alone. One common complaint was that the latrine was not readily accessible. It was at the bow of the ship on the mess deck. The latrines of the decks below had been closed. This meant that Company K men had to go up two decks and through three holds to reach it. Even those who were not yet ill slept with their clothes on; they did not want to be delayed, should they have to vomit. Several who had already vomited during the day went to an empty hold near the latrine and slept there rather than in their own compartment. A number of men strapped their steel helmet to the bar of their bunk for emergency use. The atmosphere became so unpleasant that even those who were still healthy became depressed. The sick received sympathy; the few who were unaffected had to do all the work. One man insisted that seasickness was contagious. "If I don't feel good, I go outside by myself. If I go with a bunch of guys who say they're sick, I get it sure as hell."

"We're the farthest away from the shithouse. They must have seen the Boochie names when they made the assignments. The *Kurombos* [Negroes] have the same names as the Haoles; so they didn't know."

"Don't worry. They'll pull shit too. They always get it."

Early encounters with Negro troops were unfavorable and tended to reinforce the stereotyped conceptions that most Nisei held. Initial contacts came in crap games, which started within a few hours after boarding ship. At supper it became obvious that the Negroes had been given the least desirable assignments. Many were KPs; the serving line was noisy, and the KPs threw food all over the plates, completely disregarding the sections. Some openly condemned their conduct as childish; even those who reserved judgment were resentful, since they did not appreciate their main course with dessert poured over it. The proximity of Negroes also led to anxieties about theft. The few Nisei billeted with them were pitied; they complained that the Negroes around them were eyeing their new equipment. After a few days at sea, when the compartments below the main deck began to smell of sweat, some concluded that Negroes were hopeless. Nisei were under constant pressure to take frequent showers, as uncomfortable as they were. Their compartment was being swept constantly, and it was one of the cleanest parts of the ship. Since similar norms did not hold for other troops, the comparison was unfair; but many Nisei pointed repeatedly to the contrast between their room and that above them. "The Niggers' place really stinks." When reminded that the men themselves were not to blame, most Nisei agreed, but they still felt uncomfortable about being so close to them.

The Nisei were by no means agreed in their orientation toward Negroes. The views of those who were already negative were intensified by the contacts; they felt that their beliefs were confirmed by what they saw. Some were vindictive. "When I get there, I'm going to get the Boochies to keep the Crocks out of Yoshiwara." Others, especially those with Christian backgrounds, were sympathetic and pointed out that Negroes were always being mistreated.

One morning the lone Negro officer aboard was observed eating with the enlisted men. This aroused considerable comment and indignation, even among those who were usually anti-Negro.

"Any *Kurombo* who's an officer is O.K. When you figure how much shit a *Kurombo* has to take, any guy who's good enough to get bars has got to be damn good."

"That's low, huh? Even when he's an officer, they gotta discriminate against him. What a bunch of bastards!"

Later it was learned that the lieutenant had been on duty as the officer in charge of the mess detail. However, when he was not working, he ate at a separate table in the officers' dining room. Those who saw him dining alone continued to condemn navy policy. Indeed complaints about discrimination in general mounted. The men noted that with the exception of a handful of Caucasian privates only the Nisei, the Negroes, and the Japanese prisoners of war had been assigned bunks below the mess deck. All other Caucasians, including several privates, were in quarters on the mess deck or higher.

A very unfriendly bespectacled Caucasian lieutenant was apparently in command of the Negro troops. The few Nisei who had come into contact with him found him to be a strict disciplinarian, a man very conscious of his rank who sometimes abused his privileges. He seemed unnecessarily severe, and before long the Nisei labeled him "Little Napoleon." They described him as "GI as hell" and concluded that he must be "just out of Benning." Most Nisei were contemptuous of him. They imitated his swaggering gait, laughed at his GI glasses and awkward appearance, and sneered at the amateurish manner in which he bloused his trousers over his combat boots. He looked more like a raw recruit than an officer. Several rumors about him developed. One was that when the Negro troops were en route to Fort Lawton, he forced them to double time on the train and gave them calisthenics at all the station stops.

Since the lieutenant was a disciplinarian, he had several unpleasant encounters with the Nisei. On the third day at sea the men heard that Little Napoleon had caught Private Yasui asleep on guard duty and had reprimanded him. This led to much laughter, since they knew that Yasui was lazy.

"Hey, Yasui. I hear you got caught fuckin' off last night."

"Yeah! Son of a bitch! I was havin' a good dream too."

"Who was it?"

"Aw, it was that little bastard. You know, that prick with the GI glasses."

"Little Napoleon?"

"Little Napoleon! Ha! Is that what you guys call him? That's pretty good. Yeah, that's the bastard. He's a prick, you know it?"

One rumor that aroused widespread resentment was that Little Napoleon had had two seasick Negroes arrested for failing to report for KP. Since almost everyone had been seasick, they knew how it felt. What justification was there for imprisoning a man who refused to work when he was that sick? None of the Nisei who had been unable to work because of illness had been arrested. One man insisted that he had seen Caucasians sleeping when they were supposed to be on duty; before long another rumor spread that Little Napoleon had excused a Caucasian soldier who was seasick but that he told the Negroes that illness was not an adequate excuse. He was a racist. As sympathy for Negroes increased, another rumor developed that Little Napoleon had sent a soldier with a temperature of 102° down to the brig without any blankets. Two days later a Nisei officer reported that five Negroes had been court martialed for being asleep on duty. When they pleaded that they were seasick, they were told that that did not matter. This soon spread in the form of a rumor. Although these rumors led to some anxiety, they did not affect the behavior of the Nisei. They knew that many Nisei and Caucasians had done the same thing and that nothing had happened to them. They had not even been reprimanded.

Sentiment against Little Napoleon crystallized. He was condemned largely because he was a white man taking unfair advantage of Negroes. But the men also regarded him as an incompetent officer. They resented his authority and privileges, which they felt were undeserved.

Resentment of the navy's differential treatment of ethnic minorities intensified when, on January 12, word spread among the passengers of an incident that had occurred in the newspaper office earlier that day. The Caucasian editor of the *Weigel Echo*, once an instructor at Columbia University, asked one of the Nisei staff writers to prepare an editorial on discrimination. A rather mild article was written, not mentioning the situation aboard the ship but pointing to the cost of discrimination and condemning it as undemocratic. As the issue was being readied for stenciling, an Army Transport Command officer who was responsible for the paper asked that the editorial be replaced. While he had no objection to it himself, his past experience with the navy had been such that he believed its publication would lead to trouble. The editor was outraged and took the matter to the army chaplain. The matter finally reached the captain of the ship, who ruled that the subject would not be mentioned and ordered—under threat of court martial—that the matter be dropped. The navy chaplain concurred with the captain's ruling. Indignation arose not only on the newspaper staff but also among the army officers aboard. Censorship of the newspaper was expected, but the suppression of such a mild editorial seemed unreasonable. Most Nisei accepted it as inevitable; after all, the navy had always refused to accept Nisei for service. Some noted that the Japanese prisoners of war were being treated with far greater consideration than American soldiers who happened to be black.

Since Company K men had become so accustomed to lax enforcement of regulations, the strict discipline of the navy appeared unnecessarily repressive. They felt hounded by "chicken shit," and resentment against officers was magnified with the invidious distinctions aboard. Furthermore, the treatment of Negroes made respect for the system even more difficult. Before long reaction against authority crystallized, and patterns of insubordination first developed at Fort Meade were again instituted. No conscious planning was involved; the men simply lapsed into their old habits. Orders issued to Nisei soldiers were obeyed reluctantly, if at all. Frequently the men muttered "chicken shit" before the officer was out of hearing range—letting him know how they felt. Work details were performed lackadaisically. "Disappearing" became a regular routine, and the men took pride in telling one another of ingenious tactics they had developed for getting out of work or embarrassing an officer without making themselves technically liable for punishment.

Flagrant, large-scale insubordination developed first among those assigned to guard duty. They found their work irksome and of little practical value. The war was over, and there were no enemy submarines lurking on the seas. The "lights out" regulations were no longer in effect, and smoking on deck was permitted. What was there to guard? The men felt that it was a complete waste of time for them to stand for hours in the cold wind, especially at night. Several went to sleep at their posts. When reminded by duty officers that it was a court martial offense to fall asleep while on guard duty, they began searching for hiding places. Since most of the officers' quarters were empty, they had no difficulty in finding soft beds—beds far more comfortable than their own. They also discovered that these cabins contained showers with hot, fresh water in which soap would melt. Before long most guards reported for work with soap and towel. Many did not wait for their relief shift to replace them. When they got sleepy, they left their post. If they became hungry, they went to the mess hall to eat.

The three men assigned by Nakasone as corporals of the guard were pitied. They could not ask anyone who was seasick to stand watch and sometimes had to do it themselves. Someone had to be present at posts where officers could not be avoided or where there was actually something to keep under surveillance, such as food or prisoners of war. Furthermore, they frequently had difficulty locating their men. As complaints mounted, all guards were reassigned on January 17. But the new set of guards proved equally adept in avoiding work.

Soon after white helmet liners and nightsticks were distributed to the guards, most of the equipment "accidentally" fell overboard. The clubs were an unnecessary nuisance; the POWs were guarded by armed marines as well as by passengers on temporary duty. After a few days at sea so few sticks and helmets were left on the ship that the sergeant of the guard decided against issuing any more. The Caucasian sergeant sympathized with his subordinates and neglected to report the losses. The Nisei openly admitted throwing the equipment overboard; they regarded it as a way of getting back at the "chicken shit" navy.

"If them bastards think they can push us around like they push around the *Kurochan*, they got another guess coming. We're used to gettin' pushed around, and we ain't takin' any shit."

"Some of these Haole bastards are dumb. No fuckin' sense standing out there and freezin' your ass off. And what the fuck they give us billy clubs for? Who we gonna sock with 'em? I threw mine overboard—too heavy."

"Yeah. Me too, Fuck that shit! If they treat us nice, O.K. But if they're gonna fuck us, we might as well fuck up."

Willie Yoneda became the company hero, for he was far more adept in "goofing off" than anyone else. His quick wit and pleasant disposition made him popular with most of his comrades. He had been assigned to guard duty, and at first he had encountered some difficulties in finding a safe place where he could sleep undisturbed. He could avoid officers easily enough, but he could not always evade the corporal of the guard, who was more familiar with his tactics. He used his freedom as a guard to explore the ship in detail and apparently discovered several hiding places that remained unknown to all others. He was generally asleep somewhere or playing bridge; whenever he was wanted, he was nowhere to be found. The men guffawed in merriment as Nakasone roamed through the vessel looking for Yoneda. The two were close friends, but Yoneda generally succeeded in evading the acting first sergeant. One morning a warrant officer tried to awaken Yoneda by tapping him gently and whispering, "Up and at 'em, buddy." Knowing how much trouble Nakasone and Colonel Andersen had had in arousing him at Fort Snelling, the men nearby burst out laughing. Willie did not budge; he continued sleeping peacefully. The warrant officer was puzzled; he could not understand why the others were chortling at him. Finally, he walked timidly away.

As the insubordination became more widespread, some of the more responsible men began to worry. Guard duty was taken seriously in the military, and being absent from a guard post was a court martial offense. Had the officers decided to enforce regulations, many would have been in serious trouble. One of the Nisei corporals of the guard remarked in apprehension, "Those guys are going too far. Maybe we're all Boochies and all that shit, but there's a limit to it. I went out to check the other morning about 05:00. I went to the POW hold where these guys were supposed to be guardin' the prisoners. Four guards were flat on their ass asleep, and one guy was just sittin' between two PWs just slinging the shit. Goddam! I had to chase them guys down. What if some PW did something? I don't give a shit about the rest of the boat, but the PW place is supposed to be watched. I don't want to be chicken shit, but that's going too far."

Tasks that were regarded as essential, such as keeping living quarters clean and safe, were performed effectively. Regulations against smoking in compartments were enforced by the passengers themselves. Most of them had never been on an ocean liner before, and they were fearful of being on a burning ship at sea. When a small fire broke out in one of the holds on the fourth day at sea, their uneasiness was reinforced. Those who lighted cigarettes in their bunks were immediately set upon by their neighbors, who reminded them of the regulations. Violating rules meant nothing to these men, but it provided a way of getting others to put out their cigarettes without revealing personal anxieties.

Where personal responsibility was fixed, the work had to be done. This placed considerable pressure on those who were in charge of specific tasks. They could not ask those who claimed to be seasick to work. Sometimes they had to do the work alone. The men took turns "goofing off," and on most work crews enough appeared each day out of consideration for the individual in command. The man placed in charge of keeping the latrine clean was immediately designated the "*Benjo* King." The others commiserated with him, especially after January 17, when his Nisei crew was replaced by Negroes. Since he was conscientious, he did much of the work himself; friends who were not on duty elsewhere frequently dropped in to help him.

The only members of Company K who performed their duties conscientiously were Private Nakasone, the "*Benjo* King," and Saipan. Nakasone was a good soldier; in addition he was accountable for all the work assigned to the company. His men realized that he had to do his work. Saipan had always maintained that he had a superior command of the Japanese language because of his brief residence in Japan prior to the war. When he requested permission to serve as a guard for the POWs, therefore, the corporal of the guard, who had known him before the war and tried to protect him from the others, and Nakasone agreed. Saipan was always on duty and did as he was told. He even volunteered to work when off duty; except when he slept he was always with the prisoners. The others soon noticed Saipan's devotion to the POWs. Some commented that since he looked so much like the caricatures of Japanese warloads, he must feel a special affinity to Japanese soldiers. Many wondered aloud what the Japanese prisoners thought of Saipan.

One afternoon, as the army replacements were standing about finishing their lunch, they saw Saipan enter the mess hall from the prisoners' side of the ship. A few moments later a column of prisoners entered, led by an officer. The Japanese officer approached in a military manner, snapped to attention, and saluted Saipan and the officer in charge of the mess hall. Although the lieutenant returned the salute half-heartedly, Saipan snapped to attention and saluted smartly. The observers were dumfounded. They could hardly believe what they saw. They snickered among themselves and then rushed out of the mess hall, where they burst into laughter. Even those who had vigorously opposed the tormenting of Saipan were so amused by the incident that they could not help laughing. A report of what had happened spread throughout the company within a few hours.

Some men who had previously played tricks on Saipan then told him that his work with the POWs had been so exemplary that he had been promoted to corporal. They followed him to Nakasone's room, where Saipan asked the acting first sergeant when he would receive his corporal's stripes and raise in pay. Nakasone was bewildered. Once he realized what had happened, Nakasone sought out the pranksters and rebuked them severely; in recalling the incident later, however, even he had difficulty in suppressing a smile. News of the encounter between Saipan and Nakasone spread quickly.

"I hear Saipan went up and asked Nakasone for his corporal's stripes."

"No shit?"

"They'd better take Saipan off that shift, or the PWs'll start wondering who really won the war."

Those who had befriended and protected Saipan were upset, but for the next few days their efforts were of little avail. Some announced plans to purchase a set of stripes for Saipan. A rumor developed that Saipan told Nakasone that it was a good thing the days had twenty-five hours, allowing him to sleep for one of them.

Once the ship had passed the Midway Islands, attention focused increasingly on life in occupied Japan. Articles in *Life* magazine about fraternization and the black market led to numerous discussions of the exploitation of destitute women. Some Nisei appeared excessively greedy, and those who had brought "bait" for Japanese girls frequently found themselves under attack. One rumor that was welcomed was that all Asiatic language specialists would be sent to the NYK building in Tokyo rather than to Manila. The place was described as a luxurious hotel-like structure in which enlisted men were waited on in dining halls.

"I hear it's pretty nice there."

"Well, I gotta see it. I don't believe nothin' until I see it."

"There must be a catch in it. It sounds too good."

Since the information was attributed to letters from men on duty there, the rumors were not summarily dismissed. Furthermore, it was now obvious that the ship was not stopping to disembark troops on the Philippines.

At 16:00 on January 23 an announcement was made that Yokohama was a hundred miles ahead and that land had been sighted on the port side. The passengers dropped whatever they were doing and rushed to the main deck to see. After supper many stayed on deck. Those who had gone below were brought up at 19:20 by the call: "Anchor detail, take your stations!" Some started packing. There was gaiety and singing. Those who worked in the galleys received some extra food and distributed it among their friends. It was a warm, beautiful, starry night. Shore lights could be seen from the deck. Someone started singing "Harbor Lights," and others joined him in humming the melody.

Most of the troops were up before reveille on the following morning. They wanted to get into the latrine before the rush. They knew it would be closed after breakfast, and they anticipated an inspection before anyone left the ship. Since everyone was up, for the first time the mess line was as long as it had been on the initial night aboard. After breakfast virtually everyone crowded onto the main deck. The ship was anchored in Tokyo Bay. Fujiyama could be seen in the distance. The shoreline was picturesque, and many commented on its scenic beauty. Hundreds of ships were in the bay, and U.S. Navy patrol boats were ubiquitous.

At 09:00 the passengers jammed the main deck to see the unloading of the POWs. As they watched, a small boat came along side. It had an all-Japanese crew, accompanied by one U.S. Army officer and a Nisei interpreter. The Nisei watching joked about relatives among the boat crew and prisoners.

"Oh, *aniki! Shibaraku dat-ta-na!*" [Big brother, it's been a long time.]

"Let's go! We're in the wrong section."

"Is that your cousin or Saipan's brother?"

"Hey, Nakasone. You'd better get down there. They're liable to take Saipan along too."

Several showed a special interest in the prisoners. The guards had fraternized, and some had gotten the addresses of several of them and had promised to visit. Most Nisei felt sorry for them; they knew that the Japanese regarded being captured in war as the ultimate disgrace. "It's their last square meal." One man displayed a hankerchief bearing pornographic art that one of the prisoners had given him. "Boochies go in for this stuff too, huh?"

At 10:30 two tugboats pulled along side. The public address system blared out: "All passengers clear the main deck on the starboard side." There was a rush to the superstructure so that the order could be obeyed and the unloading still be witnessed. Another order followed:

"All troop passengers lay below in their compartments." Amid groans the men reluctantly returned to their bunks.

At 14:00 the U.S.S. *General Weigel* docked at Yokohama. American soldiers were playing basketball on the pier. Nearby, Japanese laborers were loading another ship. About 14:35 someone threw a cigarette butt overboard. To everyone's dismay, about a dozen laborers dropped their work and fought for it. At first the men were amused, and soon several were throwing cigarettes down to them. All work stopped on the dock, as the laborers converged there. The first soldier to throw down an unopened pack was a Negro. The passengers continued to throw cigarettes until they were ordered to stop through the public address system. Many Nisei expressed sympathy, although it was apparent that a number of them enjoyed watching the ragged workers fight over discarded butts.

"At least we won't have to police up any more."

"*Kawaiso* [pitiful], huh? They got no shoes or nothin'."

What they saw contrasted sharply with their conception of the Japanese, and it confirmed what their friends had written about the desperate plight of the people. Watching the prisoners unload had also been a moving experience for many Nisei. "Boy, I'm glad I was born in the United States. You don't appreciate it until you go away." They became acutely conscious that they too were of Japanese ancestry. Because of their birthplace, however, they were among the conquerors.

Preparations were made for disembarking. All passengers were ordered to their respective compartments for a medical examination. Everyone understood that it would be a "short-arm" inspection, though some wondered how anyone could contract venereal disease on a ship without a single woman aboard. At 17:00 a warrant officer came down to the Nisei hold to order them to get ready. As they started putting on their gear, the Caucasian sergeant of the guard rushed down to say that they would not leave before 20:00. "Why don't they make up their fuckin' mind?" As it became obvious that no one would be leaving for some time, most men went on deck again. By 17:30 they were wandering all over the ship. At 19:10 passengers were ordered to line up by the mess hall; everyone ran down only to learn that beer was being served. Most men did not care for beer.

"We just fuck around and fuck around. The trucks ain't even here yet. Lucky if we get any sleep tonight."

As they sat in their compartments, the men became restless. From time to time someone would yell that the time had come, only to burst out laughing when the others put on their packs. After several such false alarms, they became unresponsive. But at 21:15 what was left of Company K was activated by the jolt of Private Nakasone's inimitable roar: "O.K., YOU GUYS, HIT IT!"

That was it! Without further questions they put on their packs and steel helmets. They picked up their duffle bags and one by one started up the ladder. They lined up on the mess deck. The men knew their positions on the roster, and they made room in the narrow passageway to let those ahead of them pass. By 21:30 the company was in platoon formation on the dock, ready to board the trucks. At 21:45 the company was at the Sakuragicho Station in Yokohama, aboard a special Japanese train. After fifteen days they were once again on land. They were finally in Japan.

Analytic Summary and Discussion

For a brief period the disorderly unit that had been condemned so roundly was transformed into a model company. With a change in command it once again reverted to familiar old patterns. Left largely on their own during their move to the troop transport, the men managed without incident. Once aboard, however, they became estranged from the command and again reverted to resistance. Thus, the quality of performance alternated as the troops adjusted positively or negatively to a succession of authority figures. This is not surprising, for the fate of enlisted men depends to a considerable extent on their company commander, the mediating or linking person who binds the local unit to the larger corporate body. Most soldiers view the army as a huge machine run by an anonymous "they"—people to whom they feel no personal loyalty or obligation. But they do develop beliefs and sentiments about leaders of local units. Enlisted men talk constantly about their officers—commissioned and noncommissioned—and estimates emerge from these discussions. A company grade officer is evaluated largely on the basis of decisions over which he exercises some control—local matters or emergencies not clearly covered by regulations. A company commander is seldom blamed for onerous duties imposed by battalion or regimental headquarters.

Lieutenant Catalano was exactly the kind of officer reported in other studies as winning the affection and trust of enlisted men (Homans, 1946; Bassan, 1947). He was a competent soldier. In addition, through words and deeds he repeatedly demonstrated his personal concern for the welfare of his men. He bent regulations, countermanded orders, and sometimes even participated in passive resistance against the MISLS administration. He understood the needs of enlisted men and tempered enforcement of regulations to provide the most considerate care that the circumstances permitted. He respected Nisei norms; he appreciated the obligations that his men had to one another. For example, he never asked anyone to "squeal." He was also fair. The men did not expect to get away with anything; they knew the regulations and expected some enforcement. Indeed, there was some disappointment that he did not crack down on the Hawaiian trio sooner than he did. Most members of the company genuinely liked the lieutenant; even those who did not feel personal affection felt deeply indebted (*giri*) to him. Whenever any occasion arose in which they could demonstrate their sentiments, however trivial the situations may have been, they pitched in voluntarily to contribute whatever they could. This resulted in some astonishing peak performances. In doing so, they revealed something of the aptitudes and potentialities that might have been tapped under different circumstances.

Although the reaction to Lieutenant Skinner was initially favorable, the men were soon disillusioned. Once they became convinced that he was a hypocrite, they went out of their way to be as disorderly as possible, leaving the fort in a state approximating anarchy. On the way to the troop transport, Captain Asakura turned out to be unexpectedly reasonable. Furthermore, they knew, respected, and trusted Nakasone and refrained from doing anything that would get him into trouble. As in many Nisei athletic teams before the war, they did the tasks they realized had to be done largely on their own initiative. Once aboard the transport, they reacted sharply against the distinctions between officers and enlisted men as well as the mistreatment of Negro troops. While they did what they regarded as necessary, they reverted once again to the familiar patterns of resistance.

8. «» anarchy at the replacement depot

During the autumn and winter of 1945 the once mighty U.S. Army ceased to be an effective fighting force. Most of the personnel had been drafted; it was understood that the war had to be won at any cost, and enormous sacrifices had been made. Now that the fighting was over, they wanted to go home. Although over 5 million had been released by January 1946, those who remained were impatient. They had been told that they were still needed to secure the peace, that there was much vital work yet to be done. But where was the critical work they were supposed to be doing? Thousands were sitting about day after day in idleness, in routine drilling, or in performing trivial tasks that had been created to keep them occupied. They could not understand why they were still overseas, since most of them were doing nothing that contributed to any announced objective. Virtually everyone became preoccupied with getting home.

The conduct of many officers only added to the discontent of the enlisted men. They seemed less and less concerned with the welfare of their troops and increasingly preoccupied with their personal comfort and amusement. Many left their men to their own resources. Distinctions between officer and enlisted man, minimized within the United States, were magnified overseas, where facilities for officers were often lavish. They enjoyed large supplies of liquor, better clubs, the best seats in theaters, and most of the dates with nurses and Red Cross girls. The most desirable hotels, restaurants, and resort areas were reserved for officers; and their subordinates constantly encountered signs reading "Off Limits to Enlisted Personnel." Mass meetings and protest demonstrations erupted not only in Paris but also in Manila, Honolulu, London, and Frankfurt. In them special privileges were denounced, and commanders were booed and insulted. Enlisted men collected funds to purchase space in American newspapers, and they organized campaigns for writing congressmen. The rationale that American troops were to teach democracy to people in the defeated nations failed to impress those who felt that they themselves were being treated as second class citizens. Many officers, themselves conscripts who had never cared for military life, openly sympathized with their men.

Discipline broke down everywhere. Conformity with regulations was often token and formal. Demoralization was manifested in widespread drunkenness, reckless driving, carelessness in dress, refusal to observe military courtesy, and slovenliness in performing assigned duties. Any task that was not obviously necessary was dismissed as more "chicken shit." Thus, the men from Company K found it unnecessary to alter their old habits of insubordination. Several commented that the rest of the army had finally caught up with them.

Fraternization and the Black Market

American troops in occupied Japan found themselves in a devastated area. Results of the bombing were apparent everywhere. Large sections of Tokyo had been leveled to the ground, and many people were living in makeshift shacks of wood and corrugated metal. Transportation facilities were incredibly overtaxed. Rolling stock that had survived was being utilized to the maximum, and it was not uncommon to see passengers hanging on outside the coaches, clinging to windows or balancing themselves on connecting links between cars. (See Plate 7.) The people seemed ill mannered; they shoved to get on the trains and buses, and young men were constantly trying to force their way in at the head of various lines. Many were in rags. Even those who still had clean clothing were reluctant to wear them, when so many others had so little. Most people looked hungry. Urchins, unwashed and unkempt, could be observed begging for food and sleeping in parks and railroad stations. Everywhere Japan seemed to be so dirty, so primitive, so smelly.

Initial contacts between members of the army of occupation and the conquered Japanese were dictated by necessity. Both sides were at first fearful, suspicious, and hesitant. As the early transactions proved surprisingly satisfactory, however, the apprehensions broke down. Although most contacts were casual and ephemeral, some friendships developed. Before long, people on both sides expressed surprise at how

PLATE 7 *Transportation facilities were limited.*

well they were getting along. Although linguistic barriers were formidable, joint enterprises of all kinds were carried out with the aid of gestures, dictionaries, and army language manuals. The Japanese accepted their subordinate status; the transactions were between the "gooks" and *shin-chiu-gun*. Some Americans were arrogant, but most were friendly and considerate. Many Japanese tried to anticipate the pleasures of their rulers; they tried to learn whatever they could about American tastes and customs. By the thousands they sought employment in military establishments, even when other jobs were available. Those who succeeded were envied. Soldiers off duty found themselves surrounded by children, who waved friendly greetings or asked for candy. The most extensive contacts occurred in two areas: in the black market and in the relations between American soldiers and Japanese women.

The black market in occupied Japan was the result of an unrealistic economic policy (Rundell, 1964:61–79). The income of the Japanese was restricted by law. Goods rationed by the government were sold at reasonable prices, but it was virtually impossible to live on these meager rations alone. Most families found it necessary to supplement them with purchases on the black market. American goods were at a premium. The soldiers had sugar, candy, gum, cigarettes, clothing, canned meat, and other food—all scarce items. Furthermore, the containers in which they came could be used for various purposes; galvanized and rust-proof metal was scarce. The Japanese were delighted to purchase such items, for American goods commanded high prices both in resale and in bartering with farmers for rice. The black market soon became so extensive that it was difficult to distinguish between legal and illegal enterprises. Japanese officials, themselves dependent for part of their subsistence on the black market, were reluctant to enforce the laws stringently.

American soldiers also found it foolish to abide by military regulations. The value of Japanese money had been set officially at 15 yen to one dollar, and all military personnel were paid in yen at this ratio. Enlisted men were permitted to send home postal money orders totaling up to 90 percent of their paycheck; their salaries could then be reconverted into dollars. This was referred to as "good money." Except in the PX, however, the purchasing power of the yen was not worth 15 to the dollar. Soldiers found many things on the open market that they wanted to buy—such as German cameras, silk, and cultured pearls. Furthermore, those desiring feminine companionship needed money for food and hotel bills. But they balked at paying for such items with "good money," for the cost in dollars became outrageously high. Obtaining what they wanted at more reasonable prices was simple. After the initial sale of a carton of cigarettes—purchased for 55 cents in the PX and sold for 300 yen—additional items could be purchased at the PX from the profit to sell for ten times that amount. Thus, a used German camera priced at 10,000 yen would cost over $650 at the 15:1 ratio but could be obtained at about $18 through illicit channels. Similarly, a silk handkerchief selling for 30 yen could cost $2 in "good money" or about 6 cents, and a 400-yen dinner could cost $26 or just 75 cents. Since spending "good money" in Japanese establishments was absurd, virtually all Americans engaged in black marketing of some sort during their stay in Japan. Although a few took advantage of the opportunity to acquire wealth, most men simply wanted to purchase souvenirs at reasonable prices.

Periodic raids were made, aimed primarily at "big time" operators, and the publicity given to them served to some extent as a deterrent. On the whole, however, control measures were ineffective, for the transactions were profitable for all parties. The goods in demand among Japanese constituted surplus to most Americans, who could purchase them easily at the PX or draw replacements from their company supply room. The goods wanted by Americans were items that the Japanese did not regard as essential for survival. The prices of various goods were fairly uniform throughout the country and tended to fluctuate together in response to changing conditions, as one might expect of a free market. The current price for a carton of cigarettes was used as the index. Although many Japanese stores were selling at illegal prices, they were seldom raided by the police. Nor was the U.S. Army able to control its troops; regulations were largely ignored, for its own law-enforcement personnel were almost as deeply involved as the others. That officers of all grades, MPs, and even some CID (Criminal Investigation Division) agents were engaged in black marketing during off-duty hours was a standing joke.

Although most Japanese were pleasantly surprised at the conduct of American troops, one sore point was fraternization with their women. Initial contacts came in brothels. The thought of Japanese women consorting with American men was disturbing, but such prostitution was accepted as inevitable in a defeated nation. Furthermore, commercialized vice had long been accepted as an essential part of Japanese society. Although most brothels were declared "off limits" by the army, soldiers experienced little difficulty in pursuing their erotic interests. The cabaret business boomed; night clubs sprang up throughout Japan—even in remote villages. Popular establishments, such as the *Marigold* and the *Oasis of Ginza* in Tokyo, were patrolled both by MPs and Japanese police; all patrons were searched as they entered and required to check their liquor and weapons. In addition, "pon pon" girls could be found in virtually all railroad stations, parks, and other areas in which soldiers congregated. What upset most Japanese more than anything else was the public display of affection, which struck them as coarse and vulgar. So many protests arose that in March 1946 an order was issued by 8th Army Headquarters: "Public display of affection by men in uniform toward the women of any nation is in poor taste. Particularly is this so in Japan among those who were so recently our enemies and where the people have never been accustomed to such demonstrations. . . . Such action in public is prejudicial to good order and military discipline and will be treated as disorderly conduct." This order was no more effective than the other regulations.

Although most of the contacts were casual, some stable arrangements were also established. Many romances blossomed from meetings in a cabaret. Although most patrons accepted any available partner, some developed more enduring relations with a particular woman. They went to the dance hall early and bought enough tickets to last the entire evening; for some this continued night after night. Misunderstandings sometimes arose, and it was not uncommon for soldiers to fight among themselves for the favors of a particular girl, sometimes without her even knowing about it. On occasion serious difficulties erupted, especially when one of the rivals was a Negro or Nisei and the other a Caucasian. Some girls met their boy friends after work hours; others left the establishment to "marry" Americans for the duration of their tour of duty.

Considerable pressure was placed on the women. While no organized effort was made to punish them, on occasion some man would slap a "pon pon" girl. Those who were seen walking with Americans were frequently taunted: "*Niisan ni yu yo*!" [I'll tell your big brother on you!] "*Ano sugata o mitero!*" [Look at that sight!] "*Mit-tomo nai! Haji or shiranai no ka?*" [Disgraceful! Haven't you any pride?] Combat veterans of the Japanese army, just returning after several years abroad, were especially incensed. As one young farmer recounted:

> Returning veterans are certainly angry about the behavior of Japanese women. Of course, they did worse things themselves and admit it, but they never expected Japanese women to do it. The other morning I saw a troop train come into Tokyo Station. One girl was walking with a GI, and a veteran ran after her. We had to hold him back because he was so angry. Other Japanese quieted him down. Fortunately, the American was good natured. We are used to it by now, but it hits the returning veterans hard. It makes them realize how low Japan has become. This man was crying like a baby. He told the girl that he fought his heart out for Japan, and now he was returning to find that the people were not worth his effort. I was bitter too when I first saw it, but I think nothing more of it now. Naturally I don't like it, but I have accepted it as being in the nature of things.

Despite pressures at home and from neighbors, despite continuous harassment by other Japanese, many girls continued to seek the company of Americans. The standard explanation among Japanese was that "pon pon" girls were either "factory workers" or those of "low class" who had "no education." Most Americans assumed that the women were desperate for food and acted out of necessity. To be sure, many of the streetwalkers were prostitutes or factory employees who had to supplement their meager wages. However, they were not the only women consorting with Americans. Girls who worked in military establishments and in stores catering to American trade were hired for their knowledge of the English language; they were obviously neither uneducated nor from poor families. Other women met Americans through mutual friends. Why were so many Japanese women with respectable family backgrounds and not in dire need willing to sacrifice their reputation and future to pursue relationships with Americans? It appears that for many dating Americans was part of a rebellion against the subordinate position traditional for them in Japan. Women had been condemned from birth to a life of servitude; not only were their subordinate roles fixed in custom, many were enforced by law. A "good" woman was expected to devote her entire life to rearing children and ministering to the comforts of her husband. Then suddenly, with the termination of hostilities, the legal status of women soared. Under orders from the Allied Command, women were allowed to vote for the first time and even to run for public office. Women were granted the same legal rights as men in the new constitution, and all laws discriminating against them were revised. Thus, Americans were credited with the emancipation of Japanese women (Tsurumi, 1970:276–303).

Some of the girls involved were forthright in challenging the status assigned to them. One declared defiantly:

> The lot of the Japanese woman has been a pitiful one. I will not stand for it anymore. We now have a democracy, and we have the right to do as we choose. That is why I got a job with the Americans. I do not have dates too often, but I have them occasionally. I know what my father thinks, and I don't care. The more he beats me the more I am convinced that there is something basically wrong with Japan. I have met rough Americans, and some of them have forced me to do things that I did not like. But I still prefer Americans to the Japanese.

Many of the girls misinterpreted the conduct of the Americans. Courtesies that were customary in the United States and would have been taken for granted by American women were interpreted by the Japanese as kindness or special affection. Those who were involved with Americans reported that they were considerate, that they treated women as if they were human. They not only paid handsomely for services but even brought gifts. The manner in which they were treated was a sharp contrast to what could be expected of Japanese men. One Yokohama prostitute stated her views bluntly:

> There are some Americans who are quite unreasonable. This is especially true of the younger men; they are so young they don't know what they are doing. But the older men are very kind. The Nisei are a very odd group; I cannot understand them. The Caucasians on the whole are very nice. Some of the Negro men are very rough, especially when they are drunk. But they are good natured and mean no harm. Even though they are rough, I prefer them to Japanese men. I should be ashamed to say anything like this, but the Japanese men are the least considerate and the worst of them all. I would prefer any occupation soldier to the Japanese.

Nor was the notion that Americans were considerate limited to those who were involved with them; other women who watched at a distance also noticed it. If Americans treated prostitutes so courteously, then other women must be treated with even greater consideration.

A Yokohama social worker confirmed the contentions of the girls:

> Many girls just want fun. The Americans treat them well. The Japanese woman is just as human as any other woman in the world. She craves affection even though Japanese custom does not permit any open display of affection. The Japanese woman is starved for affection, and if the American is the only one who can give her what she wants, she is willing to give up almost anything for it. They know they are throwing away their lives, but many of them have no future to look forward to anyway. They have nothing to lose, they think. They want to live now and take their chances later. Young women in Japan have thus far been virtually imprisoned. They are free for the first time. What you see is the reaction of some of those who are willing to cut off all their ties. There are thousands of other women who want to do the same thing, but they are afraid. If they were given one day in which to do anything they wanted without fear of retribution, they would flood to the Americans. The older Japanese will deny this. That is because they really do not understand how the younger generation feels. I know. I have talked to many girls, and they all say the same thing.

But even older Japanese who maintained a safe distance from Americans admitted privately that the common explantion of "pon pon" girls was inaccurate. A housewife reported:

> My husband just got back from Tachikawa where he went to visit my brother, and he said that my nieces were getting out of hand. It seems that all the young girls in that area are getting out of hand, and the parents just don't know what to do. They want to play with Americans so badly that they run away from home at night and do not come back until the next morning. Those are not bad. Some of them do not come back for two or three days. They all want to get jobs working in the airfield, and even if they don't they just go over there and stand around. The parents scold them, slap them, and threaten to disown them and kick them out of the house, but these warnings have no effect. The girls just go with their friends to meet the soldiers. What I can't understand is that my niece was a quiet and obedient girl. She is well educated; she graduated from middle school about two years ago. If the poor girls go after the soldiers, I could understand that, but I cannot see why my niece has to go. She has a home and everything. My brother is worried. He is angry, but he cannot kick his own daughter out of his house. I wish there were some solution to the problem. You cannot mistreat your own flesh and blood, but you cannot allow them to do things like that.

Although there were some unpleasant incidents and many rumors of atrocities, Americans and Japanese got along fairly well, and the occupation turned out to be an unexpected success. The unpleasantness was accepted stoically: "*Shikata ga nai; senso o maketa kara*" [It can't be helped; we lost the war]. Wherever they went, Americans were carefully watched, and many Japanese formed a favorable impression. They were frequently described as *sap-pari shite iru* [straightforward], *hogaraka* [cheerful and unreserved], or *non-biri* [relaxed]. In contrast the Japanese often described themselves as *hinekureteru* [twisted up] and bound by custom. The carefree orientation of Americans was widely attributed to their democratic training. There were countless reports of Americans giving up seats in buses for elderly women toward whom they could not possibly have amorous intentions; there were also reports of their ordering Japanese men to give up seats to women carrying infants or heavy loads. That most Americans were friendly and generous with children was acknowledged even by those who otherwise disliked them.

The initial year of occupation was a period of soul searching for many Japanese. They had been persuaded that Japan was invincible, and their traditional outlook had been shattered by defeat and the devastation of their land. Americans turned out to be very different from what they had been told. Where else had they been misled by the *Gumbatsu* [military clique]? One inevitable consequence of the black market was the immediate and widespread recognition of the incomparable superiority of American goods. This helped them to understand one reason why Japan had lost the war. Furthermore, Americans seemed to handle their possessions as if they had little value and spent money as if it had no value at all; this created the impression of immense wealth. From those employed at military bases the Japanese learned of the lax regulations of the U.S. Army and of the friendly mingling of soldiers of different ranks. Veterans of the Japanese Army could not help but note sharp contrasts. Enlisted men in the U.S. Army had unheard of privileges: they could address officers without showing deference; they played baseball with their superiors; they were never beaten; they were given passes whenever they were not needed. Knowing nothing of the demoralized state of the troops, many concluded that this was the organization to which they had lost. By the spring of 1946 many of the traditional Japanese values were being questioned, and there began a mass adoration of all things thought to be American (Kawai, 1960).

This was also a period of widespread corruption and disorder. Laws and regulations were clearly stated and widely known, but neither the American soldiers nor the Japanese paid much heed to them. In this context Nisei soldiers found themselves suddenly catapulted into positions of special importance. They were much in demand, regardless of their military occupation specialty. Commanding officers were constantly asking for them, for the presence of a single Nisei greatly facilitated the accomplishment of various tasks. Most Nisei were not unaware of their special position.

Initial Contacts with the Japanese

All army enlisted personnel aboard the U.S.S. *General Weigel* were placed on the same troop train. The unheated cars were old and dirty. Everything was covered with dust; the fingers of those who ran their hand over the upholstery turned black. Several of the windows would not open; others were boarded. A number of men commented on a peculiar odor that pervaded the area, but those who had been in Japan before assured them that they would soon become accustomed to it. After several hours the train arrived at Haramachida Station, not much more than twenty miles away. After a wild truck ride they arrived at their destination, and at 02:30 a sleepy cadreman came out to assign bunks. The Caucasians were called out first and assigned to quarters downstairs. Then, at 02:45 the assignment of the Nisei began—to two rooms upstairs and one downstairs. At 03:00 all the Negroes were still shivering outside.

The Nisei found themselves in huge rooms, each with over fifty canvas cots and six small stoves. The rooms were filthy. Some began to beat the dust out of the canvas; others ran out for wood and fired the stoves; still others went to the cadre for brooms to sweep out the dirt before unpacking. As they prepared to retire, the men saw the latrine behind the building. It was just a huge outhouse. On one side was a trough with eight faucets; the water was ice cold, and they were warned by the cadre not to drink it. Since none of the streets were paved, the depot was a quagmire of mud.

"Look at that! Just a box over a slit trench!"

"Goddam! It stinks in here! I lose fight, hey."

"My poor achin' ass! We're way the hell out in nowhere. I thought we were supposed to go to Tokyo."

"Son of a bitch! Every time we move we end up in a worse place. I thought Snelling was the shits, and then they put us on that fuckin' train with no lights. I was glad to get off that, and they put us in that dirty shithole at Lawton. I was happy as hell to get out, and then we end up in that crowded hole in the ship. Boy, I was really happy this morning 'cause I figured I wouldn't have to sleep in that damn thing again. Now look! Look at this fuckin' place! This is the worst place we've ever been in!"

"Jesus Christ! We're out at the shit-ass end of nowhere."

The replacements were dirty, tired, and sleepy. Most of them went to bed, but before long they were up to make adjustments; it was too cold. The wooden building was flimsy. Only a single, thin sheet of wood provided protection from the elements. Because of defective chimneys, smoke from all the stoves poured into the rooms, but there was no danger of suffocation. The walls were full of cracks. It did little good to pile up blankets above, for the wind blew up from below through the canvas. Most men slept, fully clothed, in their sleeping bags, with one blanket above and one below. They rolled up their field jackets for pillows. It was well past 04:30 before most of the Nisei were asleep.

On the following morning no reveille was held. Only a few went to breakfast, but by 08:30 most men were out of bed; it was too cold to sleep comfortably. Early risers had talked to the cadremen, and what they learned was passed on to the others. They were at the 4th Replacement Depot, near the village of Zama, twenty-three miles from Yokohama and thirty miles from Tokyo. The main depot was located at what had once been the Japanese War College; it contained a large Red Cross building, a theater, a fairly large but poorly stocked PX, a small gymnasium, and some office buildings. This was used for the processing of combat veterans from all parts of the Pacific theater with enough discharge points to return home. Replacements and the overflow of veterans were quartered at the annex about four miles away, what had once been the signal station of the college. Facilities there were more limited than those at the main depot. The PX was smaller and open for only a few hours a day; the mess halls were smaller and dirtier; the Red Cross was not only much smaller but was open only in the evenings. There was only one shower building, and hot water was available only during those hours when very few were using it. All the enlisted men from the ship were now at the annex, assigned temporarily to the 275th Company, 15th Battalion.

Replacements were to remain at the annex until requisitions came from various units for soldiers with their military occupation specialties. This was not a permanent station, and no one knew from day to day how long he would be there. Since the cadremen had told them that the last group of Nisei had been there for over a month, most men settled down in anticipation of a stay of similar length. Private Nakasone was relieved of his responsibilities, and the self-governing unit was dissolved. Each man was on his own.

When the men learned that they were occupying the barracks of the Japanese War College, they were astounded.

"No wonder the Boochie army is rugged. They make officers train in a dump like this. I wonder where they put the enlisted trash."

"They probably slept outside on them rocks with one blanket apiece."

"Hell, this place was probably dirtier yet. I notice there's a lot of GI stuff here; the plaster boards, the stoves, and the shithouse are all American."

"Christ! What a stinky shithouse!"

"Boy! If this is the West Point of Japan, I'd hate to see what the enlisted guys had."

At 09:00 a general formation was called, and all new replacements were ordered to an auditorium for an orientation meeting. The battalion commander, a young lieutenant colonel, addressed the group:

I'm Colonel Dickerson, your battalion commander. We've processed 30,000 men like you up to now, and we know just how to take care of you. We have work to do and need your cooperation. Any infraction of the rules will be dealt with severely. You are in the army, and this is an army post.

There are 3,000 returning combat veterans and 500 replacements now in the annex. The returnees don't give a damn. They're getting discharges. Don't be like them. When you use the showers, remember that we went to a lot of trouble to put them up for you. We didn't have to do it. We did it for you. So when you're there, don't break the windows and wreck the shower heads. They can't be replaced. Everything depends on the ships. If they don't bring any cargo, we have to do without. These facilities were put in for you. It's up to you to take care of them.

Returnees dress any way they please, but for you there will be a uniform regulation. You will wear a complete uniform. We don't care whether you wear wool ODs or fatigues, but you will wear a complete uniform.

You are restricted to this post beginning as of right now. We've had men like you here before, and we know what to expect, and we're all ready for you. There is a mound which serves as a fence around the post. That fence is off limits, and you'll be shot if you're on it. Keep away from that fence. Some men have already been shot for wandering too close to it. Remember, there are armed guards here with orders to shoot.

We're all going to work here. You will pull KP because 90 percent of the Japanese have parasites and are not fit to handle food. Some of you Nisei will work as interpreters for the Japanese labor gangs.

I want to warn you men against changing your money with the veterans. They've been around, and you're green. This Jap money is easy to counterfeit. Don't let these men who offer you a high rate of exchange hand you their fake money. We have no finance officer here at the annex, but you can get your money changed at the PX or by your CO. You can't spend much money in camp anyway. You don't need any money here.

Some of you men are planning to go out. I want to warn you that when the women of this area were inspected by our medical teams, 90 percent of them were found to be infected with VD! Think it over. All the whiskey you can get around here is bootleg stuff. You drink it at your own risk. Remember, men, *medical statistics prove that 90 percent of all women in this area have VD.*

A lot of men try to black market around here because a carton of cigarettes will bring about 200 yen. But you had better watch out. We have CID men in the area now. Some of them are Nisei dressed like Japanese. If you sell to them, they'll pinch you. So watch out.

You will be here only for two or three days. *You are all assigned right now*, and you're here only as long as it takes us to print your orders and for your unit or us to make arrangements to get you there.

Finally, men, don't talk to any of the laborers. Leave them alone, and keep away from them.

As they lined up at the warehouse to draw two additional blankets and a comforter, the replacements discussed various points in the colonel's talk. They were restricted; that meant no passes until they were transferred to another station. Those trying to sneak out would be shot. They were now overseas, and patrols used live ammunition. Although the men thought it unlikely that 90 percent of all the women in the area had venereal diseases, they agreed that 90 percent of the prostitutes might well be infected. Perhaps that was what the colonel meant. Everyone was assigned; that meant they were going to Tokyo perhaps within a few days. At least that was good news. They did not mind being restricted for a few days In general, the men felt that the colonel meant what he said. He was so young. Anyone who could attain such high rank so quickly was probably very competent. He appeared to be a strict disciplinarian, and two rumors developed to support this impression. One was that a man who was standing near the mound was placed in the guardhouse; the other was that anyone caught by the colonel himself had to pay a $10 fine.

"Boy, what a prick! He'll probably throw the book at you too."

"Yeah. No use fuckin' with guys like that colonel. He's a young bastard; so he must be tough."

At 15:45 a line started forming outside the mess hall, and before long others went out with their mess kits. By 16:15 the line was over

PLATE 8 *All brothels were off limits.*

100 yards long. After supper, as each threw his leftovers into the garbage can and washed his mess kit, the newcomers could not help noticing the large number of Japanese women and children huddled outside the mound behind the mess hall. Many of the Nisei stood staring at them. From time to time a soldier would look about carefully to be sure that no MP or officer was watching and then run out to hand his leftover food to the children. Each time someone went with his garbage the women and children fought for the scraps. The urchins were barefoot and thin; many of the women were carrying babies on their backs. Many were in torn or patched clothing. Their faces and hands were dirty, and some appeared as if they had not eaten for weeks. They did not beg

aloud but stood outside the wall gazing longingly as the American threw their leftovers into the garbage can. One Nisei joked about it, "Why don't you go out there and feed your old lady?" Most of them, however, were stunned. They had never before seen Japanese beg. They had heard of Europeans fighting over army garbage, but somehow it had never occurred to them that starving Japanese would do the same thing. They knew that the Japanese were very proud, and on the Pacific Coast many had elected to do without during the depression rather than go on relief. Some said they felt sick; others walked back to the barracks without saying anything.

"And they yell about rationing back in the states!"

"I wonder what Hiroshima's like. This is all farm land and not bombed at all, and the people here have to go through garbage cans!"

The first clash between the Nisei and the cadre came on Saturday morning, January 26, when thirty-five Company K men were assigned to KP in the 275th Company kitchen. At 05:00 a sergeant turned on the lights in one of the upstairs rooms and called them. No one responded, and he had to go down the aisle to wake up each man individually. Since his attitude was overbearing, the men reacted sharply.

"All right, soldier, get up! You're on KP today."

"What the fuck! It's only five o'clock yet. You don't need anybody yet."

"Listen here, soldier, cut that shit! When I tell you to get up, you get up."

"You gonna make me?"

"I'll give you five minutes to get up."

"Aw, shove it up your ass!"

"O.K. Just for that you'll pull shit detail all day. When you get over there, you're gonna work on the garbage detail. Get that? You're on garbage detail."

"Fuck you!"

The sergeant had no better luck with the others.

"What the fuck! Second day in Japan, and we pull KP already!"

Finally, the sergeant read off a list of names and told the men to be at the kitchen in fifteen minutes. Slowly, one by one, they began getting out of bed. As several of them muttered loudly about "beating up" the "wise guy," the sergeant left the room.

The KPs straggled into the mess hall a half hour late and found that there was nothing for them to do. As they sat about drinking coffee, they noticed a sign carved on the kitchen wall indicating that some Nisei had been there:

> Aloha from Hawaii boys of Company D
> MIS(TS), Fort Snelling, Minn.

"MIS*T*S! That's pretty good. TS is right."

"How about one for K Company?"

"Carve a big fucker."

"Watch it now. That wall's only a quarter inch thick."

After breakfast the KPs discovered that two Japanese were responsible for sorting the garbage—wet and dry. The Nisei were astonished to see each of them reaching into the trash cans and occasionally eating the things he pulled out. After the morning rush was over the KPs learned from the two that they had been employed by the army for garbage disposal. One of them, who claimed that he had been an officer in the Japanese army, stated that the previous group of Nisei had treated him harshly for not sharing the food with the women and children standing outside the mound. He contended, however, that he represented a cooperative of some 130 families in the vicinity. The waste was taken to a central plant where the coffee was steamed and evaporated to get sugar; everything edible was distributed to the families, and the rest was used to fatten pigs. He insisted that the women and children were from families that refused to join the cooperative, even though membership was open to anyone in the area. By giving them food, he argued, the Nisei were depriving the 130 families in the cooperative. He then produced papers indicating that he had been hired by the U.S. Army, and he insisted that his activities had the full knowledge and consent of the colonel.

The KPs conferred with each other and decided that the garbage should be divided. Two-thirds of it was given to the man; the remainder was taken to the mound. Guards were posted to warn of approaching MPs. KPs working inside remained alert to warn of the arrival of any officer. Although the regulations stated clearly that all garbage was to be buried, the regular mess personnel seemed completely unconcerned over what happened to it. The mess sergeant simply ordered one of the KPs to take a crew of Japanese laborers outside to bury it. Although it was obvious that the order was disregarded, the sergeant said no more.

It soon became apparent that the Japanese at the mound were well organized. As the leftovers were handed over the wall, the children took them back to the women in the rear, who had much larger containers. Each child, after relaying his share, quietly went to the end of the line. They took everything—food, boxes, scraps of wood, cans. The KPs were shaken.

"When I read about the people in Europe starvin' and beggin', I thought of how tough it must have been, but I never realized what it was like until I saw those kids scramble for the garbage here."

"Yeah, you don't know what it means when you just read about it. Did you see the look on that old woman's face when we fixed her up with some chow? She didn't say nothin', but it came from inside. You know what I mean? She really appreciated the chow."

"I hate to see people pushed down that low. People shouldn't have to live like that. The fuckin' war's over. I don't see why we have to bury this stuff. Shit, some of this untouched food is good stuff. It's silly to bury it when people outside are starving."

"That's the fuckin' army for ya. Some cocksucker sits on his ass in Washington with nothin' to do so he figures out a rule: bury all garbage. They ought'a drag his ass out here for a while and let him see what's goin' on."

The luncheon menu called for rice. Long before it was served, the KPs knew that most of it would be left over, for it had not been cooked properly. They made arrangements among themselves to distribute it. As they had anticipated, most Caucasians and Negro troops did not want any; most Nisei accepted some but refused to eat it. Three 10-gallon cans of half boiled rice were left untouched. The Japanese—the two garbage collectors and the laborers working in the area—rushed to the cans, seized huge handfuls of rice, and gobbled it down on the spot.

At this point an argument broke out among the KPs. The Nisei who had argued with the sergeant in the morning was in charge of the detail, and he was cooperating with the official garbage collector. In the meantime, however, others had learned from the laborers working near the mess hall that this man did not represent a cooperative. The workmen insisted that he was in business for himself and that he *sold* his garbage to people in the vicinity. One of the assistant cooks, who overheard the argument, declared that he had seen the man selling his wares in Haramachida. The cook added that the man had a license for a nonexistent cooperative and that the colonel probably knew of it. Although the cook had no evidence of the colonel's complicity, the men immediately recalled that the garbage collector had been an *officer* in the Japanese army. It did not seem implausible that the colonel was getting his "cut."

It was decided, therefore, to distribute the greater portion of the rice to the people outside the mound. The official collector was given a little so that he would not complain to the colonel. Within minutes after the distribution of rice began the number of people in line increased tenfold. Seemingly from nowhere women and children came running. The procession of pails was endless. MPs came by twice on a jeep. Both times the warning was sounded in time, and everyone disappeared. Once the jeep had passed, they were back again. The people thanked the KPs over and over, reiterating that it was the first white rice they had seen in years.

It was raining so hard that evening that only a few soldiers appeared for supper. The women and children came as usual, and the KPs put aside their regular work and took all the excess food to the fence. They were confident that no MPs would be riding about in such foul weather. So many people arrived that they could not be kept outside. After darkness fell even more people appeared. As the KPs tired of filling the endless pails, they appointed some of the older boys to do the distributing and devoted themselves to keeping the lines in order.

Nor were the Nisei the only ones who participated in getting food to the Japanese. A number of Caucasians walked out to the mound to give things to the children. Negroes in particular were openly sympathetic. Some of them purposely left food on one side of their mess kit untouched and then took it out to the children. Indeed, when the KPs were having difficulty carrying out the 10-gallon cans, one of the men who rushed in to help was Negro. He then stood there helping the KPs for more than a half hour. Some of the Negroes even gave up their PX rations—giving the children the five bars of candy allowed each man per week.

Ill will quickly developed between the mess sergeant of the 275th Company and the Nisei replacements. The KPs charged that he was arrogant, when all the other cooks were reasonable. Although he did not interfere with their disposing of leftovers, he did object vigorously to their serving their friends large portions. Furthermore, the food was badly prepared, and there was never enough to feed everyone in the line. Many sharp exchanges occurred, and on Sunday evening some of the men got even in a small way. The mess sergeant had called in a Japanese girl. Using a Nisei as an interpreter, he offered her ten pounds of wieners if she would spend the night with him. Instead of relaying the message, however, the Nisei told the girl that the sergeant was infected with a venereal disease and that he intended to rape her. Dissatisfied with the work of another KP, the sergeant ordered him to clean some carbon-filled stove fixtures, a dirty task. The other KPs volunteered to help him. They took each fixture and pounded it with hammers until it was no longer usable. Then they departed.

The newcomers learned about the black market from the cadre, the veterans, and especially from the Japanese laborers. They all agreed that things were so disorganized that no serious effort was being made to enforce regulations, unless a violation was so flagrant that it could not be overlooked. Indeed, it became increasingly apparent that some of the cadre, even the officers, were involved in large-scale operations in co-operation with Japanese racketeers. Supplies were provided by the Americans, and distribution was arranged by the Japanese. The men could see that the depot cadremen were in an ideal position to amass a small fortune. It was impossible to keep strict inventories; troops in transit were constantly drawing and turning in clothing and equipment. Thousands of items could be taken from the warehouses and would never be missed. Replacements still ignorant of conditions in Japan arrived daily with thousands of American dollars, which the cadre could purchase with their black market profits at a rate a bit higher than 15 to 1. Some of the cadre boasted of having saved thousands of dollars, and no one doubted what they said.

On the first night one Nisei sold a dollar bill to a cadreman for 20 yen, and with this he purchased a carton of cigarettes. He promptly sold it to a Japanese laborer for 300 yen. This he distributed to his friends, who were then able to buy their ration of cigarettes. On Saturday night, January 26, a number of men went out to Haramachida, Hachioji, and Atsugi. The order restricting replacements to the annex was obviously not being enforced, for MPs in the nearby towns did not ask to see passes. After selling more cigarettes and clothing, the men returned laden with Japanese money. This they promptly distributed among their friends. On Sunday night more than a third of the Nisei went AWOL, encouraged by the enthusiastic accounts of those who had ventured out. By the end of the first week virtually everyone in the company had several hundred yen. Those who remained in camp sold their cigarettes or ration coupons to those who left; the man who took the risk was allowed a profit of approximately 30 percent. Money also circulated through countless dice and card games. Before long it became so plentiful that 10-yen and 100-yen notes were used whenever there was a shortage of toilet paper. Some carried about rolls of 100-yen bills several inches thick. Close friends did not bother to repay small loans of 10 or 20 yen for purchasing PX items; it was not worth the trouble. As one man commented, "You don't have to worry about gettin' paid around here. They pay us five times a month—four times when they give us our weekly ration tickets and once at the end of the month. The regular pay is the smallest."

Each morning replacements who had not been placed on a KP roster were sent to the main depot for a variety of work details. Those who were assigned to warehouse duty immediately saw possibilities. Anyone who did not have enough cigarettes to sell, such as heavy smokers, had easy access to countless items in demand. It was easy to be assigned to warehouse duty; even if one were not sent there, he could walk in as if he had been ordered to report and simply go to work. As they began to appreciate just how disorganized the depot was, the men became bolder.

"That bastard went with no field jacket or nothin' this morning. Next thing I knew he was on detail with us in the warehouse. He just walked in. Shit, they didn't know the difference. All Nisei look the same to them dumb jerks. He found a set of ODs and a nice new field jacket and just put 'em on."

"You worked in the warehouse? Why didn't you get me a hood, you bastard?"

"Why didn't you tell me you needed one? Let the guys know what you want, and they'll get it for you. Somebody's bound to go in the warehouse again in a couple of days."

"Hell, they burn that stuff anyway."

"They ought to give that stuff to the Buddhaheads."

"They can't. Then they can't tell the difference between the GI stuff they gave away and the stuff the Boochies bought on the black market."

Serious discussions arose concerning the ethics of black marketing. Most men participated just to get enough spending money, and the few trying to amass a fortune were ridiculed as BTOs—"big time operators" —or as *Zaibatsu*. Few objections arose to selling cigarettes; after all, they were not necessities. But the propriety of selling clothing and food to impoverished people was questioned. When some men stole sugar from a mess hall to sell for 75 yen a pound, many felt that they had overstepped the bounds of decency. Similarly, most agreed that stealing clothing for personal use or for friends was permissible but that clothing should not be sold to people who were in rags. Those who felt very strongly about it refused to sell; they sometimes gave things away, technically a court martial offense.

"It's easy to cop clothes. All you have to do is wear a poncho. You can carry anything under that."

"Why be chicken shit? Swipe a poncho after you get there."

"Look at Uyeda. He just walked into a warehouse, swiped a duffle bag, and then filled it with long johns. He sold the uppers and lowers for 150 yen a piece."

"150 yen a set?"

"Hell, no! One piece for 150 yen."

"He's really takin' chances though."

"He said it's just as wrong to sell a pack of cigarettes as it is to sell a truckload. So why be chicken shit about it? Might as well go for broke. He just carried that fuckin' duffle bag right out of the main gate, and the guard didn't say a fuckin' thing."

"I'll be a son of a bitch!"

"Some of these guys got no shame. I don't see how they can sell stuff to guys who're dressed in rags."

"Me neither. I know that trying to stop the black market here is like tryin' to hold back a flood with a piece of straw, but I can't get myself to sell stuff to these guys. If I see some rich bastard, that's different, but not to these poor guys."

As it became increasingly apparent that it was safe, the replacements began to go AWOL in droves, and each night the returnees kept the others spellbound with tales of their orgies.

"Where'd you go?"

"Atsugi."

"Good ass?"

"Can't complain. Just 30 yen."

"Better watch out."

"Aw fuck. It's cleaner than in the States. It's a GI whorehouse. You gotta go to a pro-station for a short-arm first. Then, if you pass, they give you a ticket. It's a GI place, and you can't get in without a ticket. They got MPs out there to take care of guys who try to bust in. After you're through you go back to the pro-station again and clean up."

"Pretty safe, huh?"

"Yeah. You sign in when you take a short-arm. Then you gotta sign out again after you're through. That way they know if you didn't take your pro."

"Yeah! 30 yen each! You can get ten pieces of ass for a carton of cigarettes, and the weeds only cost 55 cents. No matter how bad it is you ain't got no kick coming. Even if you have to put a sack over her head, you ain't gettin' gypped."

"They don't seem to like Nisei though, huh?"

"Naw. I think it's because we ask them too many questions about what they're doing there. The old ones give you a hard luck story, but the new ones are *hazukashii* [bashful]."

Those who had gone to Tokyo went to the notorious *Oasis of Ginza*, and they described in exaggerated detail what happened there. Some claimed that in Tokyo Japanese girls copulated in public, but the others remained skeptical. The belief that geisha girls retained their chastity was quickly exploded.

"We hitchhiked down to Haramachida and went to a fuckin' geisha house. There was four of us. We figured a geisha house was different from a whore joint, but I guess times've changed. Goddam! I didn't know what to do. The babes came out, and pretty soon one of 'em grabbed my cock. Jesus Christ! They snuggle up and try to make ya hot. I guess the competition is pretty strong. The other guys were sayin' that if you go to a whorehouse you shell out about 50 yen, but if you take a geisha girl, you gotta pay her 50 yen and pay another 50 yen for a room. They make more money than the whores. That's why they try to make ya hot before you go to a whorehouse."

Toward the end of the second week in Zama, considerable discussion centered on a man who had apparently become infatuated with a prostitute. Since he had been a sycophant of the Hawaiian trio at Fort Snelling, he was not very popular; but the others were nonetheless concerned.

"Yamada's gonna get his ass in a sling fallin' for a babe like that. She might be a nice babe and all that; but you gotta remember that she's a whore, and she'll do anything to get money."

"She really went for him though. I was in there the other night, and she wouldn't go with anybody else. She just was there waitin' for him, and the old babes got mad. They told her she was supposed to be workin'. She won't go with nobody else any more."

"Is he gonna buy her out of the house?"

"That's what I hear. I hear he went out tonight to sell some stuff he copped at the warehouse so he could get enough money to buy her out. I hear he's tryin' to raise 5,000 yen."

Many of the younger men were without previous sexual experience. Because of tight community controls, they had been unable to seduce Nisei women they had known, and they had never patronized brothels. This was their first opportunity, and some of them quickly became obsessed with sex. Although few would admit it except to very close friends, their extreme curiosity made their virginity obvious to their older comrades. Their naiveté was often the butt of jokes, and some were finding it unbearable to go on. Some of the older men tried to protect them.

"Yeah. Some of these young guys like Junior here ya gotta watch. They never had it before, and they get so goddam excited they don't know what they're doin'."

"Fuck you!"

"That's no shit, Junior. You better watch it. The first time it's the same for everybody. You just lose your head. You get so you don't know what the fuck you're doin'. You get so you can't work or think or nothin'. All you think about is that twat. Maybe you know the babe ain't any good, but you don't care. When you get to feelin' like that, it's time somebody looked after ya."

"Yeah. That's the way it is with everybody. Everybody gets burned once, but buying a babe out of a house is too much."

"I ain't gonna buy no babe out of a house."

"Can't tell. You still got your cherry yet."

One afternoon a cadreman went through the rooms, asking if anyone wanted an American girl for "40 bucks a throw." When the men objected that $40 in American money was too much, he pointed out that American nurses, WACs, and Red Cross girls were rare in Zama. At first the corporal refused to divulge the identity of the girl for whom he was pandering. When several expressed disbelief, however, he revealed that it was one of the Red Cross girls.

"How do ya like that, huh? Pimpin' for a Red Cross girl!"

"Now that's steppin' down pretty low. I figure a Red Cross girl would get just as hard up as the rest of us, but she don't have to make money out of it."

"This is the place to do it though. Forty bucks in American dough! Wow! Somebody's goin' home a millionaire."

Zama was indeed the place for a lucrative practice. It was the only military base in Japan where so many men had American money. The replacements had still unspent dollars and were lonely. Most of the combat veterans had not seen an American woman for two or three years. Furthermore, $40 was not much in the army where, as the colonel had pointed out, one really did not need money. Since Japanese women were so readily available, however, the Nisei showed little interest.

Even interest in Rikoran waned. One evening a man who had just returned from Hachioji announced that he had heard of her whereabouts from some Japanese. Rumors then developed. One was that she was in China working on a movement of some kind; another was that some AMG (American Military Government) agents were looking for her. It was apparent that many still worshipped her, and a frequent comment was, "I feel sorry for her. I pity her when they catch her." But many others were now more interested in women of lesser beauty. Rikoran was too remote, and they were no longer content just to dream.

As the men fanned out in increasing numbers over the area surrounding the replacement depot, attention also centered on the stench that pervaded the countryside. They had started complaining about it from their first night in Japan. They had joked about "honey pots," the wooden buckets in which human excreta were transported for fertilizer, and they had related tales of embarrassing situations in which they had been trapped near them. One night two men in the company fell into a reservoir while walking through a farm on their way back to the annex. Those who were still awake when they returned related how badly they smelled, and the two were the butt of jokes for several days.

"I hear Mike and Joey fell in a shit hole last night!"

"What? Mike fell in the water again?" Several recalled that miserable night on A.P. Hill, when Mike had lost his footing while wading through the swamp and had almost drowned.

"Yeah, them guys were prowling around the fields lookin' for some pussies, and they fell into a big honey pot."

"Honey pot? Holy smokes! I thought he just fell in the water. Jesus Christ! That's really the shits! Them things really stink! That's worse than fallin' in the fuckin' GI shithouse."

"No kiddin', boy. The Boochies leave that stuff out there in them holes until they're really ripe."

The two were quickly convinced that the disaster would never be forgotten.

"Hey, Stinky, I hear you had an accident last night."

"An accident! Shit, boy, that was a catastrophe! Goddam! I heard about guys fallin' in them things, but I never thought I'd fall in."

"That's what you get for fuckin' around."

"We weren't fuckin' around. We just wanted to get into town, and we got lost. It was dark out there, and we were tryin' to find the road. I was walkin' first, and I tripped and fell in. Joey stopped in time so he only went in to his legs."

"Couldn't you smell the damn thing?"

"Sure, but everything stinks around here. Christ! How'd I know they had a damn hole right out in the field? You don't think we fell in on purpose, do you?"

"If that ain't the shits."

"You ain't shittin', boy. We couldn't believe it, you know. Goddam! We stunk so damn much we didn't come back here. We went straight to the shower and hopped in with everything on, but we still stunk. Son of a bitch! So we took off all our clothes and tried to wash it and everything, but it was no use. So we decided to throw our clothes away and go to the warehouse today to get some new ones. Joey wanted to save his boots, but they still stink. Shit, when we came back we thought we were pretty clean, but as soon as I opened the door some guy yelled, 'PHEW! WHO SHIT?' Lose fight, hey."

"Goddam! That's really somethin'. Now you got somethin' to remember Japan by."

"Goddam! We'll never live it down. Everybody in the company knows about it now. Nobody'll ever forget it either."

Seeing Japanese begging, fighting for cigarette butts, offering their bodies for candy, and eating garbage was a shocking experience for most Nisei. It contradicted what they had heard throughout their lives about the majesty of the Japanese—inherited in their blood. Ideals such as self-reliance, fortitude, pride, and cleanliness were rudely shattered. As those going AWOL dispersed over greater distances, it became apparent that the people in the immediate vicinity of the depot were not on the verge of starvation, thanks largely to the huge quantities of American goods that they handled. Almost everyone, even small children, had substantial sums of money. Those who tried selling their wares in Tokyo discovered, however, that there were many who simply did not have enough money. Furthermore, as the result of several unpleasant transactions many Nisei began to realize that not all Japanese were honest, as their parents had insisted.

"Those Japs around here are sly. You sell them something, and they pay you in 10-yen bills. Before you can count 'em all they take off with the goods."

"Japs are treacherous all right. You can never tell. The one with the most sincere face is the one that's a bastard. You can't trust 'em."

It became increasingly apparent that the Japanese people were much like everyone else—no better, no worse.

Paradoxically, the very men who identified themselves with the Japanese "race" were the ones who treated them more harshly. Some of these men, especially those who had come from the segregated communities in California, identified themselves so completely that they remarked that it was too bad that such degrading things had to happen to people of "your own blood." They apparently regarded the protection of

Japanese women from Caucasians and especially from Negroes as one of their responsibilities. Yet, when they sold things on the black market, they haggled until they had squeezed out the highest possible price. They seemed to enjoy ordering about laborers, forcing them at times to perform unreasonable tasks. This did not pass unnoticed. As one man remarked, "Some of these guys have always been squawkin' about gettin' pushed around in the States. They heard about discrimination and all that crap from the time when they were kids. Now this is the first time they're 'King Shit.' They can push guys around, and the other guys can't do a fuckin' thing. So now they take it out on the Boochies."

Deep sympathy for the plight of the Japanese people came not from these men but from the more assimilated Christian Nisei. It came from those who had always regarded America as their home, who had never felt any loyalty toward Japan, and who resented any implication that they had. These men, as well as many Caucasians and Negroes, felt sorry for the Japanese not so much because they were Japanese but because of their conviction that human beings should not be subjected to such suffering and indignity.

Contention with the Depot Cadre

As the men became better oriented to life in Zama, it became increasingly apparent that the colonel was mistaken about their already having been assigned. They made preparations for a longer stay. They took more clothing out of their duffle bags and made plans for washing their dirty fatigue clothes and underwear. This turned out to be a problem, for hot water and soap were not readily available. Only a few buckets could be obtained, even though some had gone to all the neighboring companies to steal as many as they could find. The men also gave one another haircuts. There had been no barber on the U.S.S. *General Weigel*; the shop at the main depot was small and crowded.

All the returning veterans were excused from duty. For them there was nothing to do but to wait for shipping space. They whiled away their time at various entertainment facilities at the main depot. Replacements were assigned almost daily to various tasks. All KPs for both the main depot and the annex were drawn from replacements. In addition, a few were assigned to guard duty. The rest were sent to the main depot to assist the cadre there. Most Nisei were assigned as supervisors of labor gangs of Japanese—cleaning barracks, tents, and buildings; doing construction work; digging ditches; and sorting equipment and supplies. Very few of the tasks required much effort. Guards had to walk their

PLATE 9 *Annex of the 4th Replacement Depot*

beats, and the KPs were the only ones who were actually extended from time to time.

In a few days the Nisei realized that a duty roster was kept only for KPs; their names were sent to the mess sergeants. All other duties were assigned on a random basis; men in formation were simply counted off, and the required number were sent to the various stations. This penalized those who appeared in formation regularly, and they soon began to "disappear" in large numbers. A daily game began. Knowing that work formation was at 07:30, many arranged not to be there. They timed their breakfast to linger in the dining room beyond that time. Some went to the latrine; others hid in the barracks; still others went to the shower building. In time additional hiding places were discovered. By the end of the first week most men were no longer concerned. Even if they were caught, they could jump off the trucks on the way to the main depot. On the morning of January 29, for example, almost a hundred Nisei boarded the trucks, but not more than seventy reached the depot. Each time a truck slowed down for a sharp curve several jumped off. When the trucks arrived at the main depot, cadremen were not on hand to make immediate assignments; before they arrived more than half the seventy had departed. Several more sneaked off while being marched to their place of work. Only a few were unable to escape; they worked all morning and then failed to reappear after lunch. When asked for their names, most of them gave the names of cities such as Osaka, Sendai, or Fukuoka. They laughed at the stupidity of cadremen who did not have enough sense to examine their "dog-tags." When some cadremen tried to check their charges more closely, the men gave false serial numbers. On one occasion, when a man was called by his serial number, he discovered that someone else had inadvertently invented his number! Whenever such embarrassing incidents occurred, everyone assumed an air of innocence. One man commented privately about Nisei troops: "If a man is really GI and means it, the Boochies won't fuck with him because they don't want to catch shit. But as long as they can get away with it, they'll fuck off by the numbers."

Especially at the main depot those who disappeared were difficult to locate. Since most soldiers wore fatigue uniforms, it was impossible to tell the veterans from the replacements, and to most Caucasian sergeants all Nisei looked alike. A cadreman could not walk up to a Nisei in the PX or post theater and start yelling at him; he might turn out to be a combat veteran who would take serious offense. Furthermore, most of the cadremen were fresh from the United States; on the whole they were friendly to interpreters, on whom they had to rely so heavily.

Tuesday, February 12, was a day of special confusion for the cadre. The entire company was scheduled for work. However, only 60—including 20 Caucasians—appeared at the 07:30 formation. Only 40 Nisei out of 186! The others could not be found. By the time the group arrived at the main depot only 50 were left. Most AWOLs stayed away from the annex, for they knew that no Nisei were supposed to be there that day. One group that arrived at the main depot simply lined up in a military formation behind a tall Nisei wearing khakis and marched off toward a warehouse. Cadremen, thinking that one of their colleagues was leading the group to work, did not interfere. Large numbers of Nisei congregated at the gymnasium and in the PX. Others spent the day at the Red Cross, the barber shop, and the post theater until 15:30, when they appeared at the entrucking point for a ride back to the annex.

A few days later one group experienced some difficulty in escaping. They were assigned to a company where the sergeant had gotten tired of having men vanish after a few moments of work. He therefore demanded the names of all those present, and the person he questioned unthinkingly provided the real names. When the others learned what he had done, they tormented him mercilessly; he was forced to do most of the work that morning, as punishment for his stupidity. At noon the men decided against going to lunch. They waited until all the cadre had left and then searched the desk in the orderly room for the slip containing their names. Some Japanese laborers who were taking a lunch break nearby looked on in dismay as the Nisei took the sheet and threw it into the stove.

No one, not even when serving as a KP, felt that he was being overworked. In most companies in the main depot Nisei were forbidden to do physical work. They were only to supervise the Japanese. On some occasions as many as five or six Americans—Caucasian and Nisei—would stand about supervising the work of a lone Japanese. Although the cadre-men had become accustomed to this, several Nisei complained that it was embarrassing. What they found most objectionable, however, was wasting time doing tasks that they felt were completely unnecessary. After a hall had been swept and mopped four times it seemed pointless to do it again, even though a full hour remained before break time. Some Nisei in charge of work crews simply took their men into a room, opened some K rations, smoked, and talked or slept. Some of the laborers were delighted, but others became bored and restless.

The Japanese laborers also found the work ridiculous, and most of them went through the motions listlessly. Most Nisei were surprised at the lethargy of the Japanese; they had always thought of them as industrious. One of them was quoted as saying, "In Japan men work hard, finish their task, and then they take a rest. Americans want you to move even when there is nothing to do. Therefore, it is stupid to work fast or well. That is why everyone works slowly." This made sense. But some insisted that these men were even lazier than their comrades in Company K.

"One bar of GI chocolate is good for all your laundry, 15 yen, or a piece of ass. Which would you take, Willie?"

"Laundry. Laundry's a lot of work. I'd rather jack off and get my laundry done."

"You lazy bastard! I think we were too hasty when we said those Boochies were lazier than you."

On Thursday, January 31, after six straight days of not being able to muster enough replacements to fill work quotas, the ranking sergeant of the 275th Company addressed the group upstairs: "You guys have been fuckin' around too much. I'm givin' you last warning right now. You must be in formation every morning—all of you. There'll be no more of this giving fake names shit, or you'll be digging 6-by-6s along with the Japs. If any more of you guys get caught goofin' off, you'll get permanent KP. And remember, when I blow that whistle, you guys come running down them stairs *on the double*. We won't come up here any more to call you."

The men chuckled. The sergeant must have been reprimanded. Before he could finish, someone downstairs yelled that mail had arrived. Mail! The men had heard nothing from home since leaving Seattle. As they stampeded out of the room, the sergeant was brushed aside and almost trampled.

On the following morning the sergeant blew his whistle at 09:00. He blew it again a minute later. Not a single Nisei fell out; they had decided to call the sergeant's bluff. About a fourth of them were still in bed, and the others aroused them and helped them dress. As the whistle blew for the third time, one man jumped up jokingly: "Is that the whistle?"

"Fuck 'em! Wait 'til the son of a bitch comes up here after us."

At 09:15 a corporal ran upstairs, and the sergeant went into the downstairs room to order the Nisei out. The men took their time, dragging their feet sluggishly. In a loud voice one of them mimicked the sergeant: "When I blow that whistle, I want you guys to come runnin' down them stairs *on the double*!" Everyone laughed. They continued to taunt the cadre.

"Let's all dig 6-by-6s."

"Why don't we bury some chicken shit sergeant in one?"

"Yeah. It's waste time just to dig a big hole."

Or they said things in Japanese and smiled knowingly at one another.

Another question came up. This was the second day in a row that all Nisei had been called for work details. The Negro troops had been mustered every day since their arrival. Why was it that the Caucasians were called so infrequently? No one objected to combat veterans' being excused; they deserved a rest. But why should Caucasian replacements be excused from work? Grumbling began in the ranks: "How come the Haoles don't pull shit detail? Only us and the *Kurombos*. We always get fucked." A rumor that Caucasian casuals in another company had received passes intensified the resentment. Soon men were yelling from ranks: "How come the white boys in this outfit don't pull shit detail? How come only us and the colored boys have to pull all the shit?"

The Negroes, who had been observing the encounter from their formation with considerable glee, became even more interested. As the complaints became more insistent, a corporal was sent in to order the Caucasians to join the others.

Although the men continued to take pride in their ability to "goof off," as in similar situations in the past limitations were placed on such activities. Certain duties had to be performed. A respectable minimum had to stand formation each day, and it was understood that everyone should take turns making an appearance. Those who hid one day generally reported for work on the next. On February 7, for example, when it was clear that too many were absent, roll call was taken. As friends began answering for those who were absent, the sergeant started asking for each man's last four serial numbers. This provoked much laughter, partly because this particular sergeant was well liked but even more because many hoped that those who had been "goldbricking" too often would be caught. Almost a hundred Nisei boarded trucks for the main depot that morning, but many jumped off on the way. On arrival many ran off in different directions, and the cadre could not catch them all. Finally, a sergeant took the names and serial numbers of the fifty-six Nisei who were trapped and then announced, "O.K., men. I'm sending back these names as those who showed up for work. After you leave here, I don't care where you go. I don't blame you at all." Other sympathetic cadremen came to be identified as "nice guys," and a special effort was made not to get them into trouble. The mess sergeant in the neighboring 277th Company was popular, as was the mess sergeant in the kitchen near the main depot PX, who each day fed those he knew were supposed to be on duty elsewhere.

Complaints against the 275th Company mess hall continued to mount. The men grumbled constantly of poorly prepared food, short rations, the sour disposition of the mess sergeant, and most of all of having to wait so long in line. After serving as KPs in other kitchens, they knew that none of the other mess halls were run so inefficiently. Whenever they were assigned there as KPs, they invariably had trouble with the mess sergeant. He was always concerned with running out of food and ordered the KPs not to serve too much. But the men ignored the orders, dishing out generous portions especially to other Nisei but also to Caucasians and Negroes as well.

"Why do we always get the fucked-up mess halls? Goddam! Beans, hash, stew, wienies, macaroni—all cooked like shit. Even the chicken on Sunday tastes like hell."

"Compared to guys in combat it's not bad."

"C rations are better than this shit. Anyway, that ain't the point. All the other mess halls serve better chow."

"That's true. Look at 277. They serve 2,500 guys with 12 KPs. Our mess gets 30 KPs and serves one-third that many. It's all fucked up."

"It's always the same. At Snelling all the other guys get Boochie cooks, and we have to eat in their dirty, fuckin' Consolidated Mess. But this place is worse than Consolidated—and that's really bad."

The difficulties were compounded on February 1, when an officer was assigned to supervise the work there. The regular mess personnel obviously disliked him, and the KPs found him offensive. On the second day he was on duty the KPs were drawn from Company K. Since some of those assigned were among the laziest in the unit, several others, who were off duty, walked over to the mess hall to watch them. They sneered as the officer yelled "Hubba Hubba!" at the Japanese laborers. "Well, I guess we can't call Nakasone Simon Legree anymore." The KPs performed sluggishly. Loudly, without mentioning names, they mocked the officer. "Mah word's as good as God's. When ah speaks, yo' all snap shit. Ha, ha, ha, ha, ha!" When the lieutenant ordered Nakasone, who was one of the KPs, to sweep water from the lower to the higher end of a trough, they doubled over in laughter. "Christ, the guy's dumber than he looks. We overestimated him. He don't even know about the law of gravity."

Since there was obviously not enough work for thirty KPs and fifteen Japanese laborers, the lieutenant finally excused a few men until 10:45. Almost everyone left, and as they did so they helped themselves to apples, bread, and cookies. One detail was ordered to clear off a shelf before leaving. The men took the kitchen clock along with the scrap items and threw it into the garbage can. "If they want to be GI, we can obey orders too."

Later that day other difficulties ensued. When one man was ordered to pull a portable sink from one room to another, he retorted, "Anybody with common sense can see it won't go through." When the mess sergeant ordered another man to scrub the pots, he was rebuffed, "Fuck you! I'm doin' something else."

The major complaint was having to stand in line for so long for such poor food. On February 7, for example, the company was called out for supper at 16:00. Although it was windy and snowing, the cadre refused to let the troops return to the barracks, giving them a choice of standing in line or not eating. It was not until 16:35, when everyone was thoroughly drenched, that the first Nisei reached the mess hall door. The long wait in the slush and driving snow was enough to anger most men, but what they received when they finally arrived infuriated them.

"What the fuck is this?"

"They say it's baked macaroni."

"Baked macaroni, huh? Well I'll be a son of a bitch. It looks like mashed potato and tastes like burned hot cakes. If it wasn't for them three peas down there, I couldn't tell where the macaroni leaves off and the pudding begins."

"What pudding? I just got a lot of crap."

"No shit, this ain't got no taste. A half hour in the fuckin' snow for this shit. What the hell is this anyway?"

"They must've collected the pudding from the last short-arm inspection."

"I'm hungry, goddammit, and I'd eat shit if you put pepper on it, but not this."

Individually and in groups the men got up and walked to the garbage can. Without eating, they threw the food away and scrubbed out their mess gear.

The most widely accepted explanation of the chronic food shortage at the 275th Company mess hall was that the cadre was selling food on the black market. Since supplies were sent to each mess hall on the basis of the number of soldiers assigned to the unit that day, there seemed to be no other reasonable answer. Having worked at the warehouse, everyone knew that there was no shortage of food there, and none of the other mess halls had such problems. The men recalled the mess sergeant's offering ten pounds of wieners to a Japanese girl to spend the night with him, and they concluded that he must be selling a substantial portion of his supplies. This belief was consistent with conclusions that many had formed of the company cadre. They seemed unscrupulous, selling anything—food, drinks, clothing, equipment, money, Red Cross supplies, and whatever the returning veterans had left behind. Few objected to their making some money, but they agreed that it should not be at the expense of the troops.

Since the PX was not open late enough, those who were hungry had to depend on KP duty. One man complained: "What a hell of a setup. You gotta work to get somethin' to eat!" Before long it became a common practice for those who were on KP to steal enough food to feed their friends. In some instances the mess sergeants gave them leftover food; in others, however, they just helped themselves to anything edible that could be carried.

On the night of February 7 about thirty men were called out for night KP. As they departed, they were encouraged by their fellows: "KP for you guys tonight, huh? *Shik-kari nusunde koi zo*! [Swipe a lot!]"

"What a hell of a way to send a guy off to work."

About 22:00 a dozen of the KPs returned laden with eggs, wieners, sugar, fruits, bread, and butter. They had to return to work. Knowing that very few had eaten supper, they brought the food for them. The KPs themselves were having their own party at the mess hall.

About 21:30 the following evening some KPs returned to the barracks with cans of corned beef hash, bread, and butter. A few moments later others came in with more bread and a huge chunk of meat. It was so large that the bearers staggered under the load. The meat was frozen and had to be hacked into smaller pieces and then thawed.

"We were gonna take the whole fuckin' case [of hash], but it was too heavy. We figured we'd attract too much attention carryin' it out; so you guys'll have to be satisfied with this."

"Jesus Christ! Nothin' chicken shit about you guys, is there? Why didn't you bring the fuckin' refrigerator along while you were at it?"

"Fuck you! Do ya want'a eat or don't ya?"

"Holy Christ! How in the hell we gonna eat all that?"

"Get the guys out'a bed. We need more guys to eat."

Almost everyone shared the toasted hash sandwiches and steaks.

"I hate to deprive some other guys of this chow, but goddammit I gotta eat. If them fuckin' mess sergeants keep givin' away food to get pieces of ass every night, nobody'll get chow pretty soon. Ten pounds of wienies for a piece of ass is too much. Them fuckers give away or sell food so we starve. We starve so we swipe stuff. So tomorrow somebody else has to starve. He goes on KP and swipes more stuff."

"Yeah. That's the way it goes. Hey, shall I go downstairs and tell some more guys to come up?"

"We eat good tomorrow. Everybody's on KP."

As the men became more familiar with the neighboring mess halls, they made a variety of arrangements to assure an adequate food supply. Anyone who missed a meal in one mess hall could eat elsewhere, where someone in the company was on KP.

As the men began cooking steaks and other food on the small stoves, several difficulties arose. The meat could be smelled from some distance; any inspecting officer would be able to trace it easily. Furthermore, a minor epidemic of diarrhea swept the company. Victims of the "GI dribbles" received much sympathy; each had to run all the way outside, around the building to the rear, into a cold and dirty latrine. Although some blamed the diarrhea on the unsanitary conditions in the local mess hall, others pointed out that it was impossible to wash off in cold water the grease left on utensils that had been used to fry meat in butter. Indeed, the water at the latrine was so cold that the grease simply froze and had to be scraped off. In addition, the stoves became dangerously hot, and the rooms sometimes became fogged with smoke. For some time the men had felt uncomfortable about the safety of the building. It was made of thin, dried wood, and they realized that in a flash fire they could be incinerated in a few moments. Those billeted upstairs were especially concerned; there was only one stairway, and it was narrow. Thus, when the sergeant, pointing out the fire hazard, requested politely one day that there be no more cooking on the stoves, the men consented. Some discussion had already started among them, for many had begun to feel that the home cooking was going too far.

The miseries of the 275th Company cadre were compounded on Monday, February 4, by the arrival of a large detachment of Caucasian paratroopers. In general, paratroopers were addressed with respect; they were supposed to be tough and well versed in "dirty" fighting. Before long a rumor developed that these men had had trouble with some Hawaiians at Fort Ord, and the mainlanders were delighted to hear that some Hawaiians had been thrashed in a gang fight.

"I just talked to one of the paratroopers. He wanted to know if we were from Hawaii, but I told him we weren't. He said they had a lot of trouble with the Hawaiians at Fort Ord just before they shipped out."

"I hear one guy got stabbed and so two hundred of them paratroopers went over to the Kanaka barracks and beat the hell out of 'em."

"They used their own trick on the Hawaiians, huh? Good! I'm glad they beat the shit out of them."

"I wonder if it was our boys."

"Could be! They were waitin' for their boat about the same time. The Kanakas here said they just got to Hawaii."

"I thought our boys went to L.A."

"That's what I heard. But who else could've been at Fort Ord when the paratroopers were there? G and H Company guys were ahead of us, and they're already here. L Company's just graduating about now. It must have been our boys."

"I hope Takeda and Tajima and them guys got theirs."

"Aw, them yellow guys were probably hiding. They start all the trouble, but the other guys have to do the fighting. Them guys are no fuckin' good."

"Too bad. Some of them Kanakas are nice guys."

"If the Hawaiians here hear about it, they might have more trouble."

"Let 'em. They're plenty of paratroopers here."

"This company won't back up the Hawaiians."

"Hell, no. Fuck 'em."

Members of the two units got together in the mess line on Tuesday afternoon and exchanged information. They got along easily, for they were of kindred spirits. Both groups disliked Hawaiians; both took pride in their virility; neither had any respect for the cadre or for regulations. Both realized that by joining forces they could easily terrorize the local cadre.

On the following day the company was called out at 10:35 for lunch. After standing patiently in the cold wind until 11:00, many became restless. A few isolated men started yelling for the mess personnel to open the door. Then, someone started to rattle his mess gear, and almost at once everyone up and down a line of over 150 yards followed suit. As the din of jangling mess kits reached a crescendo, someone picked up a rock and threw it on the mess hall roof. Immediately several others picked up rocks, and stones of ever increasing size began to shower on the mess hall.

"Let's go! Let's go!"

"What the fuck're we waitin' for?"

"Open the goddam door!"

No one in the mess hall dared to object to the stoning. Finally, one of the KPs opened the door. As the men went through the line, they cursed and derogated the mess sergeant and his regular staff. When the troops lined up for supper that evening, the afternoon performance was repeated. This time, after the first hail of rocks on the roof the mess hall was opened quickly. Some Japanese laborers, standing nearby awaiting transportation out of the camp, looked on in utter dismay.

A few days later, when the sergeant announced that meal tickets would be issued for the the 275th Company mess hall, the replacements were dumfounded. They agreed that the mess sergeant was flattering himself if he thought that unauthorized persons were sneaking into his mess hall.

Although military regulations were clearly stated, they simply could not be enforced under the conditions that prevailed in Zama. Black marketing was so commonplace that everyone counted his money in "G's"; anyone who did not have several thousand yen was considered poor. Soldiers went AWOL in droves. Individuals going over the mound were occasionally challenged by armed guards. After the arrival of the paratroopers, however, they went out together in such large numbers that the guards turned the other way. They walked down the streets in full view of MPs, who showed no inclination to challenge anyone. Those who went to Tokyo often stayed overnight; they knew they would not be missed, for there was no bedcheck or reveille. When firewood became scarce, someone would bring in a bench or some other furniture not immediately in use; someone else would bring out a meat cleaver stolen from a mess hall, and in a few moments they would produce kindling. The men had little concern over discipline. The penalty for missing a detail was another detail—if the man could be located. The digging of 6-by-6 holes was an empty threat that was never enforced. Cadremen who were strict were threatened with violence. Thus, ignoring or circumventing regulations became the rule rather than the exception.

The Dissolution of Company K

All officers and enlisted men classified as Asiatic Language Specialists were under the jurisdiction of the Allied Translator and Interpreter Service (ATIS). A few months after V-J Day ATIS headquarters had been moved from Manila to Tokyo—to the building that had once been the main office of the Nippon Yusen Kaisha, Japan's largest steamship company. The NYK building had miraculously survived the bombing, perhaps because of its proximity to the Imperial Palace. All linguists were processed through ATIS. Other units would send requisitions, indicating the kinds of skills required, and ATIS would send out qualified men.

Countless Nisei replacements had gone AWOL to Tokyo to visit friends at the NYK building, and from their accounts rumors developed of the facilities there. They were uniformly favorable. Enlisted men enjoyed quarters much like those of officers. Only five or six men were assigned to clean, hotel-like rooms that were steam heated. There were hot showers, porcelain wash basins, and flush toilets. Enlisted men, like officers, were given sheets, pillows, and mattresses. The building also contained a tailor shop, a laundry and dry-cleaning service, a barber shop, as well as a well-stocked PX. The tables in the dining hall for enlisted men were covered with white tablecloths; and the food served was excellent, for it was prepared by chefs who had worked on NYK ships. Passes were issued nightly, and a dance was held every Sunday night.

From Nisei friends assigned to the cadre at the main depot the men learned that the reason for their long stay at the annex was that the NYK building was filled to capacity. According to rumor, the personnel office at ATIS was having difficulty placing all the linguists available, and many competent interpreters had been assigned to routine jobs—as stenographers, medical orderlies, truck drivers, jeep drivers, and PX clerks. According to still another rumor, only the most competent were being assigned as linguists—interpreters as T/5s and translators as T/4s. According to still another rumor, anyone assigned to General Headquarters (GHQ) in Tokyo would have to remain an extra six months beyond his regular tour of duty; in spite of this, life was so luxurious in Tokyo that many capable linguists were purposely failing examinations—preferring to remain in Tokyo doing menial tasks rather than being sent to a unit in some outlying area. All this led to renewed resentment against the MISLS administration. Replacements had been rushed from Fort Snelling for thousands of miles, presumably because interpreters were needed so badly. Either the officials at Fort Snelling were unaware of the situation in Tokyo, or for reasons best known to themselves they persisted in recruiting and sending over men who were no longer needed. They had been told that Nisei linguists were so essential that discharge regulations applying to other enlisted personnel might not apply to them. And now they were sitting in Zama whiling away their time.

On February 9, two Nisei who had been stranded in Japan during the war came to visit a friend in the company. They said that they had volunteered for the U.S. Army to regain their American citizenship. When they confessed that they had been pro-Japan during the fighting, most men responded coldly. What enraged them much more, however, were rumors that "disloyal" Nisei who had been deported from the Tule Lake Segregation Center were now being employed by the U.S. Army. Since they were bilingual, they would obviously be useful, but their being accepted after what they had done was upsetting. Although these expatriates were not being hired at ATIS, they were working in other positions. They had apparently changed their minds about national loyalty; they were now claiming that they had been misled by agitators. Some of them were now fighting to regain their citizenship and were being represented by the American Civil Liberties Union. If they succeeded, they would become eligible for civil service jobs, receiving high pay for doing the same work as soldiers. A rumor that civil service interpreters were receiving $4,500 a year or better made everything seem even more unfair. Several Company K men had brothers and friends who had been killed or wounded in Italy or France; others had been in Tule Lake, where they had been threatened by those who were pro-Japan. These people would have the best of everything—escaping the perils of military service, receiving high civilian salaries for doing the same work as enlisted men, and eventually returning to the United States without penalty. They had ridiculed Nisei servicemen as "suckers," and now Caucasian liberals were reinstating them as innocent victims of "agitators." But the men knew that many of the expatriates *were* the agitators.

"Fuck them Tule Lake bastards! They fucked it up for the rest of us."

"They wanted to come here. Now let 'em make out what they can."

"They had the same choice as the guys in the 442nd who died over there. Fuck these bastards! Yellow-bellied Japs!"

"If I see any of them Tule Lake bastards, I'm gonna give 'em a bad time. I was in Tule, and I remember when they used to make speeches about how Japan was going to win the war and all that shit. One guy wanted to beat me up because I registered. If I see him, I'm gonna beat the shit out of him."

Even the more sober men were resentful: "If a guy wants to fight for Japan, that's his business. I don't give a shit about that. But when a cocksucker fights for Japan when he thinks Japan is going to win and then switches over to Uncle Sam just because we won the war, then I got no use for him. That's the way most of these bastards are."

"Yeah. I hear some of them Tule guys are trying to get their citizenship back now that America won the war. Fuck them bastards! Send 'em all over here."

"I don't care if a guy fights for his country. You can't help that. But a guy's gotta make up his fuckin' mind. You can't trust a guy that just goes where the wind blows. I got a lot more respect for a Boochie who fought for Japan and still wants to fight for Japan than I got for these cheap bastards who go around kissing the GI's ass now."

The most disturbing rumors were of possible assignments in Korea. Early in February it became known that a large contingent of Nisei at ATIS had been alerted for shipment to Korea. By common consent most American troops wanted to avoid duty there, and those in Company K concurred. Those who had been looking forward to a pleasant vacation in Japan found their daydreams rudely shattered. It was believed that Koreans hated all Americans—Caucasian, Negro, and Nisei. Furthermore, Korean women had been reported in the *Pacific Stars and Stripes* as being unfriendly to all foreigners. Other rumors were that no coal was available there and that some Nisei interpreters had been beaten for flirting with Korean girls. Concern over being sent there continued for several weeks. If not enough linguists were available in Tokyo to fill the quota, would it be necessary for some Company K men—who might be next in line—to fill in? If Nisei were being held in Zama until space became available in Tokyo, the faster their predecessors were sent to Korea—or anywhere—the sooner the group would be able to move to the NYK building. But most men agreed that they preferred remaining in Zama a bit longer to risking being transferred outside of Japan.

On January 29 a large contingent of Hawaiians had arrived at the annex. At first some feared that their old comrades at Fort Snelling had caught up with them, even though they knew that it was too early for that. It was soon established, however, that they were from Companies G and H. Those who disliked Hawaiians were tormented by the others: "O.K., watch your rough talk from now on." What disturbed many, however, was that Hawaiians were called in for classification tests only two days after their arrival—ahead of Company K men who had arrived earlier. Since they were graduates of the "regular" school, they were presumably more competent, even though they had attended classes for the same length of time. Did this mean that graduates of the oral school were to be discriminated against once more? Did it mean that the Hawaiians would be going to Tokyo ahead of them? If so, would many of them be sent to Korea? Did it mean that Company K personnel would go directly to some division without stopping at ATIS at all?

At 15:00 on February 1 replacements from Company K were ordered to their classification examination. The tests were conducted by a team of officers—Caucasian and Nisei—and a number of Nisei veterans bedecked with campaign ribbons. Only a few tried hard to score well. When they returned to their friends, they insisted that they had failed deliberately, but the others no longer believed them. Those who had been labeled "eager beavers" at Fort Snelling were no longer trusted. So few were qualified that the exercise soon became a farce. By the time the examiners reached the second half of the alphabet, they had given up hope and were asking most questions in English rather than in Japanese. They went through the routine of asking for a reading of a *kanji* text, but they were surprised when anyone recognized anything. Most men answered the simpler questions; after all, the examiners knew that they had graduated from some kind of school.

During a general meeting after the examination the officer in charge was asked why the group had to remain in Zama for so long. He replied that one reason for the delay was a controversy between ATIS and 8th Army Headquarters over who had jurisdiction over linguists. He also confirmed the rumor that so many interpreters were available that only the best would be used and that all others would be given nonlinguistic assignments.

On the following day the CIC men who had been on the transport were alerted for assignment in Korea. A list containing their names and those of a few Caucasians was posted on the bulletin board. They received little sympathy. They had held themselves aloof from Company K Nisei; only a few knew any of them by name.

"Korea, huh? That's tough shit."

"They're a bunch of snotty bastards anyway. They got it comin' to 'em."

Two days later the cadre announced that the room downstairs in which the alerted men were billeted had been quarantined. Someone had contracted measles. When warned to keep out so that the entire company would not be restricted, this became the object of joking.

"As long as I don't go to Korea I don't give a damn. I'd rather spend an extra six months in Tokyo than go to Korea."

"I guess them CIC guys are tryin' to get out of goin' there too."

"Yeah, but gettin' measles! That's goin' too far."

Wednesday morning, February 13, was bright, crisp, and frosty. It was getting warmer, and most men were out of bed early. Even those who had gone AWOL and returned late were up. "Boy, I haven't seen a nice day like this one since I left California—not since the spring of '41. Nice, huh?" Men were singing and humming as they went to wash:

Oh, what a beautiful mornin',
Oh, what a beautiful day;
I've got a beautiful feelin'
Everything's comin' my way.[1]

"Did you go to chow?"

1. From "Oh, What a Beautiful Mornin'" by Richard Rodgers and Oscar Hammerstein II. Copyright, 1943, by Williamson Music Inc.

"Hell, no! I feel good. I don't want to spoil it. Every time I go near that fuckin' mess hall, I get pissed off."

"Bacon and eggs this morning, hey."

"No shit!"

"I ain't shittin' ya."

"What happened? Mac[2] here on an inspection?"

Even the sergeant was smiling and told those washing up near him in the latrine, "You guys might ship out tomorrow because there's a hell of a lot of casuals comin' in this week, and there ain't no room for 'em unless you guys leave. So tell your buddies to hang around. We don't give a shit if you wanna go out and knock off a piece now and then, but for Christ's sakes don't be AWOL when you ship out. You tell the other guys, huh?"

The word spread quickly.

Late in the morning a few men were ordered to guard the storehouse. They were issued carbines and live ammunition—the first time any Nisei in the unit had been permitted to carry loaded weapons. This led to further rumors; one was that the storehouse had been raided the night before—by Negroes. This was greeted with skepticism: "Niggers, shit! I wouldn't be surprised if some of our boys were in there. I wouldn't hand them buggers any carbines. They might take the whole fuckin' storehouse, building and all."

As they sat about the barracks talking, word spread that many Hawaiians were assembling in front of the building with all their equipment. Since several of them claimed that they were going to the NYK building, the men watched with envy.

"Remember at Meade? They needed infantrymen so bad they were flyin' 'em over. But we sat on our fannies there waitin' and waitin'. All the Haole guys come in and left in one or two days. We waited for three fuckin' months, and the goddam war ended. Remember? Shit, we got two and a half months to sit out in this place yet."

As the first part of the Hawaiian group boarded trucks, one man lamented, "Goddam, we're really nobody's children—orphans of the fuckin' war. Nobody wants us."

"First ones here; last ones to leave. Just like Meade. Nobody wants us. We just wait and wait and pull shit detail."

"Why don't them pricks at ATIS release the rest of us for other assignments? I'm gettin' sick of this casual crap."

2. General Douglas MacArthur.

"I think Tokyo's the best deal. If we go out to the divisions, we'll probably be takin' basic training all over again."

"That's better than this shit. At least it'll do your health some good."

"I figure we ought'a be pullin' out pretty soon. I'm gettin' tired of this fuckin' place. It's too goddam dirty. Look at all the guys who had to go on sick call this morning. It's a good thing these guys are rugged, or everybody'd be sick. Look at this fuckin' room! Jesus Christ! This company's never been so sloppy."

"That's 'cause Nakasone ain't in charge anymore."

"Oh Christ! If Nakasone was in charge here, these bunks would be lined up straight as an arrow and breathin' in cadence."

Someone called attention to an article about Nisei interpreters in the *Pacific Stars and Stripes* in which a high official was quoted as saying that "quality, not quantity" was needed.

"That's no shit. I guess that leaves us out."

"Why don't we face the facts? They don't want us at the NYK. All those guys heard about Company K. They're afraid of us. You ought'a see the looks on their faces when ya tell 'em you're from Company K."

From sundown until about 19:00 the lights kept going out, and this elicited suspicion: "What the fuck! Don't tell me these cadre bastards are selling electricity to the Boochies too!"

"Well, it looks like we're gonna go through the whole fuckin' war 'Assigned, Unattached' or 'TDY' [Temporary Duty]."

"We're liabilities, you know it? It must have cost a hell of a lot to train us—$2,500 for basic, more at Meade, and a hell ofa lot at Snelling. The trip must have cost a fortune—the ship and all that equipment and chow and everything. Then you gotta figure on all the expense of keepin' up our records and bookkeepin' and all that. Then they pay us on top of that, and we get allotments. It costs over a couple a hundred bucks a month per man to keep us here in this shithole. All we do is KP, and we wouldn't have to do that if we weren't here because they'd be nobody to eat. We ain't worth shit to the government. They're gettin' fucked on the deal, and so are we."

"The thing about this outfit is that the guys ain't dumb or lazy or anything. They just don't give a damn anymore. Everybody's lost fight."

"That's just it. All the guys could be on the ball, and they will be on the ball when the blue chips're on the table."

"If they treat us good, I hate to fuck 'em up. But when I get one of them chicken shit bastards, I don't give a shit any more."

"I heard we might go as a unit to the MPs."

"Yeah? MP's about all we're good for."

"If we can all be together, I'd rather go to the MPs than get a soft job someplace else. I got used to this fuck-up outfit."

"You know, a lot of us guys've been together ever since we got drafted at Fort Sheridan back in '44. We got split up and took basic in different camps, but that was the only time we were ever on the ball. We all got together again at Meade when we got sent to A-9-3. There we really fucked up. We fucked up so much that they transferred the captain away. Then at Snelling we were all in Company K. We really fucked up there too. Now we're really fuckin' up again. We been spoiled. We got to the point where any goddam order is chicken shit. Sometimes they gotta give orders, but we call it 'chicken shit.' Nobody gives a shit anymore. Just 'go for broke' now."

At 18:15 a sergeant entered the room and announced: "Don't any of you guys go AWOL. You leave at 09:00 tomorrow."

The men yelled and jumped and pushed each other. "Hallelujah! Bivouac's over!"

"Hey, Sarge, do we all go?"

"Practically all of you guys are goin'. Wait a minute. Yeah. All you guys are goin', and most of you'll go at 9 o'clock."

The lights went out six times between sundown and 20:30, but now no one seemed to mind. They yelled and danced. Each man who entered the barracks was told the news.

"Hey, start packin'. We pull out tomorrow."

"Don't shit me."

"No! No shit! We go!"

But many questions were still unanswered. The sergeant had intimated that several orders were involved. The Hawaiians had been split, and this unit was much larger than theirs. Was each group going to a different destination? Was anyone going to the NYK building? Some went to the orderly room where a copy of the orders was available, but they were chased out before they had a chance to study them. They did learn definitely, however, that the company was being broken up and that each group was going to a different destination. Rumors emerged immediately. The T's to the Y's were going to the 24th Infantry Division. The S's were going to the 11th Airborne Division; it was presently at Sendai, but it would soon be moving to Hokkaido. The N's and O's were going to the 1st Cavalry Division. Men rushed about excitedly with each new rumor they heard. Most of them tried to pack, but the lights continued going off every ten or fifteen minutes.

"I lose fight. How in the fuck can I pack with the damn lights goin' out like this?"

"Aw, go to bed."

"We're gonna be split up, huh?"

'Yeah, I know. But I hear all the S's are going together."

"Yeah. I'm glad they went down the roster that way."

Most of the men were pleased. Although the orders had not been posted, they had been alerted officially. They assumed that they would be assigned along with their close friends, for most buddies were those whose surnames were close together in the alphabet. Friends whose names were apparently not close enough were downcast, and the others expressed their sympathies. At 21:00 the lights went out for the last time, but many talked on excitedly well past midnight. Many of these men never saw one another again.

On the following morning everyone was out of bed early, and several even shaved painfully with ice water. A rumor that one man was AWOL aroused considerable indignation; even in Company K missing a shipping order was inexcusable. At 07:15 the orders were read. The names were not in alphabetical order. Everyone was puzzled. Those who were regarded as well qualified had been grouped with others who knew virtually no Japanese. College graduates were on several different orders. What was the rationale underlying the assignments? For some reason the sergeant refused to disclose the destination of the various shipping orders. Each man was told the number of the group to which he had been assigned and what time he was to be ready to leave. Those who managed to peek over the sergeant's shoulders revealed that many units were listed: 1st Cavalry Division, 11th Airborne Division, 24th Infantry Division, 25th Infantry Division; 105th Bombardier Wing, two MP battalions, and ATIS. Some of the orders also included Hawaiians who had not left on the previous day.

Unaccountably, virtually all cliques and fast friendships were broken. In one clique of six men each member was on a different order. "I don't see how they did it. They split up every one of the asshole buddies in the whole company." Since the various groups departed at different times, those who were leaving later helped friends pack, muttering about being separated and of not being able to pack the night before because of the lights. As trucks came to the front of the barracks, each man leaving in the early orders had several friends helping him carry his gear. No one cried, but there were fond farewells to buddies who had been together through so much agony for well over a year. They promised to hold reunions when they returned home. Home addresses were exchanged. Most of the men were glad to leave Zama, but they were saddened about being separated from their friends. Some said nothing after they boarded the trucks. Others tried to be gay. As the vehicles began to move, friends ran alongside for several feet, waving and yelling.

Analytic Summary and Discussion

Unlike most other replacements, who were still disciplined, Company K personnel were better equipped to fend for themselves in the general atmosphere of corruption and chaos that prevailed in Japan during the early months of the occupation. Where so many regulations were flouted, the men found themselves at home and worked together to get what they could for themselves. This time they encountered little opposition. Those responsible for enforcing regulations did not care enough to take the trouble, and once things got out of hand there was little they could do. Since so many of the cadre were themselves involved in illegal activities, they were vulnerable; they could not send others up for court martial without implicating themselves. What happened in Zama fell short of total anarchy. Token compliance occurred when an officer was watching. Tasks that were regarded as essential were performed, and informal norms emerged to set limits on what one could do in relation to the Japanese, the cadre, and to one another. The patterns that formed initially in A-9-3 found a congenial setting in Zama.

The men trained in Company K were the most undisciplined and least competent of all the linguists in occupied Japan; yet many of them actually served as interpreters. When the large emergency shipment from the "regular" school reached Japan late in the autumn of 1945, an abundance of linguistic talent was already available. Japanese nationals who spoke fluent English, civil service workers, and veterans who had volunteered for extra duty filled all the choice positions. Hence, ATIS was forced to assign many competent linguists to menial tasks. Less than fifty Company K men reached ATIS, but soon after their arrival a change in army policy created a sudden demand for interpreters. All enlisted men who had served twenty-four months or more were released at once, creating a vacuum. Thus, some of the least competent men ever graduated from MISLS found themselves assigned as interpreters at the War

Crimes Trials in the Philippines and in Japan, as interpreters for special educational missions, as CID agents working with the Tokyo Metropolitan Police, as well as interpreters for divisional headquarters in various parts of Japan. Once released from the informal pressures of Company K, however, most of them worked conscientiously. Once on a field assignment they could no longer plead ignorance. Many made up through self-study what they had not learned at Fort Snelling. When they *had* to learn Japanese, most of them somehow managed to do so. Thus, by an ironic twist of fate a number of these men drew some of the most challenging linguistic assignments in occupied Japan.

9. «» demoralization as a social process

Hundreds of Nisei servicemen in World War II were confronted by circumstances far more trying than those faced by the members of Company K, but most of them managed to adjust without excessive rancor. Why, then, did this group fall apart when the others did not? The answer is to be found in the particular beliefs that these men formed and in the behavior patterns that they developed informally and enforced among themselves. However bizarre their antics may appear to an outside observer, once the events are viewed through the eyes of the participants, most of them become readily comprehensible. Even when one does not approve of their conduct, he can understand how soldiers sharing such convictions would act as they did. What people in demoralized groups do becomes puzzling only when their deeds are seen from the outside. This suggests that there are ways of accounting for demoralization that differ from the explanations frequently given by those who do not appreciate the standpoint of the people involved.

How did Company K develop its characteristic style of performance? Although the disorders seemingly erupted with suddenness, it is clear from the record that the demoralization of this unit was a long drawn-out affair. Many of the behavior patterns originated in contexts of which the officials at Fort Snelling knew nothing. Techniques for coping with frustrating situations developed initially in A-9-3 as a reaction to incredible inefficiency and discrimination. Success in their retaliatory tactics against Captain Larsen led to the crystallization of action patterns that were used with increasing proficiency in all succeeding situations in which difficulties arose with authorities. On encountering inconveniences on the Turkey Farm, the handful who had been in A-9-3 reacted as they had in the past. When they were able to get away with it, the practices were adopted by the others. While those who went to the "regular" school had to mend their ways at once, Company K personnel retained the same mentality and stance. They subsequently met every similar situation in the same manner. In the end the men found themselves in occupied Japan in the midst of a demoralized army. Because of their past they found it relatively easy to cope with the disorganized conditions they encountered in Zama.

The history of Company K is unique; these unusual events are not likely to recur, certainly not in this particular combination. Nonetheless, the record can serve as the basis for formulating some tentative generalizations about a class of phenomena: the process of demoralization.

Qualitative Differences in Performance

This case study contains an account of a succession of transactions executed with varying degrees of effectiveness. Most of the activities were routine, but there were occasional instances of extraordinary peak performances, many of marked sluggishness, several in which concerted action broke down, and some in which deliberate obstruction took place. This case provides data from which hypotheses may be constructed concerning the conditions under which different types of performance develop. A comparison of the extreme instances discloses clearly the importance of primary group definitions and evaluations.

In spite of the dismal overall record of the units under observation, there were several instances of clearly effective performance—both in accomplishing the recognized goals of military organization and in obstruction. These achievements, important or trivial, occasioned surprise not only among observers but even among the men themselves. Although competent performance depends on individual effort, it is the *coordination* of diverse contributions that results in efficacious accomplishment. Thus, we must focus on the conditions under which men cooperate readily. The record reveals that efficient teamwork depends on the manner in which the participants evaluate a given transaction. The key criterion is what is regarded to be in the best interests of the members of contiguous primary groups. As Shils (1957) points out, most human beings are not so much concerned with abstract values and ideals but with what is concrete and near at hand. They are more concerned with the specific people with whom they are in daily contact and with maintaining their standing in their eyes.

This does not mean that lofty ideals are irrelevant; they become important only when members of primary groups accept them and expect one another to act in terms of them. Thus, when members of a local unit believe that a given transaction is a direct contribution to a larger cause that they regard as worthwhile, execution is likely to be proficient. To be sure, grand causes are vaguely conceived and often symbolized in slogans and stereotypes; but they do matter. Although patriotic appeals regularly elicited derision, most American soldiers in World War II—unlike those in the Vietnam conflict—assumed that their cause was a just one; they saw the fighting as ugly and distasteful but something that had to be done. Furthermore, they were convinced not only of the rectitude of their cause but took it for granted that in the end the United States would win. When discussing their postwar plans, for example, all soldiers placed their fantasies within a victorious homeland. Virtually all Nisei soldiers shared these assumptions, and any transaction that was defined as a direct contribution to winning the war was evaluated as important. Another abstract value that was presupposed was the importance of survival. Thus, at Fort Meade there was little objection to any exercise in which the replacements felt that they actually learned something that would make them more proficient riflemen—not only to contribute to the war effort but also to stay alive on the battlefield. Other ideals, such as humanitarian concern with the destitute, were of importance only to some, and in Zama the men who shared them joined forces to get leftover food to the Japanese near the boundaries of the camp—at a time when they mistakenly believed that they might be shot for doing so. Obviously there must be a high degree of consensus that the larger cause is indeed a worthy one, and the connection between the goal and the transaction must be clear and plausible.

Another important cause was securing the position of the Nisei in the United States. Members of the 442nd Regimental Combat Team fought zealously for it, and other Nisei soldiers applauded and supported their effort. It was clearly understood that Nisei should conduct themselves in an acceptable manner to earn the right to remain in the United States. Furthermore, it was assumed that all Nisei—other than those who had elected to go to the Tule Lake Segregation Center—expected each person to do his share in upholding the reputation of the group. Although some objected that zealots were going to unnecessary lengths to make their point, even they complied with this expectation. In any transaction in which the men were acutely aware of their ethnic identity, especially when they perceived themselves as Nisei who were on public display, their performance became strikingly adept—regardless of the context. Even after many felt that the battle record of the combat team had largely accomplished this end, they continued to contribute to anything that could enhance the reputation of their group. When boarding the troopship at Seattle, for example, every Nisei carried his huge load all the way from the dock to his compartment without stopping, even though several almost fainted from exhaustion. The men wanted to put on an impressive showing, and no Nisei dared to fail. Had representatives of A-9-3 actually been permitted to march in the funeral parade for President Roosevelt, they would probably have put on a formidable performance.

Even when no lofty goal is involved, performance is likely to be competent when there is agreement that a transaction will bring benefits of some sort to members of the primary group itself. On the night of May 29, when B-9-3 was ordered through the jungle for the third time, the cooperation was extraordinary. Everyone was tired, and no one wanted to participate. Once trapped, however, the men realized that they had to work together as a team or their ordeal would be prolonged; furthermore, serious injuries could occur. Conscientious effort was viewed as a practical necessity. At Fort Snelling, when Company K men first saw the filthy barracks to which they had been assigned, everyone pitched in to clean them. The same thing happened on the first day at Fort Lawton and at Zama. This was obviously beneficial to all component primary groups.

If a transaction is defined as bringing advantages to parties with whom many members of primary groups identify, performance is likely to be effective. At Fort Meade most men in A-9-3 worked diligently to protect the Nisei NCOs whom Captain Larsen had threatened to demote. All peak performances in Company K were for Lieutenant Catalano. Numerous studies in industrial sociology have shown that greater productivity is achieved by supervisors who are people-oriented and permissive than by those who are task-oriented and authoritarian (Viteles, 1953:148–62; Blau and Scott, 1962:140–64). However, the actual personality of the commanding officer is not as important as the manner in which he is defined—the traits and motives that are imputed to him by his subordinates. Many members of Company K genuinely liked Lieutenant Catalano as a person; others felt a deep debt (*giri*) to him. If a leader is defined as someone who is concerned with their well being, his men will make many sacrifices for him. Company K performed impressively in each of the weekly parades, even though the men detested them. The extent to which they exerted themselves for headquarters inspections astonished all the inspecting officers. These transactions were regarded as ways of repaying the lieutenant for everything he had done for them. They felt that their conduct had hurt his military career, and they wanted to make him "look good" in the eyes of higher authorities. Very few wanted to attend the party at the Hotel Radisson; they made the effort only for the "old man."

What is decisive, then, is the extent to which the successful completion of a particular transaction, whether in support of organizational goals or not, is regarded as being in the best interests of the local unit. Once a task is defined as desirable or necessary, a diffuse social excitement tends to envelop the participants, much like an atmosphere, and gives direction to the attitudes of those who are present. Efficient coordination results from the willingness of the participants to subordinate purely personal interests to a common purpose. It is the product of selfless giving—the readiness of most participants to do their share even when it involves considerable sacrifice. In the instances cited the men did whatever they felt was necessary to bring their tasks to a successful completion. There was little concern over receiving personal credit for one's contributions, and they helped one another without even being asked. They also showed considerable flexibility, meeting with ingenuity the exigencies of changing circumstances. Efficiency arises from intelligent teamwork and frictionless acceptance of a division of labor. This is not so much a surrender of self-direction, but the free adoption of primary group goals and the identification of one's interests with those of the team. Most participants willingly, even eagerly, do whatever they can to accomplish the goal, and the few laggards are placed under heavy pressure to do their share.

If a group develops a record of consistently effective performance, its members place a high evaluation on the unit. Once this happens, efficiency tends to be self-perpetuating, for upholding the reputation of the unit becomes an additional objective within its component primary groups. It is such steady, reliable performance that is commonly called "high morale." The term refers to a sustained orientation, not to a single efficacious performance. The 100th Infantry Battalion and the vanguard of Nisei linguists in the Pacific theater were so characterized. Even in Company K competent performance in dress parades was consistent. Although it was primarily for Lieutenant Catalano, the men were proud of their infantry background, for this enabled them to dissociate themselves further from the MISLS. It was one of the few things of which they were not ashamed. Thus, the development of high morale is a cumulative process. If informal norms result in exceptional achievement, the men become aware of their accomplishment. As they develop pride in their record, maintaining that level of performance becomes defined as a worthy objective. Thus, the members expect one another to make the necessary sacrifices so that the reputation of the unit will not be tarnished. Such expectations crystallize and are enforced. In time it becomes taken for granted that everyone will do his best. In the last analysis, then, high morale becomes a matter of how the participants conceive of themselves. It rests on their regularly defining a succession of transactions as worth doing well, for it would be beneath their dignity to do otherwise.

Examination of the numerous instances of conspicuously sluggish performance reveals again that the men acted in accordance with primary group evaluations. When a transaction is unrelated to any worthy cause and when there is serious doubt that anyone in the local unit would derive any benefit from it, it is defined as a waste of time and effort. Especially where there is suspicion of corruption among leaders, as with the ranking MISLS officials, serious doubts arise over whether the unit is working for the avowed goals of the organization or merely being used to further someone's selfish interests. Most lethargic performances and failures result from the reluctance of the participants to subordinate their personal interests to a common purpose. When a transaction is evaluated as an unnecessary nuisance, the participants appear ridiculous in their own eyes. They let their friends know how they feel—that they would prefer to be doing something else that is more fruitful. New ways may be devised to pursue private concerns or amusements. Those who discover effective ways of disobeying orders without being caught are much admired; some may even become heroes. Furthermore, when separate individuals and cliques regard their respective interests as paramount, schisms are likely to develop. Each comes to feel that he must look after his own affairs and those of his close friends, and they make their own private arrangements. The few who remain committed to organizational goals are tormented as "suckers." They become objects of pity, if not of scorn or violence. As disorganization becomes widespread, even fear of punishment is muted by recognition of the impracticality of arresting everyone. It must be emphasized that the ineffectual execution of transactions does not result from individual alienation. Individual malcontents are ubiquitous; but unless they win the support of others around them, they can get into serious difficulties. Inept performances develop when local group sanctions support indolence.

Even in a demoralized unit most transactions are performed in a routine manner, for the members realize that they are part of an organization that must go on. If the primary group definition is that nothing important or desirable is at stake, members are still willing to go through their habitual movements, as long as the tasks are not too tedious. If no consensus develops in the primary groups or if there is disagreement over what needs to be done, most men perform their duties in a desultory manner rather than risk disciplinary action. Most members go through the motions with just enough involvement to make a minimum acceptable showing. A few may attempt to walk the tightrope between sedition and regulations, but for most the matter is not important enough to take the trouble. If men believe that serious punishment would be inflicted for noncompliance, that defiance would be too costly to members of the primary group, they will cooperate even if they place a low estimate on the transaction. After all, getting into serious trouble is also regarded as stupid. In such contexts differences in performance quality among various local units depend on the skills and habits of the individuals who happen to be present, especially on the activities of those enacting key roles.

If the performance of a transaction that is defined locally as worthless requires more than routine effort, however, especially if some inconvenience or pain is involved, evasion of duty becomes increasingly widespread. Absenteeism is the initial reaction. Thus, in A-9-3, on the Turkey Farm, on the troopship, and at Zama, duty formations became disorganized affairs. Varying numbers "disappeared" almost daily, and others answered roll call for them. Those unable to escape worked in a lethargic manner and ran away at the first opportunity. It is clear that compliance was induced only by fear of punishment; the men went through the motions only when they were being watched by officers. Sloppy work becomes commonplace; no one cares enough to correct mistakes. Much grumbling is evident, and activity is held up at the slightest excuse.

Consistently slipshod performance develops in a cumulative process amid countless discussions. If the cost of punishment could be great, others may initially urge the dissidents to comply, while expressing sympathy for their views. But if the others notice that these men are getting away with their dereliction, they will join them. As it becomes clear that such transgressions are difficult to punish, increasing numbers join them, and evasion of duty becomes the informal norm. Men then begin to cooperate in avoiding duties; they prearrange absences from their posts and cover for each other. Once it becomes the accepted practice in local units, individuals invent ingenious subterfuges, and they may compete with one another on this. This can develop into open insubordination; men may refuse to obey orders, even when threatened with penalties. Thus, in Division F very little time was devoted to instruction in the Japanese language. Since the men regarded studying as pointless, they wanted to be free to do other things. NCOs who tried to enforce regulations were threatened with violence.

If a group develops a record of failure and inefficiency, it places a low evaluation on itself. Not only do the members candidly agree that their performance is not satisfactory, they become acutely aware of the low estimate placed on them by outsiders. Indeed, it was their inability to meet expectations imputed to other Nisei that made so many in Company K discontented with their lot. They complained repeatedly that they had nothing of which to be proud. The most conspicuous contributions to the war effort were being made by combat troops, and many were ambivalent about having transferred out of the infantry. Even in G-2 they feared being ridiculed as "fuck-ups" and being taunted by their friends and relatives about having been unable to "make the grade" as competent linguists. Once men conceive of themselves as incompetent, such beliefs affect their definitions of specific situations. Many of the reactions found in Company K are characteristic of people who are deeply ashamed of themselves. Although some boasted of their exploits, many felt absurd; they were useless soldiers going through the motions until their time was up. One defense against the danger of appearing ridiculous is to take nothing seriously. Given their circumstances the easiest solution was not to try at all or to discredit organizational goals. The protective label of a pariah group makes any kind of effort unnecessary. Bellicose posture, obscene language, excessive consumption of alcohol, flouting authority, disrespect for the values of the outside world—all are symbolic acts that affirm that one is a certain type of person, one who is disassociated from the establishment and does not care. The low estimate the men placed on themselves is revealed further in their unwillingness to pose for a company photograph and in their refusal to bring wives and sweethearts to the company party. Their desperation is also revealed in their grasping at tiny straws to bolster their esteem, such as showing off their marching ability in meaningless parades. Some even tried to defect; they took unnecessary leaves and even reenlisted to get out.

If the members of primary groups place a low evaluation on their unit, their performance becomes consistently lethargic, for it becomes foolish to expend any effort in behalf of a worthless organization. Not only do the members not expect one another to do well; they take it for granted that no one else will try very hard. Not only is deviation from formal norms tolerated, it is demanded. When Lieutenant Skinner was absent, the cadre and acting NCOs cooperated to flout regulations concerning reveille; on the troopship the guards threw their clubs overboard and boasted of it. Definite pressures arise against effective performance; indeed individuals who comply in such contexts are viewed as a threat to the group and punished. During the early days of Division F impressive cooperation developed in administering hot foots and hot seats to eager beavers. These symbolic acts made it clear to "rate busters" that their conduct would not be tolerated. Thus, noncompliance becomes the thing to do. It is taken for granted that one will conform to organizational norms only when it becomes too costly not to do so. Should such a unit encounter a severe crisis, it is likely to collapse—as various individuals and cliques scatter to save themselves. This is what in common parlance is called "demoralization."

But the record is inconsistent. Company K was a demoralized unit, and most of its transactions were executed in a routine or lethargic manner. But there were the occasional peak performances. How is this apparent contradiction to be accounted for? The answer lies in the manner in which men define themselves while participating in various tasks. The same individuals can identify themselves in different ways while doing different things. Whenever the men defined themselves in terms of ethnic identity—as Nisei on display—their performance was always effective. Even in situations that they regarded as unreasonable—as on December 29, when Captain Asakura challenged the company's claims to virility by ordering his troops to march in formation up a steep hill with all their belongings—everyone remained stubbornly in position until one man collapsed. Whenever the men identified themselves as a unit in debt to its company commander, they did their best. When members of B-9-3 regarded themselves as a faction with a score to settle with the A-9-3 cadre, they made effective use of all their resources as infantrymen. Whenever the men saw themselves as an aggrieved party seeking vengeance, their cooperation was very efficient. But in most other transactions the same individuals identified themselves as elements of a worthless military unit engaging in stupid exercises, and they reacted accordingly.

It is apparent from this record, then, that the demoralization of a group does not result from the degeneration of individuals; it develops from cooperation on the basis of informal norms. The degree of coordination necessary to accomplish complex transactions—such as deliberately slowing down the pace of work or retaliating against an informer—cannot be attained unless the participants develop mutual expectations and live up to them. As Cooley (1909:23–50) insisted long ago, each primary group is a moral order, and most of its members feel honor bound to comply with its norms. People generally do what they believe to be the decent thing, even when it involves some risk. Even those who are not enthusiastic about a proposal feel obligated to do their share, once a decision has been reached. Many students in Division F who had no objections to learning Japanese limited their study time to a respectable minimum. When everyone else seemed to be striving to avoid duty, even those who were becoming bored doing nothing took advantage of opportunities that presented themselves to loaf. What is significant is that even individuals who disagree with the informal norms feel that they must go along, once common understandings have been established. Especially if there appears to be a high degree of consensus, it is difficult to resist expectations imputed to intimate friends. The inner conflicts reported by mainland Nisei during the Hawaiian reign of terror give some indication of the pressures felt by those who fear that they may be unable to comply. Demoralization, then, is not simply the disintegration of a unit; activities are reorganized on a different basis.

Spontaneous Resistance and Retaliation

Judging from this case study, what happens to a demoralized unit after the evasion of duties has become widespread depends largely on the reactions of authority figures—in particular the reaction of the local supervisor, the key link between the various primary groups and the larger organization. If the local leader is understanding and sympathizes with his men—as Captain Parmentier, Lieutenant Catalano, Captain Asakura, and Private Nakasone did—some kind of accommodation is worked out. The disgruntled men recognize the difficult position in which their leader has been placed and try not to get him into serious trouble. On the other hand, if the local leader makes no effort to find out what has made his subordinates so discontented and insists on enforcing regulations, resistance is likely to develop. Being forced to perform tasks defined as useless arouses resentment, for the effort seems unnecessary, and those who comply feel stupid. Their resistance may take many forms. Insubordination—disobedience of a direct command—is one of the initial reactions.

Other tactics are devised as a struggle develops between the supervisor and the personnel under him. In the beginning only a few outraged individuals challenge his authority; but as the resentment spreads, resistance takes on organization. Plans are laid ahead of time, and cooperation is prearranged. Where penalties are not severe, the men may simply talk back to their leader or send a delegation to negotiate changes. When penalties are heavy, as in military organizations, widespread resentment can develop into an organized campaign of passive resistance.

If authority figures react in ways that are regarded as unjust, they are defined as frustrating objects. This evokes aggressive tendencies, and plans are made for retaliation. When a leader is regarded as incompetent, the privileges of his rank are resented. Thus, much of the trouble started at Fort Meade in contentions over trivial matters. What made most of the men so angry was their conviction that the inconveniences were avoidable; all that was needed was better organization. When Captain Larsen was completely unresponsive to complaints and suggestions, he was singled out for blame. What the Nisei were doing was *his* fault, not theirs; *he* was the one who was incompetent; *he* was compounding his transgressions by being unfair. If authorities are regarded as corrupt, whatever they do is interpreted as an attempt to satisfy their greed. No one accused the MISLS officials of being dishonest, but many enlisted men were convinced that they were being used by Colonel Andersen and Major Kawashita as the two were furthering their personal ambitions. The officials at Zama were dismissed as thieves, and no one took them seriously. Thus, pain and inconvenience in themselves do not arouse resentment. Men become indignant when their sense of justice is violated, when they believe that they are being deprived of something that is rightfully theirs. Mainland Nisei who had quietly accepted evacuation from their homes reacted sharply to differential treatment in the army. In 1942 they realized that their loyalty was in question and that there was no way of disproving the charges against them; once they donned uniforms and risked their lives on the same basis as other soldiers, however, they felt that they were entitled to be treated like other Americans.

When a campaign of resistance and retaliation emerges, it must be remembered that the participants are mobilizing against an object that they themselves have constructed in their discussions, a conception that may be entirely inaccurate. Military life, especially for enlisted men, is full of frustrations. When aggressive tendencies arise, they tend to be focused on a limited number of objects. If members of adjacent primary groups single out an object—some individual, group, or condition—as the "cause" of their discomforts, it becomes the target of hostility and vengeance. Once responsibility is placed on someone, he is personified in ways that tend to exaggerate his negative qualities. This is a cumulative process. Anything that is even vaguely relevant to the charges against him are picked up and added to his characterization until he is viewed as an ogre. A succession of officers—Captain Larsen, Captain Schmidt, Sergeant Endo, Major Kawashita, Lieutenant Skinner—were accused of things most of them had never dreamed of doing. On hearing of some of the charges, they regarded them as so ridiculous that they did not bother to refute them. Only Sergeant Nishida took the trouble to disprove the accusations against him, and his refutation was accepted. When viewed in retrospect, it becomes apparent that many of the events that enraged the men against particular officials were just inevitable consequences of bureaucratic bungling or unfortunate coincidences. Many of the inconveniences suffered by replacements at Fort Meade probably resulted from orders emanating from sources higher than the company commander. Furthermore, there is no evidence that ranking officials of the MISLS were deliberately exploiting Nisei soldiers for personal gain. It is doubtful that MISLS officials had sufficient authority to alter the schedules of troop trains. Even if the denounced party is not in fact responsible for the repugnant conditions, however, pinpointing blame does provide an explanation for the discomforts. Such explanations are usually simplistic; but this is inevitable, for most events are too complicated for laymen to comprehend. Furthermore, enlisted men are far too busy coping with the exigencies of daily life to make careful, dispassionate analyses (Lippmann, 1922:79–156). The focusing of blame provides a handy, shortcut way of "making sense" of complex, confusing situations.

Once someone has been identified as a frustrating object, men feel that they must protect themselves and if possible strike back. If the opponent is too powerful to attack directly, other tactics are devised to get even with him. In A-9-3 the replacements manipulated formal regulations as a weapon to discredit Captain Larsen in the eyes of the regimental commander. They knew enough about military organizations to appreciate how an officer's career would be affected by inept performances by troops under his command. Thus, at some risk to themselves they launched a joint effort to get him reprimanded. They marched out of step deliberately in their "staggered formation" in the hopes of being caught by inspecting officers. They reported in large numbers to sick call to create the impression that the barracks were unsanitary. On the Turkey Farm and in Division F the intimidation of NCOs was added to the repertoire of blatant insubordination, disrespect for rank, absenteeism, and harassing officers. When the men learned of Lieutenant Skinner's duplicity, they deliberately flouted his orders; even those who had no desire to go into town in sub-zero weather did so, just to be AWOL. On the troop transport, as reactions crystallized against the navy policy of differential treatment, the men failed to report for duty, claiming to be seasick; they threw equipment overboard, made unauthorized use of officers' quarters, and intimidated those who tried to restore order. In Zama the same pattern was used again, and it became especially effective when a detachment of paratroopers joined them. Thus, a pattern of protest initially formed at Fort Meade was improved and used against a succession of adversaries.

If the opponent is viewed as having challenged the unit to a contest, a high degree of proficiency may develop in a concerted effort to defeat him. Many of the tasks ordered by him are defined as part of a sadistic effort to punish members of primary groups. The men close ranks; cooperation becomes easy, for they forget their differences to partake in a crusade against evil. A high value is placed on devising better and better ways of striking back. Those who go as far as possible are viewed as courageous; those who escape punishment are admired for their cleverness; those who comply with regulations are viewed as cowards; those who side with the opponent or inform are condemned as traitors. Some of the most efficient performances of the units under study came in attempts to outdo a foe. After Captain Larsen was removed from his command, the very men who had carried on a campaign of passive resistance became a part of a model company; by making the contrast as sharp as possible they wanted to discredit the captain. On May 30, when B-9-3 was ordered to defend a sector of A.P. Hill, men who were completely unconcerned with effective defense put into use all their infantry training—in the hopes of striking back at a few hated officers and NCOs in A-9-3. On 1 February 1946, when the Nisei in the 275th Company decided to challenge the sergeant's order to be in formation on time, they formed a united front. When the whistle blew, no one went out. Once outside, they refused to do anything until all Caucasian casuals had also been ordered into formation. Thus, when men conceive of themselves as an aggrieved party seeking justice, acts of defiance enjoy a high degree of consensus.

If collective excitement is intensified, extremists are no longer deterred by fear of punishment. The soldier who beat Sergeant Endo did not expect to go unpunished; he made no effort to escape. Angry men often condone violence that they would otherwise not tolerate; those who are caught and punished are viewed as martyrs. Both at Fort Meade and at Fort Snelling many of the men became so preoccupied with vengeance that they overlooked the possibility of tarnishing the reputation of all Nisei in the United States; they had to be reminded of this danger repeatedly, even by Lieutenant Catalano. During the final days at Fort Snelling, men who knew that being AWOL after being alerted for overseas duty was a very serious court martial offense left the company area in droves. When the lieutenant visited the restricted company on the final night, he found it virtually deserted and in complete disarray.

Since any organization has established procedures for maintaining order, recurrent outbursts of intramural violence are a symptom of demoralization. Especially in the military, it is a clear indication that some regulations are no longer enforceable. Such acts symbolize defiance of authority. Although different forms of violence are difficult to identify in terms of overt behavior patterns, it is important to distinguish two types. Some attacks on personnel, like sabotage, constitute efforts to retaliate against the organization. The beating of Sergeant Endo is to be viewed in this light. Since he lived in the company area, he was more vulnerable than others identified with the administration. Although many of the threats against Major Kawashita may be discounted as "big talk," he was hated intensely and might well have been assaulted had a suitable occasion arisen. The attacks on A Company personnel were also part of the retaliation against the MISLS administration, although the victims would have been dumfounded to learn this. The OCS men who were beaten were symbolic victims. They were not known as individuals; they were not attacked for anything that they themselves had done but simply because they happened to be in a hated category. They had been identified, albeit incorrectly, with the administration, the "cause" of all the frustrations. This type of aggression enjoys widespread support, and the perpetrators often become heroes. Disturbances of this kind are welcomed as ways of striking back at oppressors, for it is understood that unfavorable publicity would be embarrassing.

But there is a second form of violence. As effective authority breaks down, previously suppressed hostilities between factions may flare up in brutal encounters. Small numbers of disgruntled men form gangs. In this case the pattern started quite accidentally. Several drunken Hawaiians attacked an NCO in another company and got away with it; they then assaulted two mainlanders in the company and again escaped undetected. Once the gang members realized that no one would challenge or report them, they became bolder. During a reign of terror not all persons are equally vulnerable. A few are almost immune—in this case other Hawaiians and the few mainlanders reputed to be skilled pugilists. The most vulnerable targets are those individuals who are generally disliked, especially those who violate primary group norms. Eager beavers provided easy targets. Mainland Nisei who spoke fluent English and had polished manners were especially disliked by Hawaiians who were self-conscious of their speech; they felt that such "Haolefied" Nisei were being snobbish. The perpetrators of this type of violence are often viewed as hoodlums. The handful of men responsible for most of the beatings in Company K were of low standing; had it not been for their brutality, they would have gone unnoticed. They were especially disliked by their fellow Hawaiians, whose support made their attacks possible.

Even when many feel aggrieved, only a small minority actually participate in the fighting. What gives the few the freedom to act and what makes their apprehension so difficult is the tacit consent and sympathetic backing of the others—some of whom wish that they too had the courage to participate. Disgruntled bullies are ubiquitous, but they are seldom allowed to give vent to their hostilities. When a unit becomes demoralized, however, others do not care enough to take the trouble to maintain order. Just as guerrilla fighters cannot operate effectively without community support, a violent gang cannot operate for long unless others cover for them, provide encouragement and support, and tolerate the inconveniences that follow—such as collective punishment. When violence erupted in Company K, it was initially welcomed by a bored and disgusted unit. Then, many tired of it. But the terrorists were able to continue for some time, largely because others were unwilling to take the risk of challenging them. The few mainlanders who took the initiative to stop the fighting were singled out for beatings; this effectively prevented others from participating in similar measures. But once the Hawaiians in the company decided that they had had enough, the fighting within Company K stopped. Since such gangs attack only when they can do so with impunity, it is clear that intramural violence breaks out only when informal sanctions against it become inoperative.

If authority figures themselves become demoralized, a state of near anarchy may develop. Formal compliance with regulations may continue when it is unavoidable; otherwise no one cares what happens. As men are allowed to fend for themselves, each primary group improvises ways to pursue its interests. In Division F the instructors were too demoralized to care what happened. Most of them sympathized with the men; they hated Major Kawashita more intensely than their students did. Nothing much was accomplished. The disorganization in Zama is another example of what could happen.

The Emergence of Definitions and Norms

If demoralization is not the product of the depravity of individuals but of cooperation on the basis of informal norms, the question arises: how do such definitions and extemporaneous arrangements arise? Morale varies with the ways in which people who are in close, sustained contact define and evaluate the transactions in which they are participating. Perspectives are never static. The world is constantly changing. Each transaction, as it is executed, modifies the situation somewhat, and the participants continuously redefine the changing context in which they are involved. How do primary groups come to define situations as they do? How do they achieve consensus? Are there any regularities in the formation of new procedures?

Definitions that diverge from official views arise in situations that are problematic. The record reveals numerous occasions in which the men felt that their rightful interests could not be satisfied through institutional means. In A-9-3 the replacements became convinced that they were expending themselves on useless exercises and that their CO was depriving them of "free time" as part of a vendetta against them. In Zama troops assigned to the 275th Company could not get enough to eat. There were numerous occasions in which the men found themselves in quandaries. After V-E day Nisei infantry replacements were given an opportunity to choose among three assignments but had no way of ascertaining which of the alternatives would be most desirable; those who went to Fort Snelling were confronted immediately with a similar dilemma. In Division F both instructors and students were confronted by ambiguity. It was obvious that orderly instruction was impossible, but they were required to appear in classrooms and to remain there for prescribed periods. How were they to occupy themselves for four months? Even in a highly institutionalized bureaucracy unexpected events occur, and uncertainties arise in the interstices between regulations. Intramural violence is supposed to be reported to authorities, but in fact few enlisted men would do this. But when such inconveniences persist, what can be done? If there is no way of being transferred out of a unit that has become unbearable, what can one do? There were times when the men searched in vain for some regulation that would provide them with guidelines. A situation is problematic, then, to the extent that people who want to do something have no accepted procedure to guide them. Temporarily activity comes to a halt. But something has to be done. How do men typically react in such contexts? How do they mobilize their resources to carry on? A great many transactions develop in daily life that are improvised on a person-to-person basis by those who happen to be on the scene.

When confronted by such perplexities, those who are involved engage in deliberation. Anyone present may participate, but there is a self-selection of personnel. Those who believe that their vital interests are at stake take a more active part; those who are too preoccupied with other matters remain uninvolved, even though a decision may eventually affect everyone. The participants come to recognize, sometimes more intuitively than explicitly, that they share a common fate. Their differences fade temporarily into the background, and they pool their intellectual resources. Those with acknowledged competence in the particular subject matter are listened to with greater deference than the others. Whenever possible, information is sought from authoritative sources; when it is not available, relevant rumors are brought up and evaluated. Initial efforts are usually directed toward clarifying the question. Then, alternative ways of coping with the problem are considered. Various proposals are presented and evaluated. Since those with diverse interests are likely to define the situation somewhat differently, arguments may arise over the most plausible interpretation and the most practical line of action. It is in the course of such spontaneous interchanges that definitions emerge. What usually characterizes such deliberations in primary groups is candidness, for impression management is less likely to be effective among people who know one another on a personal basis.

Situations are defined by piecing together various items of information to construct plausible interpretations. Judgments, with few exceptions, are based on what the individuals really believe. The definitions may subsequently turn out to be inaccurate, but they represent the combined judgments of those who are present. Consensus develops through a process of collective selection. Few participants are excluded from the deliberation; each has an opportunity to present his views, to raise questions, and to voice objections. Only those items of information that are acceptable to a substantial portion of those who are present remain to be used in the formation of definitions. What regularities are there in this selective process?

The definitions that may be constructed are limited by the availability of relevant information from trusted communication channels. Contrary to the belief of some intellectuals and executives, rank-and-file members of organizations are not gullible on matters that are of vital concern to them. They may be uninformed or misinformed, but they are not uncritical. All items of information are examined carefully and evaluated for reliability. Suspected views are challenged. Sources of information are investigated. Was the person who provided the item in a position to know? In discussions of military matters NCOs, people who had access to the orderly room, and those with friends in the administration were questioned; on other matters college graduates were sought out for their opinions. Much depends on the reputation of the speaker. Unreliable, uninformed, and biased individuals are quickly identified, and their views are quietly disregarded. Those respected for their integrity, wisdom, and cool judgment are sought out and consulted.

The key selective factor in the construction of definitions is *plausibility*. What survives from the welter of comments and suggestions are the items that are consistent with what is already accepted as true (Shibutani, 1966:68–94). To the extent that there is continuity of personnel, the formation of the outlook of a primary group is cumulative. Past definitions that have stood the test of time form an ordered perspective, which in turn becomes the basis for constructing new definitions. The culture of any group is the product of the succession of collective adaptations made to the peculiar life conditions it encounters, the product of its history. Initially, beliefs are verbalized, but in time they become widely accepted and taken for granted. Then, these unspoken assumptions are repeatedly reaffirmed in daily interaction. Many of the conversations recorded, for example, are predicated on acute awareness of the anomalous position of the Nisei in World War II. There is also a remarkable faith in the United States government; whenever the men encountered differential treatment, they blamed it on some local official. They remained convinced that once the facts were brought to the attention of higher authorities, the undesirable condition would be rectified. Similar unstated beliefs are reinforced daily by the acceptance of statements predicated on them. Once a set of beliefs crystallizes, all succeeding situations—official announcements as well as other events—are interpreted in these terms. It is in this manner that each unit develops a distinctive style of performance. Each unit develops a unique combination of assumptions, its own language, its special symbols, and some distinctive practices. The perspective shared in each circle is the cumulative product of the peculiar combination of personalities, views, and experiences that happen to be represented. Each primary group, then, develops a distinctive outlook, though it may be similar in many respects to that of its neighbors. This perspective serves as a filter through which each new situation is viewed.

Another selective factor in the formation of primary group definitions is collective excitement. Emotional reactions tend to constrict the perceptual field, lower critical ability, and make men suggestible in a limited direction. Just as an angry person becomes selectively responsive to possibilities of injuring the frustrating object, a frightened person becomes highly sensitized to possibilities of escaping a dangerous one. But the excitement of a few does not affect group definitions, for their remarks are neutralized by others who are not similarly upset. If a common mood develops, however, standards of plausibility tend to be relaxed. Thus, when intense collective anger developed against Sergeant Endo, the MISLS administration, and Lieutenant Skinner, most of the men became so preoccupied with vengeance that they overlooked the possible consequences of violence. The few who called attention to possible damage to the reputation of the Nisei were brushed aside. Similarly, when several other companies were alerted for overseas duty soon after V-J Day, considerable excitement developed in Company K. Many felt that important decisions would have to be made quickly, and they entertained seriously several rumors that under other circumstances would have been rejected as implausible. It should be noted, however, that even in these contexts some individuals remained coldly detached and continued to caution their friends (Shibutani, 1966:95–128).

Collective deliberation generally leads to the establishment of some measure of consensus, but the degree and type of consensus attained varies from one situation to the next. In considering various ways in which consensus is related to concerted action, a distinction made by Scheff (1967) concerning different uses of this concept is especially helpful. He notes that for some investigators consensus refers merely to agreement among individuals over a statement. For others consensus designates coorientation—the emergence and reaffirmation in serial communication of a shared definition of the situation in which the participants are under the impression that all others view the circumstances in a similar manner and thus feel constrained to comply. It is quite possible for such a common orientation to develop among people who are not necessarily in agreement about what line of action should be pursued.

When both coorientation and agreement develop in the deliberation of interrelated primary groups, problematic situations are resolved and coordinated action follows, sometimes at a high degree of effectiveness.

When Captain Larsen threatened to demote all Nisei NCOs in A-9-3, the replacements developed a common orientation (their friends were in jeopardy) and agreement that they should protect them. At Fort Lawton, once the men knew definitely that they were leaving, they shared a common definition (that the task was unreasonable) and decided among themselves to carry up the heavy bags piecemeal on their own time and then took turns standing guard. Countless illegal transactions, such as feeding the Japanese in Zama, were executed in this manner.

If a common orientation emerges but people still disagree about what to do, arguments persist. If the differences are not resolved, efforts are made to negotiate a settlement. What should be done to repay Lieutenant Catalano? Most men knew that army regulations prohibited giving him a gift, but several argued that other Nisei units in basic training had purchased gifts for their officers and had encountered no difficulties. It was the cadre that suggested giving a gift to the lieutenant's wife, and this proposal was quickly accepted and implemented. If coorientation develops in the absence of agreement, several independently operating individuals and coteries arise; they act on their own, and the mounting of any kind of coordinated effort becomes difficult. Although mainlanders terrorized by the Hawaiian trio shared a common definition of their plight, they were unable to mobilize their resources to defend themselves. Nothing happened.

Although exchanges among members of primary groups tend to be quite candid, there are occasions in which people do not feel free to speak their minds. Sanctioned communication may interfere with frank interchanges. Communicative behavior, as everything else human beings do, is subject to social control. Certain kinds of statements enjoy group approval; others elicit hostility or deprecation and are in effect censored. Thus, when Private Shindo started complaining about ratings when the others were enraged at Sergeant Endo, he was told bluntly to shut up. Similarly, as interethnic tension developed between Company K and Company A, Nisei who felt no hostility toward Caucasians were unable to speak up in defense of the OCS men, creating the impression that there was a united front against Caucasians. What people avow openly, even to those with whom they are in intimate contact, many differ from what many of them believe privately. Because of such limitations two instances of pluralistic ignorance developed, and both were important in the history of Company K. When confronted at Fort Meade with a choice between infantry and intelligence service, men who were unimpressed with infantry values were not as free to speak as were those who wanted to stress their manliness. During the period preceding the classification examination, those who wanted to qualify for the "regular" school did not dare say so, thereby creating the impression that everyone was going to fail on purpose. In both instances there appeared to be a high degree of consensus among members of adjacent and overlapping primary groups, when in fact many actually disagreed both in the definition of the situation and in what they regarded as the appropriate course of action.

Thus, occasions arise in which consensus is either superficial or only apparent. In the two contexts in question most of the men said much the same sorts of thing, virtually all remarks favoring infantry service. Some actually meant what they said; others were only pretending. Since exchanges among friends are assumed to be trustworthy, most men believed what their comrades were saying. The few who became suspicious rarely made their challenges openly; therefore, disagreements were not disclosed in discussion, and the surface appearance was not penetrated (Glaser and Strauss, 1964). Most people acted in terms of definitions that apparently enjoyed consensus, and many acted contrary to what they actually believed to be in their best interest. But others took advantage of opportunities not to comply. At Fort Meade the interviews were conducted in private, and each person had the opportunity to express himself away from his friends. The record subsequently revealed that many had lied. The classification examination was held in the open, and dissidents had to do their work quickly. Few had the temerity to defy the others blatantly, but apparently many were able to write enough. In both instances effective, concerted action could not be mounted in spite of apparent agreement. Many were disillusioned, and several friendships were terminated.

Whenever formal procedures are found wanting, attempts are made to improvise what people regard as being more sensible ways of getting things done. If soldiers become convinced that some task they are required to perform is a stupid waste of time, they find less tedious ways of occupying themselves. If they feel that they are being cheated of something that is rightfully theirs, they seek other ways to get what they believe they deserve. Most informal norms evolve gradually through trial and error, with the adoption of attempts that prove successful. The practice of "goofing off" developed in this manner in A-9-3. At first the adventurous few "disappeared" on their own initiative. When others saw that they could get away with it and that the cadre was unable to pinpoint responsibility, they tried it. When disciplinary action still was not taken, the practice became even more widespread. Similarly, the "staggered formation" developed by accident, when soldiers trained in different IRTC camps found themselves stepping on one another's heels. Instead of making adjustments, those who were looking for trouble continued to march out of step. The procedure for passing time in Division F also developed in this manner. The students discovered on the first day that the instructors were easily intimidated. At first hot foots were given to anyone who fell asleep from boredom, but attention became focused more and more on eager beavers. Techniques for administering hot seats were improved step by step until they reached a danger point, and then they were abandoned by common consent. The same was true of repaying Lieutenant Catalano in dress parades. At first the men happened to outperform the other companies simply because of their superior infantry training. When they saw how pleased the lieutenant was, they began to work harder at it. After that, each opportunity that arose to please him aroused a spontaneous outpouring of support. In Zama the resourcefulness of the men reached its peak in efforts to assure an adequate food supply. In the beginning a few returned with food from mess halls in which they had served as KPs; most of these items were gifts from the mess sergeants. This was so thoroughly appreciated that others did the same thing—bringing food in ever larger quantities and stealing whatever they thought would be useful. Once they realized how easy it was, they began bringing uncooked food that could be prepared in the barracks whenever needed. They even stole utensils necessary for cooking. Informal arrangements were made to feed anyone in the company who was still hungry. Thus, informal norms of all kinds may be regarded as solutions to problematic situations.

If the problematic situation is recurrent, a workable solution, however it may have originated, is repeated until mutual expectations become fixed. Thus, elaborate procedures developed to cover for those who were "goofing off." Furthermore, common understandings emerged that a respectable number would appear at each formation to enable men to take turns being absent. Those who disappeared too often were criticized. A new pattern cannot be sustained without cooperation; hence, other supporting norms develop. It is clear that a distinction must be made between mere disorder and an organized pattern of deviant behavior.

Once procedures are established and reaffirmed through continued use, ideological justifications arise to explain them. What happened in A-9-3 was justified as a suitable response to gross incompetence and discrimination. What happened at Fort Snelling was vindicated in terms of everyone's having been "Shanghaied." Even when the battalion sergeant major had refuted this charge, the excuse was reiterated outside the company. What happened on the troopship was explained in terms of the discriminatory policy of the navy, especially the invidious distinctions between officers and enlisted men. What the men did in Zama was justified quite easily; everyone else was doing the same thing. Thus, in collective behavior we find support for what Freud said of individuals: human beings frequently do not know why they do what they do, but they can provide good reasons for it if called on to do so. Nor are such rationalizations necessarily accurate. Many members of Company K were deeply upset about being double-crossed by their erstwhile comrades who went off happily to the "regular" school. But little was said of this, for it was too painful and humiliating. Instead, attention was directed to the inferior facilities, the poor food, discrimination, and their bleak future. Thus, this case study supports the theory of the formation of folkways enunciated long ago by Sumner (1906:2–3) and similar explanations proposed by Park (1927) and by Malinowski (1945).

Although local enforcement is rarely needed, primary groups do administer their unwritten codes through informal sanctions. Dissidents are brought into line and heroes elevated through spontaneous expressions of approval and disapproval, such as the show of displeasure through sarcastic remarks or changes of facial expression, ridicule, gossip, refusal to reciprocate favors, and sometimes ostracism. The capacity of coteries to enforce emerging norms is not to be underestimated. In this case study many frightened men participated in antics that could have gotten them into serious trouble. In Division F some became so bored that they wanted to study just to have something to do, but they did not dare do so until many others indicated that they felt the same way. In Company K those who were most adept in "goofing off"—men who would have been scorned in other Nisei units—enjoyed considerable prestige. In virtually all companies a few individuals develop goldbricking into a fine art, aided and abetted by sympathetic comrades who cover for them. In this company the goldbricks became heroes, and at times they competed with each other to see who could get away with the most flagrant violations. Of course, reasonable allowances are made for the idiosyncrasies of each individual. Saipan was not held to the same standards as the others.

Not all primary groups are able to enforce their norms with equal effectiveness, especially when compliance may turn out to be costly. The relative effectiveness of informal controls depends on the manner in which primary group members are oriented toward one another. Mutual trust is of decisive importance. Conformity becomes more a matter of personal obligation, not to a general principle but to particular individuals who are viewed with affection. The solidarity of a primary group is seriously affected whenever members become suspicious of one another's sincerity. When hypocrisy is detected, men become disillusioned and begin to play safe. A conspicuous contrast is provided by the difference between Hawaiian and mainland Nisei. The Hawaiians saw themselves as a minority within a minority and were acutely aware of being outnumbered by people who were different from themselves. They were convinced that they had to stick together as a defensive measure, and they felt honor bound to go to one another's assistance. Not only were mainland Nisei not constrained by such a code, but in Company K many had good reasons for not trusting one another to the extent of risking injury. Hence, in spite of their claims to virility they were never able to mobilize their resources to defend themselves against a small gang. The mainlanders could not count on one another, and the trouble went on until the Hawaiians decided to stop it. These data support the hypothesis that the effectiveness of informal norms varies with the extent to which individual members are dependent on their primary group (Short and Strodtbeck, 1965).

Demoralization in any organization is generally a long developmental process in which a succession of typical definitions arise and are enforced within overlapping primary groups. It cannot be explained in terms of simple "causes," nor can it be attributed to a handful of "agitators." It is a gradual breakdown that consists of an increasing reluctance on the part of participants to comply with institutional norms. Once a unit begins to deteriorate, individuals become more and more aware of the views of their comrades. The wavering of faith is intensified by continued irritations and sustained deprivations. Typical definitions that emerge are that most formal tasks are a waste of time, that most regulations are unjust, that people who comply with them are stupid. There is usually a transitional period of misunderstanding and ill will. For a time many continue to comply with regulations, from habit or from fear of punishment. Factionalism frequently occurs, and those in each segment begin to question the motives and sincerity of the others. As mutual trust breaks down, considerable time and effort is devoted to watching one another, and wholehearted endeavor becomes impossible. With dissension comes increasing individualism. Each starts thinking in terms of private advantages, and local interests replace organizational aims as the basis for making decisions. As the unit develops a record of consistently dismal performances, it is scorned by outsiders; the members become ashamed of belonging to it and see little point in extending themselves in its behalf. Demoralization becomes clearly discernible when the unit encounters an obstacle. When a situation requiring coordinated effort under stress arises, total collapse may occur.

A Practical Implication of the Study

With increasing industrialization throughout the world, more and more activities previously carried out in smaller units are being accomplished in large formal organizations. In spite of the many obvious advantages of bureaucratization, the growth of large organizations has not always led to greater productivity. Meeting production quotas has been a problem both in capitalist and socialist nations. Furthermore, it is widely recognized that the fate of communities may be seriously jeopardized by the breakdown of morale in some of its component organizations. Indeed, situations may become critical, especially if law enforcement agencies become demoralized. Police officers must exercise discretion, for all laws cannot be enforced at all times. They often work alone, and on noticing an offense each must form a judgment— deciding which offender to arrest, which to caution, and which to ignore. How effectively laws are enforced in a given community, then, depends on the morale of its police force. Furthermore, in times of severe crisis military units are called on to intervene in civil affairs. Should this happen, certain questions invariably arise. Which unit is reliable? Which party will it serve? Astute administrators—whether they be politicians seeking power or benevolent despots ruling in behalf of those whose interests they profess to know—are becoming increasingly concerned with problems of control. How can high morale be maintained? How can demoralization be avoided?

But effective performance may develop both in executing the tasks for which an organization is established and in retaliating against it. From the standpoint of the organization, *morale is high to the extent that its avowed objectives are supported by the definitions spontaneously formed in its component primary groups.* The beliefs that emerge in local units and the commitments that rank-and-file members form toward those with whom they are in direct contact are of decisive importance. Thus, military morale depends on the soldiers' acceptance of the legitimacy of organizational demands (Moskos, 1970:147; George, 1971:307).

Various attempts have been made to manipulate morale, with varying degrees of success. One point appears certain: high morale cannot be compelled. If it cannot be coerced with the authoritarian controls of a military organization, how can it be commanded elsewhere? Many efforts have been made to induce high morale through persuasion. If people can be convinced that their personal interests or those of others with whom they identify are being served, they will not only do their part but will urge others to do theirs. It becomes a matter of personal pride to contribute one's share. But if the undertaking does not serve the interests of the participants, difficulties are likely to arise. People can be fooled for a while by clever promotional measures, but in time the truth has a way of becoming known. As long as the participants have their own communication channels, they will define situations for themselves, perhaps in ways that are inconvenient to those in charge. In the long run tricks are exposed for what they are. If outward compliance is required, people will make acceptable statements in public; but among close friends they will speak their mind. This suggests that there are definite limitations to the extent to which human behavior can be manipulated and controlled through formal regulations.

Some observers have lamented that in modern mass societies the ordinary person is trapped in a succession of gigantic transactions over which he has no influence. Yet the results of this study suggest that people are not robots, mere cogs in a machine. If they believe that their just interests are not being served, their displeasure is reflected in the manner in which the machine works—as in a slowdown campaign. Under some conditions they may even bring the machine to a grinding halt—as in wildcat strikes or in outright rebellions. Human beings think for themselves. Although some may be prevented from speaking for a time, no one can be forced to believe something that is not plausible to him. Furthermore, it is doubtful that conversations among intimate friends can be controlled. This suggests that high morale, as Hocking (1941:308) contended, is necessarily democratic. It can only be offered by those who make the contribution. High morale, like affection and respect, is something that has to be earned.

glossary

AGCT. Army General Classification Test

AJA. Americans of Japanese ancestry (used primarily in Hawaii)

Aloha. (Hawaiian) Greeting; Love; Compassion

AMG. Allied Military Government

Articles of War. Code of laws for the government of the U.S. Armed Forces

As you were. A military command rescinding a previous order that has not yet been carried out.

Assembly center. Temporary camps in which Japanese evacuees were housed pending construction of the more permanent relocation centers.

At ease. A military command to maintain silence

ATIS. Allied Translator and Interpreter Service

AUS. Army of the United States—to distinguish conscripts from the all-volunteer Regular Army

AWOL. Absent without leave

Baka. (Japanese) Fool; blockhead; simpleton

Banzai. (Japanese) A cry of salutation: Live Forever!

BAR. Browning Automatic Rifle

Benjo. (Japanese) Toilet

Boochie. (Nisei slang) Japanese

Buddhahead. (Nisei slang) Japanese

Bushido. (Japanese) Code of the samurai

Cadence. The rate of march in steps per minute

Cadre. A group of officers and enlisted men organized to establish and train a new unit

Casuals. Unassigned military personnel

CBI. China-Burma-India theater of operations

Chicken shit. (Military argot) Excessive adherence to rules on matters regarded as trivial

CIC. Counter-Intelligence Corps

CID. Criminal Investigation Division

CO. Commanding officer

Company punishment. Limited penalties imposed by the company commander

CQ. Charge of Quarters—an NCO in charge of organizational headquarters for a particular period of time, usually during off-duty hours

Dat kind. (Hawaiian pidgin) This or that (used to refer to objects one cannot name specifically at the moment)

Dayroom. Recreation room of a unit

Detail. (also, Fatigue detail) A group of men assigned temporarily to a particular duty

DO. Duty Officer—an officer in charge of organizational headquarters for a particular period of time, usually during off-duty hours

Dog tag. (Military argot) Identification tags, bearing each soldier's name, serial number, and blood type

Dry run. A practice operation, generally with blank ammunition

ERC. Enlisted Reserve Corps

Fuck-up (Military argot) One who is unsuccessful in adjusting to military requirements

G-2. Intelligence officer or intelligence section

Geta. (Japanese) Wooden clogs

GI. (Military argot) (a) Any item that is of government issue; (b) an officer or NCO who adheres too strictly to regulations; (c) to scrub barrack floors; (d) term used to refer to fellow soldiers

GI dribbles or GI shits. (Military argot) Diarrhea

Gig. Punishment for failure to maintain GI equipment or regulations

Goldbrick. (Military argot) A soldier who frequently and successfully shirks his duties

Gumbatsu. (Japanese) The military clique that ruled Japan during World War II

Hakujin. (Japanese) White people

Haole. (Hawaiian; *lit.* "stranger") White people

Head. (Naval terminology) Latrine

Imin. (Japanese) Emigrant

Inu. (Japanese; *lit.* "dog") Informer

IRTC. Infantry Replacement Training Command

Issei. ((Japanese; *lit.* "first generation") Immigrants from Japan

JACL. Japanese-American Citizens League

Judo. (Japanese) A refined form of jiu-jitsu—hand to hand combat based on movement, balance, and leverage

Kaiwa. (Japanese) Conversation

Kamikaze. (Japanese; *lit.* "divine wind") Japanese pilots assigned to make a suicidal crash on target with planes loaded with explosives

Kanaka. (Hawaiian; *lit.* "human being") Term used by Nisei to refer to anyone from Hawaii

Kanji. (Japanese) Chinese ideographic characters

Kendo. (Japanese) Art of fencing

Keto. (Japanese; *lit.* "hairy barbarian") Derogatory term for white people

Kibei. (Japanese; *lit.* "returned to America") American-born children of Japanese immigrants who had been educated in Japan

Kotonk. (Hawaiian pidgin) Derogatory term for mainland Nisei

KP. Kitchen police—temporary duty as cook's assistant

Kuro-chan. (Nisei slang) Though not hostile, a condescending term for Negro

Kurombo. (Japanese slang) Derogatory term for Negro

Lose fight. (Nisei slang) Deflation felt when conscientious effort leads to failure because of someone's error

Luna. (Hawaiian) Overseer; foreman

Manini. (Hawaiian pidgin) Self-centered; stingy

M-1. The Garand rifle

Mess. An Army meal or place where such meals are served

MISLS. Military Intelligence Service Language School

Mochiron. (Japanese) Of course; without doubt

Monku. (Japanese) Complaint; writhe in agony (in Nisei slang "No *monku*" designates "don't complain")

MOS number. Military occupation specialty number

MP. Military Police

NCO. Non-commissioned officer

Nihonjin. (Japanese) Japanese people

Nisei. (Japanese; *lit.* "second generation") American-born children of Japanese immigrants

No lie. (Nisei slang) Is that so? You're not lying, are you?

OCS. Officer Candidate School

OD. Olive Drab—winter uniforms

Off limits. Areas that soldiers are forbidden to enter

Orderly room. The office of a company or similar organization

Pon-pon (Slang used by Japanese and American occupation troops) Sexual intercourse

POE. Port of embarkation

Police up. Clean up an area

POW. Prisoner of war

PX. Post exchange—an army store

Rating. Grade of a noncommissioned officer

Relocation center. Camps at which Japanese evacuated from the Pacific Coast were incarcerated

Retreat. Evening ceremony during which the colors are lowered for the day

Reveille. The first daily formation at which the presence of soldiers is checked

Ronin. (Japanese) An unemployed samurai

Samurai. (Japanese) Warriors in feudal Japan

Section 8. Psychiatric discharge

Seppuku. (Japanese) Ceremonial suicide by disembowelment. Commonly known as hara-kiri outside of Japan

Shelter half. Half of a pup tent (each tent provides shelter for two men)

Shimpai. (Japanese) Concern; solicitude (in Nisei slang "no *shimpai*" designates "don't worry")

Shin-chu-gun. (Japanese) Army of occupation

Short-arm inspection. Genital examination by medical personnel

Snafu. (Military argot) Situation normal, all fucked up

SOP. Standard operating procedure

Sosho. (Japanese) Cursive writing of Chinese characters

Statement of Charges. A form signed by a soldier testifying to the loss or damage of government property for which deductions from his pay will be made

Sumo. (Japanese) Wrestling

Tanoshimi. (Japanese) Pleasure; a delight to be anticipated

TDY. Temporary duty

TS. (Military argot) Tough shit (used in situations in which nothing can be done to alleviate a soldier's misfortune, leaving him with no alternative to accepting it with resignation)

Urusai. (Japanese) Annoying

WAC. Women's Army Corps

Wahine. (Hawaiian) Woman

Waste time. (Nisei slang) Anything that is ridiculous and not worth the required effort

WRA. War Relocation Authority, a civilian agency charged with operating the relocation centers and with facilitating resettlement

Yamato damashii. (Japanese) The Japanese spirit of loyalty and patriotism

Zaibatsu. (Japanese) Financial clique that built Japan's military strength before and during World War II

bibliography

Adams, Romanzo C., 1934. The unorthodox race doctrine of Hawaii. Pp. 143–60 in E. B. Reuter (ed.), *Race and Culture Contacts*. New York: McGraw-Hill.

———, 1937. *Interracial Marriage in Hawaii*. New York: Macmillan.

Allen, Gwenfread, 1950. *Hawaii's War Years, 1941–1945*. Honolulu: University of Hawaii Press.

Ardant du Picq, Charles, 1921. *Battle Studies: Ancient and Modern Battle*. J. N. Greely and R. C. Cotton (trans.). New York: Macmillan.

Barnard, Chester I., 1938. *The Functions of the Executive*. Cambridge: Harvard University Press.

Bassan, Morton E., 1947. Some factors found valuable in maintaining morale on a small combatant ship. *Bulletin of the Menninger Clinic*. 11:33–42.

Benedict, Ruth F., 1946. *The Chrysanthemum and the Sword*. Boston: Houghton Mifflin.

Blau, Peter M. and W. Richard Scott, 1962. *Formal Organizations*. San Francisco: Chandler.

Blumer, Herbert, 1943. Morale. Pp. 207–31 in W. F. Ogburn (ed.), *American Society in Wartime*. Chicago: University of Chicago Press.

———, 1969. Collective behavior. Pp. 65–121 in A. M. Lee (ed.), *Principles of Sociology*. New York: Barnes and Noble.

Bogardus, Emory S., 1947. Japanese return to West Coast. *Sociology and Social Research*. 31:226–33.

Broom, Leonard and John Kitsuse, 1955. The validation of acculturation: A condition of ethnic assimilation. *American Anthropologist*. 57:44–48.

Butler, Judson R., 1946. Military rank and social climate as determinants of social attitudes. *Journal of Personality*. 14:184–98.

Cantril, Hadley and Mildred Strunk, 1951. *Public Opinion, 1935–1946.* Princeton: Princeton University Press.

Carr, Elizabeth, 1960. A recent chapter in the story of the English language in Hawaii. *Social Process in Hawaii.* 24:54–68.

Caudill, William, 1952. Japanese-American personality and acculturation. *Genetic Psychology Monographs.* 45:3–102.

Cooley, Charles H., 1909. *Social Organization.* New York: Charles Scribner.

Cortright, David, 1975. *Soldiers in Revolt: The American Military Today.* New York: Doubleday.

Daniels, Roger, 1962. *The Politics of Prejudice.* Berkeley: University of California Press.

Fisher, Anne M., 1952. Debt of dishonor. *The Reporter.* 6 (February 5): 21–23.

Frazier, E. Franklin, 1957. *Black Bourgeoisie.* Glencoe, Ill.: Free Press.

Fuchs, Lawrence H., 1961. *Hawaii Pono: A Social History.* New York: Harcourt, Brace and World.

George, Alexander L., 1971. Primary groups, organization, and military performance. Pp. 293–318 in R. W. Little (ed.). *Handbook of Military Institutions.* Beverly Hills, Calif.: Sage.

Glaser, Barney G. and Anselm L. Strauss, 1964. Awareness contexts and social interaction. *American Sociological Review.* 29:669–79.

———, 1967. *The Discovery of Grounded Theory.* Chicago: Aldine.

Greene, Stephen, 1946. Nisei: Ears for the government. *Common Ground.* 7: 17–20.

Grinker, Roy R. and John P. Spiegel, 1945. *Men under Stress.* Philadelphia: Blakiston.

Griswold, A. Whitney, 1938. *The Far Eastern Policy of the United States.* New York: Harcourt, Brace.

Grodzins, Morton, 1949. *American Betrayed.* Chicago: University of Chicago Press.

———, 1956. *The Loyal and the Disloyal.* Chicago: University of Chicago Press.

Heinl, Robert D., 1971. The collapse of the armed forces. *Armed Forces Journal.* 108 (June 7):30–38.

Hocking, William E., 1941. The nature of morale. *American Journal of Sociology.* 47:302–20.

Homans, George C., 1946. The small warship. *American Sociological Review.* 11:294–300.

Hormann, Bernhard L., 1960. Hawaii's linguistic situation: A sociological interpretation in the new key. *Social Process in Hawaii.* 24:6–31.

Iga, Mamoru, 1957. Japanese social structure and the source of mental strains of Japanese immigrants in the United States. *Social Forces.* 35:271–78.

Jacobson, Alan and Lee Rainwater, 1953. A study of management representative evaluations of Nisei workers. *Social Forces*. 32:35–41.

Janowitz, Morris, 1959. *Sociology and the Military Establishment*. New York: Russell Sage Foundation.

Kawai, Kazuo, 1960. *Japan's American Interlude*. Chicago: University of Chicago Press.

Kimura, Yukiko, 1943. Some effects of the war situation upon the alien Japanese in Hawaii. *Social Process in Hawaii*. 8:18–28.

Kitano, Harry, 1969. *Japanese-Americans: The Evolution of a Subculture*. Englewood Cliffs, N.J.: Prentice-Hall.

Klapp, Orrin E., 1948. The creation of popular heroes. *American Journal of Sociology*. 54:135–41.

Kuykendall, Ralph S. and A. Grove Day, 1961. *Hawaii: A History*. Englewood Cliffs, N.J.: Prentice-Hall.

Lang, Kurt, 1972. *Military Institutions and the Sociology of War*. Beverly Hills, Calif.: Sage.

Lasswell, Harold D., 1933. Morale. Vol. 10, pp. 640–42 in the *Encyclopaedia of the Social Sciences*. New York: Macmillan.

Leighton, Alexander H., 1946. *The Governing of Men*. Princeton: Princeton University Press.

———, 1949. *Human Relations in a Changing World*. New York: E. P. Dutton.

Lind, Andrew W., 1938. *An Island Community: Ecological Succession in Hawaii*. Chicago: University of Chicago Press.

———, 1946. *Hawaii's Japanese: An Experiment in Democracy*. Princeton: Princeton University Press.

———, 1955. *Hawaii's People*. Honolulu: University of Hawaii Press.

Lippitt, Ronald, 1939. Field theory and experiment in social psychology: Autocratic and democratic group atmospheres. *American Journal of Sociology*. 45:26–49.

Lippmann, Walter, 1922. *Public Opinion*. New York: Harcourt, Brace.

Little, Arthur W., 1936. *From Harlem to the Rhine: The Story of New York's Colored Volunteers*. New York: Covici-Friede.

Lord, Walter, 1957. *Day of Infamy*. New York: Henry Holt.

MacDonald, Alexander, 1946. *Revolt in Paradise: The Social Revolution in Hawaii after Pearl Harbor*. New York: Stephen Daye.

McGovney, Dudley O., 1947. Anti-Japanese land laws of California and ten other states. *California Law Review*. 35:7–60.

Malinowski, Bronislaw, 1945. *The Dynamics of Culture Change*. P. M. Kaberry (ed.). New Haven: Yale University Press.

Mandelbaum, David G., 1952. *Soldier Groups and Negro Soldiers*. Berkeley: University of California Press.

Marshall, Samuel L.A., 1947. *Men against Fire*. New York: William Morrow.

Masuoka, Jitsuichi, 1936. Race preference in Hawaii. *American Journal of Sociology*. 41:635–41.

Matsumoto, Toru, 1946. *Beyond Prejudice: A Story of the Church and the Japanese-Americans*. New York: Friendship Press.

Mauldin, Bill, 1945. *Up Front*. New York: Henry Holt.

Morimoto, Patricia T., 1966. "The Hawaiian Dialect of English: An Aspect of Communication during the Second World War." Unpublished master's thesis, University of Hawaii.

Moskos, Charles C., 1970. *The American Enlisted Man*. New York: Russell Sage Foundation.

Murphy, Thomas D., 1954. *Ambassador in Arms: The Story of Hawaii's 100th Battalion*. Honolulu: University of Hawaii Press.

Nakahata, Yutaka and Ralph Toyota, 1943. Varsity victory volunteers: A social movement. *Social Process in Hawaii*. 8:29–35.

National Opinion Research Center, 1946. "Attitudes toward the Japanese in our Midst." Denver. Report #33. Mimeographed.

Newsweek, 1971. The troubled U.S. army in Vietnam. 77 (January 11):29–37.

Nishi, Setsuko M., 1963. *Japanese-American Achievement in Chicago: A Cultural Response to Degradation*. Unpublished doctoral dissertation, University of Chicago.

Nitobe, Inazo O., 1913. *Bushido: The Soul of Japan*. Tokyo: Teibi.

Norbeck, Edward, 1959. *Pineapple Town, Hawaii*. Berkeley: University of California Press.

Packer, Peter and Bob Thomas, 1966. *The Massie Case*. New York: Bantam Books.

Park, Robert E., 1927. Human nature and collective behavior. *American Journal of Sociology*. 32:733–41.

Petersen, William, 1966. Success story, Japanese-American style. *New York Times Magazine*. January 9, pp. 20–21, 33–43.

Reinecke, John E., 1938. "Pidgin English" in Hawaii: A local study in the sociology of language. *American Journal of Sociology*. 43:778–89.

——— and Aiko Tokimasa, 1934. The English dialect of Hawaii. *American Speech*. 9:48–58, 122–31.

Roethlisberger, Fritz J. and William J. Dickson, 1939. *Management and the Worker*. Cambridge: Harvard University Press.

Rostow, Eugene V., 1945. The Japanese-American cases: A disaster. *Yale Law Journal*. 54:489–533.

Rundell, Walter, 1964. *Black-Market Money: The Collapse of U.S. Military Currency Control in World War II*. Baton Rouge: Louisiana State University Press.

Sabagh, George and Dorothy S. Thomas, 1945. Changing pattern of fertility and survival among the Japanese-Americans on the Pacific Coast. *American Sociological Review*. 10:651–58.

Scheff, Thomas J., 1967. Toward a sociological model of consensus. *American Sociological Review*. 32:32–46.

Schmid, Calvin F. and Charles E. Nobbe, 1965. Socioeconomic differentials among nonwhite races. *American Sociological Review*. 30:909–22.

Schneider, David M., 1946. The culture of the army clerk. *Psychiatry*. 9:123–29.

Shibutani, Tamotsu, 1966. *Improvised News: A Sociological Study of Rumor*. Indianapolis: Bobbs-Merrill.

——— and Kian M. Kwan, 1965. *Ethnic Stratification*. New York: Macmillan.

Shils, Edward A., 1950. Primary groups in the American army. Pp. 16–39 in R. K. Merton and P. F. Lazarsfeld (eds.). *Continuities in Social Research: Studies in the Scope and Method of "The American Soldier."* Glencoe: Free Press.

———, 1957. Primordial, personal, sacred, and civil ties. *British Journal of Sociology*. 8:130–45.

——— and Morris Janowitz, 1948. Cohesion and disintegration in the *Wehrmacht* in world war II. *Public Opinion Quarterly*. 12:280–315.

Shirey, Orville C., 1946. *Americans: The Story of the 442nd Combat Team*. Washington, D.C.: Infantry Journal Press.

Short, James and Fred L. Strodtbeck, 1965. *Group Processes and Gang Delinquency*. Chicago: University of Chicago Press.

Smith, Madorah E., 1945. Changes in the causes of feelings of inferiority since the outbreak of world war II as listed by students at the University of Hawaii. *Journal of Social Psychology*. 22:23–30.

Social Process in Hawaii, 1945. Islander's reflections on mainland Japanese. 9:39–49.

Spencer, Robert F., 1950. Japanese-American language behavior. *American Speech*. 25:241–52.

Stouffer, Samuel A. et al., 1949. *The American Soldier*. Vols. I and II. Princeton: Princeton University Press.

Sumner, William G., 1906. *Folkways*. Boston: Ginn.

tenBroek, Jacobus, Edward N. Barnhart, and Floyd W. Matson, 1954. *Prejudice, War, and the Constitution*. Berkeley: University of California Press.

Thomas, Dorothy S., 1952. *The Salvage*. Berkeley: University of California Press.

——— and Richard S. Nishimoto, 1946. *The Spoilage*. Berkeley: University of California Press.

Time, 1973. Blackboard battlegrounds: A question of survival. 101 (February 19):74.

Tsurumi, Kazuko, 1970. *Social Change and the Individual: Japan before and after Defeat in World War II*. Princeton: Princeton University Press.

U.S. Department of the Army, 1948. *The Procurement and Training of Ground Combat Troops*. Washington, D.C.: Government Printing Office.

———, 1954. *The Personnel Replacement System in the United States Army*. Washington, D.C.: Government Printing Office.

U.S. House of Representatives, Select Committee Investigating National Defense Migration, 1942. *Fourth Interim Report*. Washington, D.C.: Government Printing Office.

U.S. Military Intelligence Service Language School, 1946. *MISLS Album*. Fort Snelling, Minn.

Viteles, Morris S., 1953. *Motivation and Morale in Industry*. New York: W. W. Norton.

White, Ralph K. and Ronald Lippitt, 1960. *Autocracy and Democracy: An Experimental Inquiry*. New York: Harper.

Wirth, Louis, 1941. Morale and minority groups. *American Journal of Sociology*. 47:415–33.

index

Absenteeism (AWOL). *See also* Insubordination, as indication of demoralization
at Fort Meade, 103, 108, 122, 135, 167
at Fort Snelling, 181–182, 186, 219, 225, 239, 242, 243, 275, 305, 306, 325, 327–328, 332, 333, 423, 424
norms about "goofing off," 93, 124, 183, 341, 357–358, 394, 431–432
on U.S.S. *General Weigel*, 355, 358
in Zama, 382, 384, 388, 390–391, 400
Acculturation and assimilation, 70, 79
charges of "Haolefication," 78, 81, 425
in Hawaii, 30, 79
on the Pacific Coast, 27, 30
variation in, 53, 58, 78–79, 99, 139, 197, 216, 220, 284, 388–389
A Company (Fort Snelling), 88, 199, 246–247, 248–259 *passim*, 271, 273. *See also* Conflict: Nisei hostility toward Caucasians, between Nisei and OCS men, resolution of between Nisei and OCS men, *and* violence, interethnic; Officer Candidate School, Fort Snelling
attitudes toward K Company within, 273, 276, 280
attitudes toward MISLS administration within, 280, 281–282, 283, 285
Adams, Romanzo C., 21
Agitation, anti-Japanese. *See also* Evacuation of Japanese; Legislation, anti-Japanese
before World War II, 21–24
during World War II, 36, 43, 44–45, 50, 64–65
Agitation, pro-Japanese, 52–53, 55, 58
Aliens. *See* Enemy Aliens
Allen, Gwenfread, 33, 36
Allied Translator and Interpreter Service (ATIS), 400–404 *passim*, 408, 409
Aloha spirit, 21, 28, 36, 82, 83
American Civil Liberties Union (ACLU), 45, 51, 67, 401
American Friends' Service Committee, 57
American Legion, 50, 65–66
Andersen, Col. Karl, 86, 209, 216, 260, 261, 263, 264, 265, 272, 296, 298, 310, 312, 317, 421
A-9-3 (Fort Meade), 102–150 *passim*, 167–168, 186, 219, 409–430 *passim*
reputation of among Nisei, 143, 154
Ardant du Picq, Charles, 10
Army installations
Fort Lawton, Wash., 332, 339–347, 430
Camp McCoy, Wis., 38, 80
Fort Meade, Md. (*see* Fort Meade, Md.)
Camp Savage, Minn. (*see* Military Intelligence Service Language School)
Camp Shelby, Miss., 39, 40, 45, 61, 81, 83, 84, 105
Fort Snelling, Minn. (*see* Military Intelligence Service Language School)
Zama, Japan (*see* Zama)
Asakura, Capt. James, 331–345 *passim*, 363, 419, 420
Asian Exclusion League, 21–22, 24
Assembly centers. *See* Evacuation of Japanese

Barnard, Chester I., 3, 13

Bassan, Morton E., 362
Benedict, Ruth F., 72, 73, 75, 76
Biddle, Atty. Gen. Francis, 41, 42
Black market, 288, 342, 344, 345, 359, 366–368, 372, 376, 382–384, 388, 389, 396, 399–400, 409
Blau, Peter M., 414
Blumer, Herbert, 6
B-9-3 (Fort Meade), 150–156 *passim*, 168, 170–177 *passim*, 423
  reinstitution of discipline in, 151–152
Bogardus, Emory S., 68
Buddhist church, 25
Bushido spirit, 25, 71, 72, 94
Butler, Judson R., 68

Cadre
  at Fort Meade, 105, 108–168 *passim*
  Nisei cadre, 81, 91, 93, 105–106, 147–148, 149, 152, 182, 185, 186, 200, 207, 226, 237–238, 239, 303, 304–305, 430 (*see also* Endo, Sgt. William)
  at Fort Snelling, 303, 325, 328–329, 337, 422–423 (*see also* Endo, Sgt. William) at Zama, 378, 381–382, 389–400 *passim*, 409
California Joint Immigration Committee, 24, 44
Cantril, Hadley, 50, 51
Carr, Elizabeth, 29
Catalano, Lt. Joseph, 186, 198, 205–334 *passim*, 362–363, 414, 416, 420, 424, 430
  attitudes toward, 307–309, 313, 315, 363, 414–415
Caudill, William, 57–58
Chinese
  in Hawaii, 20, 27, 29, 34
  Nisei attitudes toward, 293–294
Citizenship, renunciation of U.S., 54
Clinton, Lt. John, 334, 345, 346
Commissions, refusal of
  by Division F instructors, 324
  by OCS men, 280, 282, 283, 323–324
Communication channels, 13, 285, 427–428. *See also* Primary groups, informal communication channels within
Conflict
  between "eager beavers" and "fuck-ups," 213–214, 215, 218, 241, 284, 302, 425
  between Hawaiian and mainland Nisei, 81–83, 105, 139, 208, 223–224, 226–236, 275, 277–278, 284, 305, 424–425, 434
  intergenerational, 30–31
  between Nisei and depot cadre, 106–168 *passim*, 225, 237–238, 239–246 *passim*, 306, 319–334 *passim*, 378, 381–382, 389–400 *passim*, 422–423
  Nisei hostility toward Caucasians, 89, 94, 98, 115, 130, 139–141, 205, 250, 252–255, 271, 277
  Nisei hostility toward Negroes, 83
  between Nisei and OCS men, 1, 2, 223, 247–252, 254, 256–258, 272, 430
  resolution of between Nisei and OCS men, 279–284, 285
  between students and MISLS administration, 259–264 *passim*, 265–271 (*see also* Kawashita, Maj. Robert)
  violence among demoralized, 7, 8
  violence, informal sanction of, 237, 242, 244–245, 271, 284, 423–424, 425
  violence, informal sanctions against, 236–284, 425
  violence, interethnic, 1, 223, 256–257, 271, 273, 276, 284, 424
  violence, intramural, 2, 223, 226–237 *passim*, 273–274, 284, 424–425, 426
  violence against Japanese, 34, 42–43, 66–67
  violence outside K Company (Fort Snelling), 237, 273–275, 276, 279, 284, 285
  violence against NCOs, 225, 237, 243–246, 423 (*see also* Endo, Sgt. William)
  violence in relocation centers, 52
  violence against returning evacuees (*see* Resettlers, violence against)
Consciousness of kind. *See* Ethnic identity; Group solidarity
Consensus, 413, 426, 429–431
  as agreement, 429
  in appearance only, 430–431
  as coorientation, 429
Consolidated Mess Hall (Fort Snelling), 204, 256, 258–259, 275, 330
Cooley, Charles H., 12, 420
Co-operation, conditions of effective, 412–415
Cortright, David, 7, 8
Culture, definition of, 72

Daniels, Roger, 22
Day, A. Grove, 21, 30
Definition of the situation. *See* Primary groups, emergence of definitions within; Deliberation, collective
Delaney, Lt. George, 109, 110, 113, 116, 118, 119–120, 121, 123, 164, 165
Deliberation, collective, 53, 427–428, 429
  self-selection of personnel in, 427
Demonstrations, protest

as indication of demoralization, 8
in military units, 350, 364–365
in relocation centers, 52
Demoralization (low morale), 2–3, 9, 103, 124–126, 146, 147, 280, 285, 411. *See also* Absenteeism (AWOL); Insubordination; Passive resistance; Petitions, protest; Retaliation; Sabotage
among OCS men (Fort Snelling), 248, 251, 280, 285
among American occupation troops, 250, 364–365, 372, 412
conditions and indications of, 2, 6, 7, 8, 284, 365, 417–419, 420, 424
definition of, 5–8, 419
as a process, 412, 416–419, 419–420, 434
Desegregation. *See* Segregation, ethnic
Deviant behavior
of "eager beavers," 216 (*see also* Conflict, between "eager beavers" and "fuck-ups"; Nisei norms; Primary groups, informal norms within; Shindo, Pfc. Raymond)
from Nisei norms (*see* Nisei norms; Primary groups, informal norms within; Onishi, Pfc. John)
DeWitt, Gen. John L., 45, 50. *See also* Western Defense Command (U.S. 4th Army)
Dickerson, Col. James, 375–376
Dickson, William J., 10
Dies, Rep. Martin, 44. *See also* Un-American Activities Committee (U.S. House of Representatives)
Differential treatment. *See* Discrimination, ethnic (charges of); Discrimination between officers and enlisted men
Discipline, breakdown of. *See also* Absenteeism; Insubordination; Sabotage
among occupation troops, 365
at Fort Meade, 112–150 *passim*
reinstitution of at Fort Meade, 151–152
at Fort Snelling, 203–286 *passim*, 325–334 *passim*
at Zama, 390–400 *passim*
Discrimination, ethnic (charges of), 96, 428. *See also* Segregation, ethnic; Legislation, anti-Japanese
in U.S. Army, 3, 89
in Hawaii, 27–28, 30, 277–278
at Fort Meade, 127–143 *passim*, 168*n*, 411, 432–433
against Negroes, 54, 353, 355, 363, 393–394 (*see also* Negro troops)
on Pacific Coast, 20–26 *passim*, 64–67 *passim*
at Fort Snelling, 89, 173, 200, 247, 251, 252–255, 272, 282, 285
at Zama, 393–394
Discrimination between officers and enlisted men
in U.S. Army, 349
in U.S. Navy, 348–350, 363, 433
overseas, 364–365
Disloyalty. *See* Loyalty, national

"Eager beavers." *See* Conflict, between "eager beavers" and "fuck-ups"; Deviant behavior, of "eager beavers"; Shindo, Pfc. Raymond
E Company (Fort Snelling), 179–196, 219. *See also* Turkey Farm
reputation of in Nisei community, 184
Emmons, Gen. Delos, 36, 38, 39
Endo, Sgt. William, 198, 237–246 *passim*, 284, 304, 422, 423, 429, 430
Enemy aliens. *See also* Evacuation of Japanese; Loyalty, national
arrest of, 37, 41, 43–44
regulations governing, 35, 42, 43–44, 47
"English Standard" schools (Hawaii), 29. *See also* Pidgin English
Enlisted Reserve Corps (ERC), 80, 95
Ethnic identity, 25, 27, 53, 72–74, 77, 78, 79, 106, 388–389, 419. *See also* Group solidarity; Ethnic pride, Nisei
feelings of inferiority to Caucasians, 77, 78
Ethnic pride, Nisei, 77, 100, 129, 168. *See also* Ethnic identity; Group solidarity
Ethnic stereotypes. *See* Stereotypes, ethnic
Evacuation of Japanese (from coastal areas), 3, 37, 47–70 *passim*, 94, 421. *See also* Relocation centers
agitation for, 43, 44
compensation of evacuees, 69
exclusion orders, 45
exclusion orders and regulations, rescinding of, 57, 64, 66, 68
Executive Order 9066, 45
internment of Japanese citizens and Nisei, 47–49
justification of as "military necessity," 47, 51
obliteration of distinction between citizen and alien, 45
public opinion concerning, 44–45, 50
regulations governing Japanese citizens, 47
regulations governing Nisei, 47
voluntary evacuation, reactions to, 47
Evasion of duty. *See* Insubordination
Executive Order 9066, 45

Factionalism, 6, 7, 139, 143, 213–214, 233, 284, 416, 424–425, 434. *See also* Conflict, between "eager beavers" and "fuck-ups" *and* between Hawaiian and mainland Nisei
Federal Bureau of Investigation (FBI), 35, 42, 43, 45
Fisher, Anne M., 69
Formal organizations, 4, 435
  bureaucratic problems in, 422
  goals of, 5, 7, 412, 413, 416, 435
  norms of, 434
  importance of primary groups within, 13
Fort Meade, Maryland, 101–168 *passim*, 414, 424, 430. *See also* A-9-3 (Fort Meade); B-9-3 (Fort Meade)
442nd Regimental Combat Team. *See* Nisei servicemen, 442nd Regimental Combat Team
Fraternization, 287–288, 289, 290, 359, 366, 369–371, 384–386
Frazier, E. Franklin, 77
Fuchs, Lawrence H., 20, 27, 28, 29, 35, 77

George, Alexander L., 11, 435
Germany, surrender of. *See* V-E Day
Glaser, Barney G., 15, 431
Greene, Stephen, 85*n*
Grinker, Roy R., 4
Griswold, A. Whitney, 23
Grodzins, Morton, 45, 54
Group solidarity, 434. *See also* Ethnic identity; Ethnic pride
  among Nisei, 30, 84–85, 106, 138, 139, 167–168, 286

Hammerstein, Oscar II, 404*n*
Haoles (Hawaii), 21, 27–28, 29, 30, 34, 77
Hawaii. *See also* Immigration, Japanese; Japanese in America, history of, in Hawaii
  attack on Pearl Harbor (*see* Pearl Harbor)
  attitudes toward various ethnic groups in, 27–28, 34
Hawaiians, native, 20, 27, 29, 34
Hawaiian Territorial Guard, 35, 37
Heinl, Robert D., 6, 8
Hocking, William E., 14, 436
Homans, George, 362
Hormann, Bernhard L., 78

Ickes, Interior Secretary Harold, 64, 67
Iga, Mamoru, 75
Immigration, Japanese, 20–24 *passim*
  "Gentlemen's Agreement" concerning, 23
  to Hawaii, 20–21
  Immigration Law of 1924, 24
  McCarren-Walter Act, 69
  naturalization rights for Issei, 69
  opposition to, 20, 21, 22
  to Pacific Coast, 20, 21–24 *passim*
Infantry Replacement Training Command (IRTC), 97, 103
  desegregation in, 98
Insubordination, 6, 8. *See also* Absenteeism; Passive resistance; Retaliation; Sabotage
  as seen in black-market activities, 367–368
  as indication of demoralization, 418, 420, 423
  at Fort Meade, 108, 122, 123, 138, 167, 168
  among American occupation troops, 365
  at Fort Snelling, 2, 178, 181, 187, 211, 219, 239, 240, 259, 263 266–267, 270, 325, 333, 418, 423
  on U.S.S. *General Weigel*, 355–358 *passim*
  at Zama, 378, 390–400 *passim*
Intelligence service. *See* Military Intelligence Service Language School
Intermarriage, 24, 29, 30, 39, 70
Internment of Japanese. *See* Evacuation of Japanese
Issei, 30. *See also* Immigration; Enemy aliens; Evacuation of Japanese; Japanese in America, history of; Japanese culture; Loyalty, national
  attitudes toward war of, 40–41
  legal status of, 35

Jacobson, Alan, 58
Janowitz, Morris, 10, 11
Japan
  occupation duty in, 287, 294, 297, 359
  occupied, 359, 365–373, 409–410
  occupied, food shortages in, 377–381
  postwar corruption and disorder in, 372–373, 409
  surrender of, 67, 221
Japanese in America, history of, 20–23. *See also* Immigration, Japanese
  in Hawaii, 20–21, 27–30, 31–32
  on Pacific Coast, 22–24, 30
Japanese American Citizens' League (JACL), 33
Japanese-Americans. *See* Nisei culture; Nisei norms; Nisei servicemen
Japanese American Student Relocation Council, 55

Japanese culture, 31, 72–76. *See also* Bushido spirit
some characteristics of, 57, 72, 74–75, 79
popular beliefs about, 214, 289, 369–371
status of women in, 213, 289, 369–371
stereotypes of Japanese "race" (*see* Stereotypes, ethnic)
values of, 25, 72, 73–76, 77, 79, 213, 372
Japanese linguists
classification of as essential, 296, 298
classification examination (Zama), 403–404
need for in occupied Japan, 294, 295
publicity about, 406
use of in War Crimes Trials, 409–410
Japanese motion pictures, 25, 212–213, 289–293
Japanese prisoners of war, 348, 351, 355–360 *passim*
Japanese women
discussions about (*see* Fraternization)
emancipation of, 370
status of in Japanese culture, 213, 289, 369–374
Justice, U.S. Department of, 41, 45

Kawai, Kazuo, 372
Kawashita, Maj. Robert, 260–281 *passim*, 294, 315, 316, 421–426 *passim*. *See also* Military Intelligence Service Language School (MISLS), Fort Snelling, attitudes of E and K Company personnel toward administration
K Company (Fort Snelling), 196–409. *See also* Military Intelligence Service Language School; E Company (Fort Snelling); Primary groups; Zama
company commander. *See* Catalano, Lt. Joseph
dissolution of, 407–409
linguistic assignments of K Company personnel, 409–410
norms within, 216 (*see also* Nisei norms)
reputation of personnel among Nisei, 219–220, 286
self-conceptions within, 196, 220, 419
transformation of performance in, 287, 305, 308, 317, 362, 363
violence in (*see* Conflict)
Kibei, 31, 35, 78–79, 96, 97
Kimura, Sgt. Edward, 237–238, 240, 241, 243, 246, 247
Kimura, Yukiko, 37, 41
Kishi, James, 262, 281, 315, 316
Kitano, Harry, 76
Kitsuse John, 26
Klapp, Orrin E., 69
Knox, Secretary of the Navy Frank, 34
Korea
attitudes toward duty in, 402
attitudes toward Koreans, 34, 403
Kuykendall, Ralph S., 21, 30
Kwan, Kian M., 79

Lang, Kurt, 11
Larsen, Capt. Vernon, 116–117, 123, 136–154 *passim*, 164, 167–168, 327, 412, 414, 421–430 *passim*
Lasswell, Harold D., 6
L Company (Fort Snelling), 236, 237
Legislation, anti-Japanese, 22–24, 45, 50–51, 54, 65. *See also* Agitation, anti-Japanese; Evacuation of Japanese
Leighton, Alexander H., 53
Lind, Andrew W., 20, 28, 30, 33, 34, 36, 37, 39, 40
Lippitt, Ronald 10
Lippmann, Walter, 422
Lord, Walter, 33
Loyalty, national, 2, 3, 19, 51, 52, 53, 54. *See also* Issei
changes in public opinion about, 63, 64, 67–68
declarations of disloyalty, reasons for, 53–54
efforts by those convinced of loyalty of Japanese, 51–52
formal determination of, 52–55
Nisei attitudes toward disloyal, 58, 356, 401–402
questions concerning Issei loyalty, 22, 31–34, 44–45
questions concerning Nisei loyalty, 31–34, 35, 36–38, 44–45, 86, 95
renunciation of U.S. citizenship by Nisei, 54
segregation camp for "disloyal" (Tule Lake), 53–54
service in army as refutation of charges of disloyalty, 39, 40, 56, 79, 83–84, 86, 96, 413

McCarren-Walter Act, 69
MacDonald, Alexander, 28, 35, 40, 41
McGovney, Dudley O., 24
Malinowski, Bronislaw, 433
Mandelbaum, David G., 11, 285
Marshall, Samuel L. A., 11
Mass hysteria, 50
Massie case (Hawaii), 255, 255*n*
Masuoka, Jitsuichi, 78

Matsui, Sgt. Yukio, 152, 159, 163
Matsumoto, Toru, 45, 56
Mauldin, Bill, 90
Military Intelligence Service, 63. *See also* Military Intelligence Service Language School
Military Intelligence Service Language School (MISLS), 86. *See also* Military Intelligence Service Language School, Fort Snelling
  history and organization of, 85–90, 259
  at San Francisco Presidio, 85
  at Camp Savage, 85–87, 259
Military Intelligence Service Language School (MISLS), Fort Snelling, 1, 9, 88–89, 170–334, 411, 414
  attitudes of administration toward Nisei, 88, 259, 280–281, 282
  attitudes of E and K Company personnel toward, 184, 185, 334
  attitudes of E and K Company personnel toward administration, 259–263, 265–271, 280, 281–283, 285, 297, 298, 300, 315, 346, 401, 416, 421, 424, 429 (*see also* Kawashita, Maj. Robert)
  attitudes of OCS men toward administration, 280, 281–282, 283, 285, 297
  "conversational course," 187, 189, 196, 220
  language instructors at (Division F), 203–204, 210, 211–212, 216, 261, 262, 266–268, 270, 297, 425–426
  reputation of, among Nisei, 87, 89–90, 170, 173, 174, 178, 180
  reputation of, and attitudes about, E and K Company personnel among administrators, 266, 321
  reputation of E and K Company personnel among Nisei, 185, 219–220, 286
"Military necessity," 170–171. *See also* Evacuation of Japanese (from coastal areas); Pearl Harbor, explanation of
Morale
  definition of, 3–5
  as seen in effective performances of K Company, 14, 206, 209, 309–313, 414, 415–416
  high morale, characteristics of, 3, 6, 84
  high morale, definition of, 415–416, 435
  high morale, indications of, 415–416
  low morale (*see* Demoralization)
  manipulation of, 436
  variations in, 10, 362, 412, 426
Morimoto, Patricia T., 78, 85
Moskos, Charles C., 435
Mueller, Sgt. Walter, 117, 134, 135, 138, 151, 153
Murphy, Lt. Col. Melvin, 205, 207, 208, 209, 262, 263, 265, 272, 275, 321, 330, 334
Murphy, Thomas D., 36, 81, 85

Nakahata, Yutaka, 38
Nakanishi, Pfc. James, 129, 156, 157, 166, 197
Nakasone, Pfc. Henry "Cyclone," 161, 303, 311, 312, 331, 320–363 *passim*, 374, 395, 406, 420
National Opinion Research Center, 50, 64, 67, 68
Native Sons of the Golden West, 51
Negro troops, 54, 54–55*n*, 286–287, 347, 348, 351, 353, 354, 355, 363, 373, 393–394
  Nisei attitudes toward, 83, 252–253, 358, 363
Nisei cadre. *See* Cadre, Nisei cadre
Nisei culture, 26, 71, 76–77
  communication channels, 106, 184, 196 (*see also* Communication channels)
  norms (*see* Nisei norms)
  self-conceptions, 72, 79, 93
  values of, 76, 77, 79, 91–94 *passim*, 100
Nisei norms, 80–81, 91, 92, 93–94, 97–98, 99–100, 106, 168, 184, 197, 207, 239, 263, 413–414, 418. *See also* Primary groups, pressures to conform within
Nisei servicemen. *See also* Military Intelligence Service Language School; Military Intelligence Service Language School, Fort Snelling; K Company (Fort Snelling); Primary groups; Selective Service System
  before World War II, 38, 80
  casualties among, 40, 59–60, 62–63, 84, 90, 97, 101
  442nd Regimental Combat Team, 2, 14, 40, 61, 62–63, 81, 83, 84, 97, 101, 170, 413
  interest in, among Japanese, 49, 61–62
  media coverage of, 62–64, 66, 69, 84, 97, 98
  Nisei NCOs (*see* Cadre, Nisei cadre), 286, 406 (see also *Pacific Citizen*)
  Nisei officers, 91, 92
  norms governing (*see* Nisei norms)
  100th Infantry Battalion, 38, 40, 62–63, 79, 83, 84, 87, 97, 331
  reputation of, 2, 61, 62, 63, 80–81, 87, 89, 99, 100, 178
  reputation of E and K Company per-

sonnel, 185, 219–220, 286
segregation of (*see* Segregation, ethnic)
service in army as refutation of charges of "disloyalty," 39, 40, 55, 79, 83–84, 86, 96, 413
stereotype of "combat man" among, 90, 91
Nishi, Setsuko, 75
Nishida, Sgt Teruo, 172, 187, 189, 264, 422
Nishimoto, Richard S., 48, 51, 53
Nitobe, Inazo O., 25
Nobbe, Charles E., 70
Norms
formal, 43
informal (*see* Primary Groups, informal norms within; Nisei norms)

Obstruction, 141, 412. *See also* Passive resistance
Office of Strategic Services (OSS), 86, 88
Office of War Information, 88
Officer Candidate School (OCS), Fort Snelling, 86, 88. *See also* A Company (Fort Snelling)
admission of K Company men to, 300
100th Infantry Battalion. *See* Nisei Servicemen, 100th Infantry Battalion
Onishi, Pfc. John, 124, 157, 166, 183
Organizational goals. *See* Formal organizations, goals of

Pacific Citizen, 59, 60*n*, 94, 97, 98, 169, 172
Pacific Stars and Stripes, 403, 404
Packer, Peter, 255*n*
Park, Robert E., 433
Parmentier, Capt. Thomas, 152, 153, 171, 420
Participant observation (as a research procedure), 16–17, 18
Passive resistance, 421, 423
Pearl Harbor, attack on, 19, 33–34
anti-Japanese agitation following, 36
explanation of, 33–34, 41
ensuing regulations on Japanese, Kibei and Nisei (Hawaii), 35–36
Petersen, William, 69
Philippine Islands, 288
Filipinos in Hawaii, 20, 21, 29
hostility of Filipinos toward Japanese and Nisei, 34, 42, 43, 288–289
Pidgin English
definition of, 29*n*
sensitivity concerning use of, 81, 224
use of, in Hawaii, 28–30, 78, 79
use of, in Nisei army units, 85
Portuguese (in Hawaii), 20, 27, 29
Prejudice, ethnic, 21. *See also* Discrimination, ethnic; Segregation, ethnic; Stereotypes, ethnic
Primary groups, 5, 8, 10, 11, 412–436 *passim*. *See also* Group solidarity; Ethnic pride
authority figures and leadership within, 10, 362, 414–415, 420–421, 425
definition of, 11–12
emergence of definitions within, 13, 14, 15, 411, 412, 416, 417, 418, 421–422, 426, 427–430, 435, 436
evaluation of unit, 415, 416, 418, 419
importance of evaluations arising within, 13, 14, 412, 415, 416, 418, 426
factionalism within (*see* Factionalism)
goals of, 415
informal communication channels within, 13, 139, 430, 436
informal norms within, 10, 12, 13, 409, 411, 416–426 *passim*, 431–434
informal sanctions within, 433
interpersonal relations within, 10, 12, 433–434
pressures to conform within, 81, 93–94, 97, 98, 100, 106, 141, 142, 168, 174, 216, 419, 420, 430–431
feelings of resentment concerning Language Classification Examination (Fort Snelling), 194–195, 196–198, 431, 433
self-conceptions within, 416, 419
solidarity of (*see* Group solidarity; Ethnic pride)
structure of performance, 4
style of performance, 4, 5, 6, 13, 411, 428–429
Problematic situations
definition of, 426–427
informal norms as solutions to, 432
Problem minority, 19, 20, 70. *See also* Discrimination, ethnic; Segregation, ethnic
Publicity
about 442nd Combat Team (*see* Nisei Servicemen, media coverage of)
about Tule Lake demonstrations, 58
Public opinion (concerning evacuation), 44–45, 50
shifts in American public opinion concerning Nisei, 63, 64, 67–68

Rainwater, Lee, 58
Reenlistment, discussions about, 300–302
Registration of evacuees, crisis surrounding. *See* Loyalty, national, formal determination; Relocation centers
Reinecke, John E., 29*n*, 78

Relocation centers, 11, 48, 60, 81, 185. *See also* Evacuation of Japanese
closing of, 68, 217
draft resistance in, 60–61
Gila River, Ariz., 50
Granada, Colo., 61
Heart Mountain, Wyo., 60, 61
Poston, Ariz., 60
registration crisis in, 52–55
Tule Lake Segregation Center, Calif., 53, 62, 287, 401, 402, 413
Resettlement program, 55–57, 58, 68, 95. *See also* Evacuation of Japanese
informal norms governing resettlers, 58
participation of resettlers in war effort (nonmilitary), 61–62
reception of resettlers, 56, 57, 66, 98
violence against resettlers, 66–67
Retaliation against organization
as indication of demoralization, 421, 422
at Fort Meade, 114, 122–123, 141–142, 153, 154, 168, 411–412, 422, 423, 424
at Fort Snelling, 325, 424
during Vietnam War, 8
Rikoran, 290–293, 386–387
Rodgers, Richard, 404*n*
Roethlisberger, Fritz J., 10
Roosevelt, Eleanor, 50
Roosevelt, Pres. Franklin D., 39, 47, 55, 65
Executive Order 9066, 45
funeral for, 114–115, 116, 156
Rostow, Eugene V., 69
Rundell, Walter, 366

Sabagh, George, 23*n*
Sabotage
as indication of demoralization, 8, 423
in Language Classification Examination, 188–194 *passim*
during Vietnam War, 8
aboard U.S.S. *General Weigel*, 356, 423
at Zama, 383, 400
"Saipan," 230, 231, 257, 314, 345, 358–359, 360, 433
Samurai code. *See* Bushido spirit
Scheff, Thomas J., 429
Schmid, Calvin F., 70
Schmidt, Capt. Alan, 179, 185, 186, 422
Schneider, David M., 220
School Battalion (Fort Snelling) Oral Language School (Division F), 179, 200–201, 419, 420, 426, 432. *See also* K Company
Scott, W. Richard, 414
Segregation, ethnic, 11. *See also* Discrimination, ethnic (charges of)
in the U.S. Army, 38, 54, 61, 95, 98
as facilitating internal conflict, 285
in Hawaiian schools, 29–30
at Fort Meade, 101, 151, 153, 168
at MISLS, 88
on the Pacific Coast, 22–23, 24, 26, 79
rationale for, in U.S. armed forces, 39
Selective Service System
reinstitution of for Nisei, 40, 58–59, 61, 90, 95, 169
resistance to, 60–61
suspension of for Nisei, 49, 86
termination of, 259
Shibutani, Tamotsu, 17, 79, 428
Shils, Edward A., 10, 11, 412
Shindo, Pfc. Raymond, 232, 265, 430. *See also* Conflict, between "eager beavers" and "fuck-ups"; Deviant behavior, of "eager beavers"
Shirey, Orville C., 63, 84
Short, James, 434
Skinner, Lt. Donald, 319, 320, 322, 323, 346
resentment against, 324–327, 328–331 *passim*, 332–334, 337, 338, 363, 419, 421, 423, 429
Smith, Madorah E., 78
Social Process in Hawaii, 83
Spencer, Robert F., 26
Spiegel, John P., 4
Stereotypes, ethnic
of AJAs, 83, 91, 105, 223–224, 229–230
of Caucasians (by Nisei), 7, 98–99, 247
of "real Japanese" (by Nisei), 72, 361
of Japanese "race," 20, 22, 44, 214, 388
of "kotonks," 82–83, 91, 223, 224
of Negroes (by Nisei), 352–353
of Nisei, 57–58
transformation of, 62–64 *passim*, 68, 69–70
Stimson, Sec. of War Henry, 65–66
Stouffer, Samuel A., 5, 90, 91, 170, 285
Strauss, Anselm L., 15, 431
Strodtbeck, Fred L., 434
Strunk, Mildred, 50, 51
Sumner, William G., 433

Tajima, Pfc. Henry, 226, 230, 231, 232, 234, 235
Takeda, Pfc. Noburu, 226, 230, 231, 235, 257, 313, 314, 325
tenBroek, Jacobus, 22, 23, 45, 69
Time, 7, 63

Thomas, Bob, 255*n*
Thomas, Dorothy S., 23*n*, 30–31, 48, 51, 53
Tokimasa, Aiko, 29*n*
Toyota, Ralph, 38
Transaction, social
  definition of, 16
  as unit of analysis, 16, 18, 412
Treasury Department, 42
Tsuboi, Pfc. Jiro, 226, 230, 231, 235, 314
Tsurumi, Kazuko, 370
Tule Lake Segregation Center. *See* Relocation centers
Turkey Farm (Fort Snelling), 88, 180, 182, 185, 219, 264, 412, 417, 422

Un-American Activities Committee (U.S. House of Representatives), 31, 44, 50, 51
U.S. Department of the Army, 18, 95, 102, 177, 219, 264
U.S. House of Representatives, 22, 24
U.S. Supreme Court, 51, 68–69

Varsity Victory Volunteers, 38, 39
V-E Day, 135, 140, 169
Vietnam War, 6, 7, 8, 413
Violence. *See* Conflict
Viteles, Morris S., 414

War Crimes Trials, 409–410
War Department, 45, 49, 54, 83, 85, 169, 261, 272
War Relocation Authority (WRA), 47, 50, 51, 52, 55, 64, 183, 217
U.S.S. *General Weigel*, 347–361, 373
Western Defense Command (U.S. 4th Army), 45, 51
White, Ralph K., 10
Williams, Lt. James, 109, 116, 132, 134, 145
Wilson, Gen. Jonathan, 260, 323–324
Wirth, Louis, 4–5

Yoneda, Pfc. William, 263, 264, 348, 392

Zama (4th Replacement Depot), Japan, 373–410, 412, 414, 423, 426, 433